T0213665

Springer Actuarial

Editors-in-Chief

Hansjoerg Albrecher, University of Lausanne, Lausanne, Switzerland

Michael Sherris, UNSW, Sydney, NSW, Australia

Series Editors

Daniel Bauer, University of Wisconsin-Madison, Madison, WI, USA

Stéphane Loisel, ISFA, Université Lyon 1, Lyon, France

Alexander J. McNeil, University of York, York, UK

Antoon Pelsser, Maastricht University, Maastricht, The Netherlands

Ermanno Pitacco, Università di Trieste, Trieste, Italy

Gordon Willmot, University of Waterloo, Waterloo, ON, Canada

Hailiang Yang, The University of Hong Kong, Hong Kong, Hong Kong

This is a series on actuarial topics in a broad and interdisciplinary sense, aimed at students, academics and practitioners in the fields of insurance and finance.

Springer Actuarial informs timely on theoretical and practical aspects of topics like risk management, internal models, solvency, asset-liability management, market-consistent valuation, the actuarial control cycle, insurance and financial mathematics, and other related interdisciplinary areas.

The series aims to serve as a primary scientific reference for education, research, development and model validation.

The type of material considered for publication includes lecture notes, monographs and textbooks. All submissions will be peer-reviewed.

More information about this series at http://www.springer.com/series/15681

María del Carmen Boado-Penas · Julia Eisenberg ·
Şule Şahin
Editors

Pandemics: Insurance and Social Protection

 Springer

Editors
María del Carmen Boado-Penas ⓘ
Department of Mathematical Sciences
University of Liverpool
Liverpool, UK

Julia Eisenberg
FAM
TU Wien
Wien, Austria

Şule Şahin
Department of Mathematical Sciences
University of Liverpool
Liverpool, UK

ISSN 2523-3262   ISSN 2523-3270   (electronic)
Springer Actuarial
ISBN 978-3-030-78336-5   ISBN 978-3-030-78334-1   (eBook)
https://doi.org/10.1007/978-3-030-78334-1

JEL Classification Code: C02, G22, H84, I13, I18

Mathematics Subject Classification: 60, 62, 91B15, 91B30, 91B82, 91G70

© The Editor(s) (if applicable) and The Author(s) 2022. This book is an open access publication.
Open Access This book is licensed under the terms of the Creative Commons Attribution 4.0 International License (http://creativecommons.org/licenses/by/4.0/), which permits use, sharing, adaptation, distribution and reproduction in any medium or format, as long as you give appropriate credit to the original author(s) and the source, provide a link to the Creative Commons license and indicate if changes were made.
The images or other third party material in this book are included in the book's Creative Commons license, unless indicated otherwise in a credit line to the material. If material is not included in the book's Creative Commons license and your intended use is not permitted by statutory regulation or exceeds the permitted use, you will need to obtain permission directly from the copyright holder.
The use of general descriptive names, registered names, trademarks, service marks, etc. in this publication does not imply, even in the absence of a specific statement, that such names are exempt from the relevant protective laws and regulations and therefore free for general use.
The publisher, the authors and the editors are safe to assume that the advice and information in this book are believed to be true and accurate at the date of publication. Neither the publisher nor the authors or the editors give a warranty, expressed or implied, with respect to the material contained herein or for any errors or omissions that may have been made. The publisher remains neutral with regard to jurisdictional claims in published maps and institutional affiliations.

This Springer imprint is published by the registered company Springer Nature Switzerland AG
The registered company address is: Gewerbestrasse 11, 6330 Cham, Switzerland

"The more I see, the less I know for sure."

—John Lennon

"Not even the most tempting probability is a protection against error; even if all the parts of a problem seem to fit together like the pieces of a jig-saw puzzle, one must reflect that what is probable is not necessarily the truth and that the truth is not always probable."

—Sigmund Freud

Preface

Dear Reader,

At the beginning of 2020 we, like many other people and probably like you, were taken by surprise at the severity and the speed of spread of the novel coronavirus COVID-19.

Like everyone else on the planet, we were watching the virus devastating countries and jumping from country to country and from continent to continent. Reading the reports of the OXFAM (a confederation of independent charitable organisations focusing on the alleviation of global poverty) and the World Bank, we realised that the virus was killing not only directly but also through hunger and collateral damages. The definition of vulnerable became vague and sometimes a condition of survival.

COVID-19 unsheathed once again and in a cruel clarity the gap between rich and poor countries. Whilst the rich countries supported their citizens during lockdowns and bought vaccines, some of the poor countries were facing famines or an epidemic with no lockdown, no masks and no vaccine.

The question arises how can governments provide the needed social protection to the population of their countries or at least to the most vulnerable population strata in a quick manner? Could an insurance product provide the necessary protection?

Having had months of discussions with different experts on this topic, we realised that many researchers and practitioners might profit from the combined expertise of actuaries, mathematicians, statisticians, sociologists, virologists and jurists/lawyers.

This triggered the idea to put together a book explaining the facets of possible pandemic insurance.

In this sense, the present book is not just a collection of loose contributions on different topics. It is thought to be a combination of jigsaw parts that can be put together to create individual or collective insurance products aimed to supplement the social protection most governments promise to their citizens. We hope that the interdisciplinary nature of the contributions will help to build bridges between different disciplines and provide a platform for new collaborations.

Liverpool, UK
Vienna, Austria
Liverpool, UK
May 2021

María del Carmen Boado-Penas
Julia Eisenberg
Şule Şahin

Acknowledgements

The editors would like to take this opportunity to heartily thank the contributors, who both supported the idea of the book and wrote high-quality chapters within an extremely tight deadline.

We would like to express our utmost appreciation to A. D. Wilkie for his invaluable comments and contribution to the introductory chapter and additional thanks to Dominic Calleja for his English support in the chapter.

Most sincere thanks go to Rick Cosstick and Rachel Bearon, (University of Liverpool), for their willingness to help us when needed. Thanks also to the Vice Rector for Research and Innovation at TU Wien, Johannes Fröhlich, for inspiring the interdisciplinary idea of this book.

We are also grateful to the University of Liverpool for the financial assistance for the open access of the book. Julia Eisenberg expresses her gratitude to the Austrian Science Fund (FWF) for the funding of her research, grant number V 603-N35.

Finally, special thanks go to Hansjörg Albrecher for his encouragement, continuous support and trust.

María del Carmen Boado-Penas
Julia Eisenberg
Şule Şahin

Contents

6 Risk-Sharing and Contingent Premia in the Presence of Systematic Risk: The Case Study of the UK COVID-19 Economic Losses

Hirbod Assa and Tim J. Boonen

7 All-Hands-On-Deck!—How International Organisations Respond to the COVID-19 Pandemic

María del Carmen Boado-Penas, Gustavo Demarco,
Julia Eisenberg, Kristoffer Lundberg, and Şule Şahin

Contributors

Matthew Aldridge School of Mathematics, University of Leeds, Leeds, UK

Hirbod Assa Kent Business School, University of Kent, Kent, UK

Nuria Badenes-Plá Instituto de Estudios Fiscales, Av del Cardenal Herrera Oria, Madrid, Spain

María del Carmen Boado-Penas Department of Mathematical Sciences, University of Liverpool, Liverpool, UK

Tim J. Boonen Amsterdam School of Economics, University of Amsterdam, Amsterdam, The Netherlands

Jonathan P. Caulkins Heinz College, Carnegie Mellon University, Pittsburgh, PA, USA

Gustavo Demarco Pensions Global Solution Group – Social Protection and Jobs, The World Bank, Washington, D.C., USA

Sherry A. Dunbar Luminex Corporation, Austin, TX, USA

Julia Eisenberg Institute of Statistics and Mathematical Methods in Economics, TU Wien, Vienna, Austria

David Ellis School of Mathematics, University of Bristol, Bristol, UK

Gustav Feichtinger Institute of Statistics and Mathematical Methods in Economics, TU Wien, Vienna, Austria

Runhuan Feng Department of Mathematics, University of Illinois Urbana-Champaign, Urbana, IL, USA

Peter Filzmoser Institute of Statistics and Mathematical Methods in Economics, TU Wien, Vienna, Austria

José Garrido Department of Mathematics and Statistics, Concordia University, Montreal, QC, Canada

Dieter Grass International Institute for Applied Systems Analysis (IIASA), Laxenburg, Austria

Richard F. Hartl Department of Business Decisions and Analytics, University of Vienna, Vienna, Austria

Rachel Hillier Partner at Capital Law, Cardiff & London, UK

Petar Jevtić School of Mathematical & Statistical Sciences, Arizona State University, Tempe, AZ, USA

Longhao Jin Department of Mathematics, University of Illinois Urbana-Champaign, Urbana, IL, USA

Peter M. Kort Tilburg School of Economics and Management, Tilburg University, Tilburg, The Netherlands

Sooie-Hoe Loke Department of Mathematics, Central Washington University, Ellensburg, WA, USA

Kristoffer Lundberg Ministry of Health and Social Affairs, Division for Coordination and Support -- Policy Analysis Unit, Stockholm, Sweden

Alexia Prskawetz Institute of Statistics and Mathematical Methods in Economics, TU Wien, Vienna, Austria

Luca Regis ESOMAS Department, University of Torino and Collegio Carlo Alberto, Torino, Italy

Christopher Rieser Institute of Statistics and Mathematical Methods in Economics, TU Wien, Vienna, Austria

Şule Şahin Department of Mathematical Sciences, University of Liverpool, Liverpool, UK

Frank Schiller Deutsche Aktuarvereinigung (DAV), Cologne, Germany

Andrea Seidl Department of Business Decisions and Analytics, University of Vienna, Vienna, Austria

Yi-Wei Tang Cepheid/Danaher Diagnostic Platform, Shanghai, China

Gary Venter Actuarial Science, School of Professional Studies, Columbia University, New York, NY, USA

A. D. Wilkie InQA Limited and Heriot-Watt University, Horsell, Woking, UK

Stefan Wrzaczek International Institute for Applied Systems Analysis (IIASA), Laxenburg, Austria

Linfeng Zhang Department of Mathematics, University of Illinois Urbana-Champaign, Urbana, IL, USA

Chapter 1
COVID-19: A Trigger for Innovations in Insurance?

María del Carmen Boado-Penas, Julia Eisenberg, and Şule Şahin

Abstract This chapter gives an overview of the consequences of the novel coronavirus, COVID-19 on the insurance branch. The main problems caused by the pandemic on the commercial insurance, and in particular, on the business interruption and possible innovations are discussed. The aim is to prepare the reader for the following chapters specifically by demonstrating connections between different aspects of modelling a pandemic. These models are necessary to create new insurance products supplementing governments' actions in response to a pandemic.

1.1 Introduction

An actuary's almost knee-jerk reaction to a pandemic is to model it. However, this temptation has to be moderated and an actuary must consider the implications of other important factors in planning for future pandemics. On the one hand, the legal aspects of insurance policies might outweigh any actuarial considerations. On the other hand, developing models for future pandemics based on the most recent one bears the risk of preparing for the last war. The next pandemic will most certainly be different.

Being a relatively rare event in the Western world, epidemics are a growing public health threat in Africa and Asia. The World Health Organization (WHO) considers the zoonotic diseases, those caused by pathogens transmitted from animals to humans, as the dominant cause of epidemics and pandemics. Coronaviruses (CoV) are a large family of zoonotic viruses that cause illnesses ranging from the common cold to more severe diseases such as the Severe Acute Respiratory Syndrome coronavirus

M. C. Boado-Penas (✉) · Ş. Şahin
Department of Mathematical Sciences, University of Liverpool, Liverpool, UK
e-mail: carmen.boado@liverpool.ac.uk

Ş. Şahin
e-mail: sule.sahin@liverpool.ac.uk

J. Eisenberg
FAM, TU Wien, Wien, Austria
e-mail: jeisenbe@fam.tuwien.ac.at

© The Author(s) 2022
M. C. Boado-Penas et al. (eds.), *Pandemics: Insurance and Social Protection*,
Springer Actuarial, https://doi.org/10.1007/978-3-030-78334-1_1

(SARS-CoV) and Middle East respiratory syndrome coronavirus (MERS-CoV). The outbreak of SARS-CoV, started in China in 2002 and was defeated by disease prevention and control systems (Deng and Peng 2020). MERS-CoV was first reported in Saudi Arabia in 2012 and has since spread to several other countries. Although most of coronavirus infections are not severe, more than 10,000 cumulative cases have been associated with SARS-CoV and MERS-CoV in the past two decades, with mortality rates of 10% and 37% respectively (Huang et al. 2020; Sohrabi et al. 2020; Zhu et al. 2020).

COVID-19—the novel coronavirus pandemic declared as such by the World Health Organization on the 11th of March 2020—has quickly reached an incomparable dimension, and every individual, every government has been caught by surprise fighting against the crisis caused by COVID-19. The pandemic has triggered what is likely to be the deepest global recession since World War II.

The COVID-19 pandemic has strikingly proved that "we are only as safe as the most vulnerable among us".[1] Considering nearly 55% of the world's population do not have access to any sort of financial social protection, and many countries rely on market-based solutions to fill the gap, the recent pandemic demonstrates that the situation does not only hurt the poorest and most vulnerable, it threatens the well-being of the entire global community. Those who are unable to quarantine themselves because of precarious financial situations not only endanger their own lives but also the lives of others. It is clear that if one country does not contain the virus, others are bound to be infected and re-infected (ILO 2021). Therefore, developed countries should provide assistance, not only for altruistic motivations but for self-protection. Similar considerations hold true for the case of total vaccination.

1.2 Discussions from the Perspective of Insurance and Social Protection

1.2.1 Commercial Insurance

Insurance can be defined as a contract under which an insurer or the government agrees to offer a promise of coverage in the event of a specified loss, injury, sickness, or death in exchange for the payment of a specified premium. Whilst injuries are minor and occur in groups of individuals or companies, commercial insurance does an outstanding job of distributing damages that are unforeseen individually but anticipated collectively. Life insurance, pensions, insurance for motor accidents, burglaries, household fires, marine insurance, etc. are examples which fit the definition perfectly. However, significant catastrophic losses that affect large populations are single events which cannot be pooled. War casualties, epidemics/pandemics, coastal flooding triggered by rising sea levels, major tsunamis, earthquakes, big volcanic

---

[1] https://www.ilo.org/global/about-the-ilo/newsroom/news/WCMS_739678/lang--en/index.htm.

activities are some examples of these catastrophic events which might affect entire societies. Then, it becomes the responsibility of the society to share the losses, which must be managed and performed by governments.

The *limits* of the commercial insurance are derived by both the severity and the prevalence of the damages. Considering the extent of losses, insurance companies might balance one type of extreme risk against others, such as forest fires in some place versus earthquakes in others. Additionally, it might be possible to introduce innovative insurance products, or adjusting the current ones to serve the aim to cover specific parts of the *uninsurable losses*. COVID-19, in this regard, tested the extent and the effectiveness of the current insurance products. Although it might not be the aim or duty of the commercial insurance to protect societies from possibly unbounded damages, it could still contribute to pandemic response (i.e. social protection provided by governments and international organisations) by dealing with some aspects of the damages caused by the pandemic. Some possible innovative responses might include life insurance products being adapted or converted to include deaths caused by the pandemic, or occupational sickness insurance might be extended to include new disease and conditions that have been caused by SARS-CoV-2 infections, etc. In the case of the COVID-19 pandemic, in particular, applying these innovations retrospectively in the existing insurance products may not necessarily mean an overall increase in costs to the insurance industry. For example, an increase in death claims might be compensated by a decrease in future pension/annuity claims.

A tsunami-like amount of business interruption, travel, and medical treatment claims crashed over the insurance sector in response to the COVID-19 pandemic. Whilst some insurance companies had already added pandemic exclusion clauses to their policies following the SARS-CoV epidemic in 2003, others did not incorporate a clear defined list of possible diseases to be covered, leaving some ambiguity about individual extent of cover.

Insurance associations all over the world declared that pandemic coverage had been optional and most policyholders chose to save the money and did not purchase this type of business interruption insurance. In France, the financial regulator supervising both banking and insurance, Autorité de contrôle prudentiel et de résolution (ACPR), on the 22nd June 2020 made public that 93.3% of insurance policies did not cover the pandemic, 2.6% did and 4.1% were unclear. For instance in Germany, many pre-COVID-19 insurance policies merely referred to the Infectious Diseases Protection Act, Infektionsschutzgesetz (2001) (IfSG), and the diseases listed therein. This ambiguity created by the fact that COVID-19 was not explicitly mentioned in IfSG led to a number of court cases. On the 1st of October 2020, the Munich Regional Court ruled that the insurance company Versicherungskammer Bayern had to pay out 1.01 million euros business interruption insurance to Augustinerkeller, a famous restaurant in Munich. However, after the ruling the Berlin-based German Insurance Association (GDV) stated that the Munich decision would have no implications for other pending cases.

The refusal of some insurance companies to pay pandemic-related claims has eroded trust in the sector in general. Numerous court proceedings followed as claimants sought retribution in numerous countries. The lawyers, representing the interests of the insurance sector, insisted that businesses could not claim for losses resulting from nationwide lockdowns as it would be catastrophic for the industry.[2] For instance, Michael Crane, a lawyer for insurance company QBE QBE.AX, stated during one of the hearings that a pandemic had been foreseen, however a government's response in the form of introducing a nationwide lockdown had been an "inconceivable" measure before 2020. This means in particular, that the actuarial equivalence principle (expected future premium payments should be equal to expected future benefits) does not work here. The premia charged by insurance companies did not contain the possibility of protracted lockdowns, i.e. the customers did not pay for the risk of business interruption to the extent that was widely experienced during COVID-19 pandemic. In this way, COVID-19 has shifted the insurability question from the actuarial to the legal sphere.

Even one year after the start of the COVID-19 pandemic there is still no clear line of jurisprudence on the gray areas of contracts containing a list of diseases in which COVID-19 is not mentioned. As a consequence, from South Africa to USA legal decisions have been taken in favour of both insurers and policyholders. And as of March 2021, it is far from clear which legal trend, if any, will eventually prevail in this battle.

It is not surprising, now in light of COVID-19, that new contracts engaged after the beginning of the pandemic often contain a pandemic/epidemic exclusion clause. Many insurance companies are not ready to undertake the risk of a pandemic. The reason is that when this rare event happens losses occur for everybody—pandemics do not respect geographical borders—therefore collective risk sharing and balancing over time are not working for a pandemic. As an example, the Wimbledon Tennis tournament had a business interruption policy with a pandemic insurance clause which costed them around £1.5 million per year. This annual premium had been paid for 17 years before a claim occurred. Small and medium-sized businesses may not be able to afford such-a-high premia over a substantial number of years. For instance in the UK, before COVID-19 many small businesses had business interruption policies that enabled them to claim up to a maximum of between £50,000–£100,000 in case of a pandemic and lockdowns. This cap essentially reduced the premia. However, as of March 2021 the actual losses in most cases exceed these amounts by a multiple. Besides, some insurance companies are reportedly trying to reduce their losses and to pay claims as quickly as possible by offering very low settlement or interim payments. The news organisation Reuters reported on a café in East London getting a settlement offer totalling £13.[3]

There is a clear demand for insurance coverage for the case of a new epidemic/pandemic. Thus, insurance companies are confronted with the challenge of developing innovative policy structures and mitigation strategies for both public and private sectors.

---

[2] In this case, the contracts are subject to moral hazard since governments might influence the claim payments through national lockdowns.

[3] https://www.reuters.com/article/uk-britain-insurers-idUSKCN2AT3B6.

Partnerships and collaborations between governments and (re)insurance companies are needed to enable insurance protection for pandemic risks that would be otherwise uninsurable. In this regard, a parametric pandemic insurance design for governments has been introduced by Boado-Penas et al. (2021). As for the practical examples, in the UK Flood Re is a joint initiative between the government and insurers to include flood cover in household insurance policies in an affordable way. In practice, every insurer that offers home insurance in the UK must pay a fee into the Flood Re Scheme and can choose to pass the risk to Flood Re for a fixed price. This keeps the premia down for consumers, and protects insurance providers from very large exposures.

1.2.2 The Role of the Governments and Social Protection

Governments must act promptly to make rapid progress toward collectively financed, comprehensive, and permanent social-protection systems which are already wretchedly inadequate at safeguarding the lives and livelihoods of their citizens. Having access to health insurance, unemployment, and sickness benefits is crucial to protect vulnerable groups and thus the whole community (ILO 2021).

During the COVID-19 pandemic, the economic shutdown and the subsequent business losses led to an unprecedented rise in the number of unemployed people. In general, Eurostat (2020), the estimated income losses at the EU level represent around 5% of total earnings and its distribution is very unequal. This inequality is present both between nations, and within them, with the greatest effect realised by the most vulnerable sub-groups of the working population. Emergency legislation in some countries made significant concessions to increase the capacity of their health systems and provide relief to those citizens and sectors that are particularly impacted by the coronavirus crisis. Spain's government, for example, launched a monthly basic income scheme up to €1,015 for the most vulnerable households in June 2020. The programme supported around 850,000 households. In the UK, social measures such as the Coronavirus Job Retention Scheme[4] or the Self-Employment Income Support Scheme were introduced so that a portion of usual monthly wage costs was paid for the time the employee is on furlough (Machin 2021).

As a result of the economic recession caused by the coronavirus crisis, most major economies lost at least 2.4% of their GDP over 2020. In developing nations (excluding China) the pandemic crisis led to a fall in nominal US dollar GDP of 10% while the private finance dropped by $700 billion in 2020 (OECD 2021).

Governments usually include several financial programmes in their budget which target vulnerable social groups, i.e. those who are disproportionately exposed to risk. Due to prolonged and strict lockdowns which will be discussed in Chaps. 7, 8 and 9 in various dimensions, unemployment benefits have become vital for millions to survive

---

[4] Through the Coronavirus Job Retention Scheme, the UK Government have committed to reimbursing 80% of employees' wage cost up to £2,500 per worker.

during the COVID-19 pandemic. Persons who have not been considered vulnerable at the start of a pandemic may be pushed to the edge of poverty or even beyond by the loss of their jobs, illness and expensive medical treatment. Consequently, governments are facing a challenge of identifying the vulnerable depending on the current situation and preparing beforehand feedback response strategies, see for instance The Lancet Editorial (2020).

At the same time, international organisations are working closely with global experts and governments to provide advice to countries on measures to protect health and bolster economic recovery. For the most vulnerable countries, the World Bank, as we can see in Chap. 7, has approved some financial emergency support to urgent needs in the wake of the pandemic. Also, the World Bank together with the International Monetary Fund urged G20 to establish the Debt Service Suspension Initiative, so that emerging countries concentrate their resources on fighting the pandemic and safeguarding the lives.

1.3 Listening to the *Wind of Change*

The 21st century with its urbanisation, internationalisation and overpopulation has created the optimal conditions for novel infectious diseases to multiply, and spread. The increasing threat of experiencing pandemics more often in the near future will force many institutions, the insurance sector being one of the pioneers, to propose path-breaking solutions.

It is not a coincidence that the origin of actuarial modelling goes back to 14th century and the Black Death, an outbreak of medieval plague which was believed to kill 30–50% of Europe's population. That is when the City of London started recording the deaths and produced regular statistics of mortality with the aim of recognising the patterns and use past data to predict the future. Seven centuries have passed, and yet not much has changed concerning the data as it once again became the driving force of actuarial modelling in a pandemic.

On the other hand, a core change has occurred to the perception of the number of pandemic victims. Whilst in the 14th or even in the 20th century with no antibiotics the general attitude towards widespread death was rather fatalistic, in the 21st century death is considered more and more a technical problem, see for instance The Guardian (2020). The number of COVID-19 deaths affects many people not least because they strongly feel that these deaths have been preventable. Therefore, it is not surprising that there is a high demand for a more structured and more extensive additional social protection during events like COVID-19.

This book provides a collection of interdisciplinary scientific studies that can be used to develop epidemic/pandemic response strategies for both the commercial insurance sector and government provisions (social protection). By putting together innovative mathematical, statistical, actuarial, legal and social academic contributions, along with a review of existing realities, we have listened to the early breezes of the *winds of change* triggered by COVID-19.

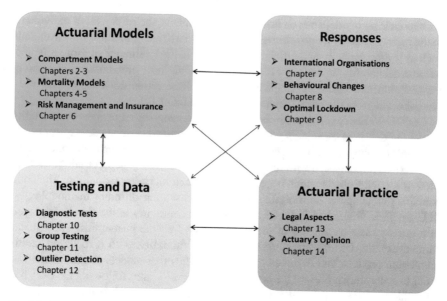

Fig. 1.1 The structure of the book

There are direct and indirect connections between all chapters, as can be seen from the contents presented below. Figure 1.1 lists and collects the chapters of the book under four main parts—*actuarial models, responses, testing and data, actuarial practice.*

Below, we give a summary of each chapter and indicate potential links between them. The book starts with actuarial mathematical modelling of pandemics for two branches—compartment models and mortality models.

In Chap. 2, R. Feng et al. bridge the gap between epidemiological and actuarial models and present insurance product designs to provide healthcare coverage during a pandemic. This chapter starts with an extensive description of the main compartmental models—commonly used in the medical literature—characterised by a system of differential equations in the case of deterministic models or transition probabilities for stochastic models. Then, the authors apply actuarial techniques to COVID-19 data and calculate premia to be paid continuously from (healthy suscepti-ble) policyholders and actuarial reserves for three epidemic models. This chapter also discusses the application of epidemic models for contingency planning and resource allocation.

In Chap. 3, A.D. Wilkie introduces an actuarial model for infections such as COVID-19. The chapter presents variations of an actuarial multiple state model which considers the duration of infection of the newly infected individuals. This is a main distinguishing feature of these models compared to SIR models. The chapter presents empirical results based on the UK data whilst emphasising the possible problems of the use of a model for prediction purposes. The prediction accuracy of such models highly depends on actions of governments, the responses of individuals to the measures taken by governments, and the disease itself, medical improvements such as testing capacity and efficiency, advances in the treatments of those affected, vaccine availability as well as efficacy, and possible new mutations with different transmission and virulence characteristics. All these dimensions are discussed in several chapters of the book (Chaps. 7, 8, 9, 10, 11 and 12).

COVID-19 has sparked research dramatically in many different areas but mortality modelling deserves significant attention considering the heterogeneous effect of the pandemic on population. The recent experience has proved that the pandemics might have various impacts on the mortality of different sexes, age groups, ethnic and socio-economic backgrounds which necessitates advanced mortality modelling. The book contains two chapters on mortality modelling presenting different methodologies. In Chap. 4, L. Regis and P. Jevtić discuss the discontinuity in the trends displayed in mortality rates as a result of the shocks caused by the pandemics. The chapter summarises the current literature on stochastic mortality, with a focus on multi-population models, and explores the characteristics that models should possess in order to accurately represent the behaviour of mortality rates following the COVID-19 pandemic. The authors also introduce a general framework using affine jump-diffusive processes for multi-population models with continuous-time jumps.

Statistical analysis shows that mortality models are often missing systemic risk elements which could capture the impact of the extreme events. In Chap. 5, G. Venter introduces a mortality model for contagious events including pandemics by adding annual jumps to capture both tiny and catastrophic risks. The chapter describes how to model mortality based on parametric regression by fitting smoothing splines across the age, period, and cohort variables in Markov Chain Monte Carlo (MCMC). Furthermore, the chapter examines the Bayesian shrinkage methodology for smoothing as well as the predictive benefits of such smoothing. The analyses have been illustrated using French male and female mortality data.

Insurance is the transfer of risks from individuals or corporations who cannot bear a potential unexpected financial catastrophe. When the number of individuals/contributors is high the insurers spread the financial risks from expensive claims (risk pooling) and can offer a reasonable level of premia. In the unlikely event of a pandemic, losses will happen at the same time for everybody, and consequently, the risk pooling and balancing over time principles are not working. Thus, for the macro level events, like a pandemic, insurance seems to be a suboptimal solution to mitigate risks. In Chap. 6, H. Assa and T. Boonen discuss three risk management setups: risk-sharing, insurance and market platform. They explore the efficiency of insurance schemes in the presence of a macro risk event with significant impact. They come to the conclusion that a social insurance scheme in the form of "Insurance-by-Credit" (no premia payments before the losses occur) outperforms standard insurance by changing the ex-ante view to ex-post: borrowing from the future instead of the past. This risk-sharing concept turns out to be optimal if one neglects credit risk and moral hazard.

Since the World Health Organization declared the COVID-19 outbreak as a Public Health Emergency of International Concern, governments across the world have implemented a variety of policies and strategies to contain the spread of the virus and its negative effects on their citizens. International organisations have supported these efforts through policy and best practice analyses, as well as evidence based policy recommendations. In Chap. 7, M.C. Boado-Penas et al. give an overview of the responses of international organisations, in particular of the World Bank and the EU, to the COVID-19 pandemic. Special attention is given to the guidance of these organisations towards vulnerable groups through changes in social insurance and pension plans.

Chapter 8 by N. Badenes-Plá focuses on changes in individuals' behaviour in different countries arising from a pandemic. While the virus spreads worldwide, the strategies to defeat it cannot be designed without consideration of cultural values and political organisation. This chapter presents an overview of the response and the degree of acceptance of citizens to government interventions to stop the spread of COVID-19 pandemic. The author analyses the behavioural characteristics of the citizens of different countries, toughness of measures, lockdown fatigue, and public trust in their government on the extent of compliance to pandemic measures. Changes in behavioural patterns due to isolation and/or social distancing are described in detail indicating long-term consequences that might affect the pricing of insurance products. For instance, unhealthy habits acquired during lockdowns, or newly acquired or exacerbated mental-health problems may impact on the quality of the remaining life expectancy of individuals.

Once a pandemic happens, it is too late to start planning social protection actions. Governments need to follow the proverb "Repair your cart in December, in July your sledge remember". Harsh suppression measures—that also include the social distancing of the entire population, using Personal Protective Equipment (PPE), closure of schools, leisure and hospitality sectors as well as non-essential retail have been introduced during the COVID-19 pandemic in many countries. However, due to the unprecedented surge in COVID-19 cases and fatalities, after already a few months, most countries were forced to increase the intensity of the lockdown by restricting the suppression rules to limit the spread of the virus. In Chap. 9, J.P. Caulkins et al. consider the problem of optimising the start and the duration of a lockdown, with fixed or variable intensity, considering the more virulent strains of the SARS-CoV-2. One of the important features of the considered model is the recognition of lockdown fatigue. At some point, people start breaking rules no matter how obedient they have been at the beginning of the lockdown, see Chap. 8 for details. The decision to begin or to end a lockdown is always a trade-off between the economic prosperity of a country and the saving of lives. The optimal strategy turns out to be extremely sensitive to the assumptions of the model. The duration of a lockdown depends on its start, and entering a lockdown after a certain number of days since the beginning of the pandemic will feature a different strategy. One can even get the so-called Skiba points, meaning that starting a lockdown at a particular day of the pandemic might provide several completely different optimal strategies.

Before the COVID-19 pandemic, the general public was not familiar with PPE and may not have given sufficient importance to hand hygiene. Since at least the spring of 2020, everyone learnt the new terminology around the SARS-CoV-2 outbreak, see for instance Yale Medicine (2020). Droplet transmission, incubation period, reproduction number—the COVID-19 virus has brought epidemiological language and modelling literally to our living rooms as telework has become the every day reality for many in 2020 and 2021. In 2021, one can recite like a prayer that seven of the known coronaviruses, whose name comes from the crown-like spikes, can infect people, that social distancing, masks and handwashing are the best methods to "flatten the curve". Chapter 10 by S. Dunbar and Y.-W. Tang provides a biochemical overview of the testing procedures necessary to understand and monitor the course of an epidemic. Different biomarkers and possible laboratory specimen for identification of COVID-19 are presented and explained. Furthermore, this chapter discusses the lessons learnt from COVID-19 that would help to speed up the response to a future pandemic. In particular, preventing the high numbers of deaths will require an earlier detection of the disease by using specific biomarkers, targeted treatments, and appropriate triage of patients, particularly those who are susceptible to the most severe course of the disease.

At the beginning of a pandemic, even if the biochemical procedures to follow are clear, the question arises of how to test: individually or in groups. When the resources are scarce and the prevalence level (the ratio of the already infected to the entire population) is still comparatively low, pooled testing, also called group testing, may provide better results than individual testing. Firstly, pooled testing has the potential for very large resource-saving and second, it requires less time than individual testing. In Chap. 11, M. Aldridge and D. Ellis discuss the mathematics behind some one- and two-stage pooling strategies under perfect and imperfect tests, and consider the practical issues in the application of such protocols. The pool testing procedures can be used for instance for surveillance purposes or to monitor the prevalence of the new variants of a disease, which is particularly important if the new variants start to threaten the success of vaccination programmes.

Data collection and analysis play a crucial role in decision-making processes. Defective or deliberately forged data can have fatal consequences. For instance, an underestimation in the number of needed tests can lead to a new upward spiral of a pandemic and, consequently, to more excess deaths. In this line, Chap. 12 by C. Rieser and P. Filzmoser introduces outlier detection techniques applied to COVID-19 pandemic data from different countries. In many applications, outliers are considered the most interesting subject for analysis, because they suspiciously differ from the data majority and might indicate a "contamination" of the given data sets. The data (for instance, the number of newly infected or dead) are regarded as compositions, where the compositional parts are treated as multivariate smooth functions. Here, only relative information expressed in terms of log-ratios between the compositional parts is considered as relevant in the analysis. The presented outlier detection method focuses on the evolution of the data over time rather than on the absolute values. If the evolution of one data set steps out of line compared to similar other data sets (for instance by analysing several different infection testing stations) this clearly indicates a problem with the data cleanliness. Considering the COVID-19 publicly available data from different countries, Chap. 12 explores which countries might have "contaminated" data sets.

COVID-19 has evoked legal challenges regarding the traditional indemnity insurance to protect people and businesses from the losses caused by pandemics. Discrepancies between the expectations of insurers and insureds considering the coverage of the policies seem to be the origin of the disputes as mentioned earlier. The recent evidence, once again, proves that indemnity-based pandemic insurance is obsolete and leads to long delays in payments. In Chap. 13, R. Hillier discusses the legal challenges of insuring against a pandemic. The chapter builds upon the insurance indemnity principle (the insurers cover just the actual loss) and illustrates the pandemic-related problems of the traditional insurance schemes by several court cases that occurred during the COVID-19 pandemic. The author states that a possible solution against business interruption caused by a pandemic could be a parametric insurance, where a pre-agreed payout is made if pre-defined event parameters (triggers) are met. This type of insurance would provide immediate help without a time-consuming loss assessment. Parametric insurance appears to be a simple method of providing quick financial support in combination with the governmental economic packages in the wake of a pandemic. Observing that a parametric design has challenges in terms of defining a robust trigger, the chapter opens a room for possible innovative hybrid insurance products combining indemnity and parametric features.

Last but not least, the closing chapter, Chap. 14 by F. Schiller, analyses the methods and ideas proposed in this book along with their feasibility in times of a pandemic from an actuary's perspective. The chapter discusses the insurability and risk management of extreme events and pandemics in particular and reflects on the potential future consequences of COVID-19 for the insurance sector. The lessons learnt will help the insurers to better adjust and response to the future extreme events. However, the crisis caused by the COVID-19 pandemic has highlighted that the capacities of the financial and (re-)insurance markets are limited, and governmental help in "dark times" is one the whales on whom the world rests. A global disaster cannot be dealt with single-handedly—neither by states nor by insurance companies, no matter the size. Just acting together in a determined and concerted manner can help to tackle the problem.

References

M.C. Boado-Penas, J. Eisenberg, Ş. Şahin, COVID-19: a social reinsurance design. Submitted (2021)

S.Q. Deng, H.J. Peng, Characteristics of and public health responses to the coronavirus disease 2019 outbreak in China. J. Clin. Med. **9**(2), 575 (2020)

Eurostat, Impact of COVID-19 on employment income-advanced estimates, 2020 (2020)

The Guardian, Will coronavirus change our attitudes to death? Quite the opposite. Y.N. Harari, 20 April, 2020. https://www.theguardian.com/books/2020/apr/20/yuval-noah-harari-will-coronavirus-change-our-attitudes-to-death-quite-the-opposite

C. Huang, Y. Wang, X. Li, L. Ren, J. Zhao, Y. Hu, . . ., B. Cao, Clinical features of patients infected with 2019 novel coronavirus in Wuhan, China. Lancet **39**(10223), 497–506 (2020)

Infectious Diseases Protection Act, 2001. https://www.rki.de/DE/Content/Infekt/IfSG/ifsg_node. html

ILO News, COVID-19: social protection systems failing vulnerable groups, 2021. https://www.ilo. org/global/about-the-ilo/newsroom/news/WCMS_739678/lang--en/index.htm, 25 March 2021

The Lancet Editorial, Redefining vulnerability in the era of COVID-19. Lancet **395**(10230), 1089 (2020). https://www.sciencedirect.com/science/article/pii/S0140673620307571

R. Machin, COVID-19 and the temporary transformation of the UK social security system. Critical Social Policy, February 2021

OECD, Global Outlook on Financing for Sustainable Development. OECD, Paris, 2020 (2021). https://www.oecd-ilibrary.org/development/global-outlook-on-financing-for-sustainable-development-2021_e3c30a9a-en

Oxfam Media Briefing, The hunger virus: How COVID-19 is fuelling hunger in a hungry world, July, 2020. https://oxfamilibrary.openrepository.com/bitstream/handle/10546/621023/mb-the-hunger-virus-090720-en.pdf

C. Sohrabi, Z. Alsafi, N. O'Neill, M. Khan, A. Kerwan, A. Al-Jabir, . . ., R. Agha, World Health Organization declares global emergency: a review of the, novel coronavirus (COVID-19). Int. J. Surg. **76**(71–76), 2020 (2019)

Yale Medicine, Our new COVID-19 vocabulary—what does it all mean?, 7 April 2020. https://www.yalemedicine.org/news/covid-19-glossary

N. Zhu, D. Zhang, W. Wang, X. Li, B. Yang, J. Song, . . ., W. Tan, A novel coronavirus from patients with pneumonia in China. New England J. Med. **382**(8), 727–733, 2019 (2020)

Open Access This chapter is licensed under the terms of the Creative Commons Attribution 4.0 International License (http://creativecommons.org/licenses/by/4.0/), which permits use, sharing, adaptation, distribution and reproduction in any medium or format, as long as you give appropriate credit to the original author(s) and the source, provide a link to the Creative Commons license and indicate if changes were made.

The images or other third party material in this chapter are included in the chapter's Creative Commons license, unless indicated otherwise in a credit line to the material. If material is not included in the chapter's Creative Commons license and your intended use is not permitted by statutory regulation or exceeds the permitted use, you will need to obtain permission directly from the copyright holder.

Chapter 2
Epidemic Compartmental Models and Their Insurance Applications

Runhuan Feng, José Garrido, Longhao Jin, Sooie-Hoe Loke, and Linfeng Zhang

Abstract Our society's efforts to fight pandemics rely heavily on our ability to understand, model and predict the transmission dynamics of infectious diseases. Compartmental models are among the most commonly used mathematical tools to explain reported infections and deaths. This chapter offers a brief overview of basic compartmental models as well as several actuarial applications, ranging from product design and reserving of epidemic insurance, to the projection of healthcare demand and the allocation of scarce resources. The intent is to bridge classical epidemiological models with actuarial and financial applications that provide healthcare coverage and utilise limited healthcare resources during pandemics.

2.1 Introduction

The COVID-19 pandemic has affected the insurance industry in many ways. Some notable impacts include a surge in insurance digitisation, volatile capital markets, and disruption in supply chain. These issues have prompted researchers to think beyond the standard actuarial framework to find ways to tackle them. While there has been extensive research on the transmission dynamics in the epidemiology and

R. Feng · L. Jin · L. Zhang
Department of Mathematics, University of Illinois Urbana-Champaign, Urbana, IL 61801, USA
e-mail: rfeng@illinois.edu

L. Jin
e-mail: longhao2@illinois.edu

L. Zhang
e-mail: lzhang18@illinois.edu

J. Garrido (✉)
Department of Mathematics and Statistics, Concordia University, Montreal, QC H3G 1M8, Canada
e-mail: jose.garrido@concordia.ca

S.-H. Loke
Department of Mathematics, Central Washington University, Ellensburg, WA 98926, USA
e-mail: sooiehoe.loke@cwu.edu

© The Author(s) 2022
M. C. Boado-Penas et al. (eds.), *Pandemics: Insurance and Social Protection*,
Springer Actuarial, https://doi.org/10.1007/978-3-030-78334-1_2

13

medical literature, traditional actuarial work has largely focused on mortality and morbidity rates using classical frequency and severity analysis. Actuarial life tables and mortality models lack flexibility and robustness to describe the rapidly changing environment during a pandemic. To that end, we explore the epidemiology literature and combine some of the commonly used models with actuarial methodologies.

In the past decade, some developments in the actuarial literature have intended to fill the gap between these two fields. A recent survey article by Feng et al. (2020) offers an overview of several approaches for infectious disease modelling such as compartmental, network, or agent–based models, and discusses their applications to epidemic and cyber insurance coverages. These novel applications represent efforts to integrate medical modelling with actuarial techniques.

This chapter focuses on compartmental models commonly used in the medical literature and their applications in the context of epidemic insurance and pandemic risk management. The organisation of this chapter is as follows. Section 2.2 introduces some important compartmental models, such as the celebrated SIR and SEIRD models. These are characterised by a system of differential equations, in the case of deterministic models, or transition probabilities for stochastic models. Section 2.3 presents an overview of some designs of epidemic insurance plans using compartmental models from the preceding section. Common actuarial concepts including annuities, benefits, and insurer's reserve levels are explored. The section concludes with case studies using the actual COVID-19 data set. Section 2.4 illustrates another application of compartmental models in the subject of allocation of resources during a pandemic. In particular, we project the demand for critical medical resources and use actuarial concepts of capital allocation to optimally stockpile resources prior to a pandemic and to ration limited existing resources during the pandemic.

2.2 Compartmental Models in Epidemiology

The modelling of epidemics has a rich history and dates back at least to Aristotle's work; see Brauer and Castillo-Chavez (2012) for a historical account. A popular framework is the so-called compartmental modelling, where the entire population is segregated into multiple compartments which correspond to different stages of a disease. Then the dynamics in the system is studied via a system of differential equations.

2.2.1 SIR Model

Consider a population of size $N(t)$ which is indexed by time t. Based on the seminal work of Kermack and McKendrick (1927), the SIR model considers the following three compartments: **S**usceptible, **I**nfected, and **R**emoved. There are many variations of the SIR model; in what follows we lay out the assumptions for the most basic SIR model. First, there are no births or immigration, i.e. no one is added to the susceptible compartment, and the population mixes homogeneously. Second, the transmission dynamic is driven by the *law of mass action*, which means that the rate of secondary infection depends on both the size of the susceptible class and that of the infected class. The rate β represents the number of contacts per individual per time unit to transmit the disease. Hence $\beta I(t)$ can be interpreted as the rate of contagious contacts. Since the disease is only transmittable in a contact between infected and susceptible individuals, then $\beta I(t)S(t)/N$ is the actual rate of transmission. Third, the size of the infected class is subject to exponential decay. The rate α represents the proportion of the infected class being removed and hence $\alpha I(t)$ is the rate of decrease in the infected class. Lastly, this simple model does not distinguish causes of removal, which may include recovery, immunity, or death. Once in the removed class, the $R(t)$ individuals can no longer return to the susceptible or the infected classes. The following figure summarises some of the assumptions in the model:

Based on the assumptions, the population size is fixed, that is $N(t) = N$. Moreover, the assumptions lead to the following system of differential equations

$$S'(t) = -\beta I(t)\frac{S(t)}{N}, \tag{2.1}$$

$$I'(t) = \beta I(t)\frac{S(t)}{N} - \alpha I(t), \tag{2.2}$$

where $S(t)$ and $I(t)$ are the number of susceptible and infected individuals, respectively. The number of removed individuals is thus $N - S(t) - I(t)$.

A main concern in epidemiology modelling is whether a disease will spread upon its introduction. Observe from (2.2) that the disease spreads (i.e. $I'(t)$ is positive) if $\beta S(t)/N - \alpha > 0$, and if $\beta S(t)/N - \alpha < 0$ the disease dies out. This explains the relevance of a key quantity called the effective (or general) reproductive number,

$$R_t = \frac{\beta}{\alpha}\frac{S(t)}{N},$$

which represents the average number of secondary infections due to a single infectious individual at a given time t. It is also worth noting that $1/\beta$ is the average time between contagious contacts and $1/\alpha$ is the average time until removal. Then it is easy to see that β/α is the average number of contacts by an infected person with others before removal. When $t = 0$, R_0 is known as the *basic reproduction number*, a common measure used to determine if a disease will spread out during the early phase of the outbreak. If $R_0 > 1$, the disease will start to spread, but not if $R_0 < 1$. Zhao et al. (2020) gives a preliminary estimate of R_0 for the coronavirus pandemic in China (from Jan. 10 to Jan. 24, 2020) to be between 2.24 and 3.58.

2.2.2 Other Compartmental Models

Here is a brief description of additional epidemic models that are commonly used in disease modelling. Since mortality analysis is based on ratios instead of absolute counts, consider the deterministic functions $s(t)$, $i(t)$ and $r(t)$ that denote, respectively, the fraction of the population in each class S, I and R. Dividing equations (2.1)–(2.2) by the constant total population size N yields

$$
\begin{aligned}
s'(t) &= -\beta i(t)\, s(t), && t \geq 0, \\
i'(t) &= \beta i(t)\, s(t) - \alpha i(t), && t \geq 0, \\
r(t) &= 1 - s(t) - i(t), && t \geq 0,
\end{aligned}
\tag{2.3}
$$

where $s(0) + i(0) = 1$.

These ratio functions can be interpreted as the probability of an individual being susceptible, infected or removed from infected class, respectively, at time t. Note that movements between compartments depend on their relative sizes. Hence these probabilities correspond to mutually dependent risks in the SIR model, as opposed to the usual independent multiple decrements in life insurance models (see, for example, Chap. 8 of Dickson et al. (2013)).

To simplify the notation, the time variable will be dropped whenever it is not strictly necessary. Also, we will use the convention that uppercase letters like S, I and R denote the number of individuals in the compartments whereas lowercase letters like s, i and r refer to the proportion of the population in these compartments.

2.2.2.1 SIS Model

In the SIR model, compartment R represents removal, either by achieving full immunity or by death. Some diseases, such as AIDS, have no cure, and subsequently the infected individuals who have recovered are susceptible to the disease again. This phenomenon motivates a simpler class of models, called SIS, with only two compartments: S and I.

In the most basic SIS model, due to Kermack and McKendrick (1932), no births or immigration can occur, so that the total population $N = S(t) + I(t)$ remains constant. Denoting the infection rate by β and the recovery rate by α, as above, the corresponding differential equations are given by

$$
\begin{aligned}
s' &= -\beta s i + \alpha i, \\
i' &= \beta s i - \alpha i.
\end{aligned}
\tag{2.4}
$$

Since $s + i = 1$, this yields a logistic differential equation

$$
i' = (\beta(1 - i) - \alpha)i,
\tag{2.5}
$$

hence, unlike the SIR model, the SIS admits explicit analytical solutions to the system of equations in (2.4):

$$s(t) = \frac{\left(\frac{1-\frac{\alpha}{\beta}}{i(0)} - 1\right)e^{-(\beta-\alpha)t} + \frac{\alpha}{\beta}}{1 + \left(\frac{1-\frac{\alpha}{\beta}}{i(0)} - 1\right)e^{-(\beta-\alpha)t}},$$

$$i(t) = \frac{1-\frac{\alpha}{\beta}}{1 + \left(\frac{1-\frac{\alpha}{\beta}}{i(0)} - 1\right)e^{-(\beta-\alpha)t}}. \tag{2.6}$$

2.2.2.2 SEIRD Model

To predict the evolution of critical cases, Hill (2020) develops an app based on a variation of the SIR model, called SEIRD. It includes seven mutually exclusive compartments, namely, the susceptible (S), exposed (E), mildly infected (I_1), those infected with hospitalisation (I_2), infected with intensive care (I_3), recovered (R), and the deceased (D). This SEIRD model is characterised by a set of ordinary differential equations that describe population flows among all aforementioned compartments:

$$
\begin{aligned}
S' &= -(\beta_1 I_1 + \beta_2 I_2 + \beta_3 I_3)S, \\
E' &= (\beta_1 I_1 + \beta_2 I_2 + \beta_3 I_3)S - \gamma E, \\
I'_1 &= \gamma E - (\delta_1 + p_1)I_1, \\
I'_2 &= p_1 I_1 - (\delta_2 + p_2)I_2, \\
I'_3 &= p_2 I_2 - (\delta_3 + \mu)I_3, \\
R' &= \delta_1 I_1 + \delta_2 I_2 + \delta_3 I_3, \\
D' &= \mu I_3.
\end{aligned}
\tag{2.7}
$$

All the parameters in this system of equations admit a clinical interpretation; β_i, $i = 1, 2, 3$, is the transmission rate to the infected class I_i; $1/\gamma$ is the average latency period; $1/\delta_i$, $i = 1, 2, 3$, is the average duration of infection in class I_i, before recovery to the class R; then p_i, $i = 1, 2, 3$, represents the rate at which conditions worsen and individuals require healthcare at the next level of severity; μ is the transition rate from the most severe cases in class I_3 to the deceased class D.

The system of ordinary differential equations above represents a decomposition of the instantaneous change in the population into those in each compartment. For example, the first equation shows that the instantaneous rate of reduction in the number of susceptible, $-S'$ matches the sum of the rates of infection due to contacts with the infected in all classes, $\beta_1 I_1 S + \beta_2 I_2 S + \beta_3 I_3 S$. The products are due to the law of mass action in biology. For example, the rate of secondary infection by the mildly infected, $(\beta_1 N)I_1(S/N)$ can be interpreted as the number of "adequate" contact each infected individual makes to transmit the disease $\beta_1 N$, multiplied by the number of infectives I_1, multiplied by the percentage of contacts with a susceptible, S/N. All other equations can be explained in a similar way.

2.2.2.3 Stochastic SIR Model

Stochastic compartmental models form another popular framework in epidemiology modelling. These are natural extensions of the deterministic models presented above. Here we assume a continuous time Markov chain framework. Another approach is to add a Brownian perturbation to each compartment (see e.g. Sect. 4 of Allen (2017)), but it is not discussed here.

For an arbitrary time interval $[t, t + dt]$, the probability of an infection and the probability of recovery are given by:

$$P((S(t + dt), I(t + dt)) - (S(t), I(t)) = (-1, 1)) = \frac{\beta}{N} S(t)I(t) \, dt + o(dt),$$
$$P((S(t + dt), I(t + dt)) - (S(t), I(t)) = (0, -1)) = \alpha I(t) \, dt + o(dt) \tag{2.8}$$

Figure 2.1 shows the dynamics in compartments S and I, respectively, comparing the deterministic and stochastic SIR models.

Another random variable of interest is the duration of the epidemic, defined as:

$$T = \inf\{t > 0 : I(t) = 0\}. \tag{2.9}$$

In other words, this is the first instant when there are no more infectives in the population (or the time of the disease–free state). Other related random variables are the final size of the susceptible population ($S(T)$) and the area under the trajectory of the stochastic processes $\int_0^T S(u) \, du$ and $\int_0^T I(u) \, du$. For a thorough discussion of stochastic epidemic models and methods for their statistical analysis, see e.g. Andersson and Britton (2012).

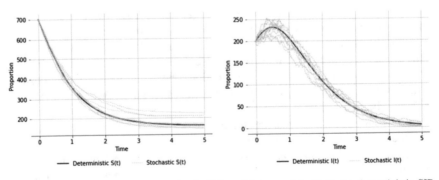

Fig. 2.1 Ten sample paths of the stochastic SIR model (orange) in (2.8), the deterministic SIR model (black) in (2.1) and (2.2) for $\beta = 3$ and $\alpha = 1.5$

2.2.2.4   More Compartmental Models

Several other compartmental models have appeared in the epidemiology literature, but are not covered in this chapter. For example, the multi–group model separates the population into different groups, allowing for varying infection and recovery rates along groups. It can be applied to bordering counties, states or countries. Another model pertinent for COVID-19 applications is the quarantine–isolation model; it describes the effect of isolating susceptible individuals from the infected individuals. When no vaccine is available, this is perhaps the only measure available to governments to contain the contagion. Interested readers can refer to Hethcote (1978) and Brauer and Castillo-Chavez (2012) for a plethora of sophisticated epidemic models.

2.3   Epidemic Insurance

The idea of designing an insurance coverage against the financial impact due to infectious diseases is similar to what motivates coverage against other contingencies, such as accidental death or destruction of properties. Where it differs significantly from property and casualty insurance, at least from an actuarial point of view, is in the time–varying reference groups, such as the number of policyholders bearing the premiums and the number of policyholders eligible for compensation, which evolve quickly over time through an epidemic.

To illustrate these differences, this section first reviews some of the insurance policies and models proposed in Feng and Garrido (2011), using the basic SIR model, and subsequently quantifies infection risk by combining epidemiological and actuarial methodologies. The reserve level of an epidemic insurer is also studied, and using historical COVID-19 data, several case studies are presented.

2.3.1   Annuities and Insurance Benefits

Assume that an infectious disease insurance plan collects premiums continuously from susceptibles, as long as they remain healthy and susceptible. In the meantime, medical expenses are paid continuously to infected policyholders during the whole period of treatment, or until death.

Using the *Equivalence Principle* to determine level net premiums, that is

$$\mathbb{E}[\text{present value of benefits}] = \mathbb{E}[\text{present value of benefit premiums}] \qquad (2.10)$$

and defining the actuarial present value (APV) of a t-year annuity of benefit payments $\bar{a}_{\bar{t}|}^{i} \triangleq \int_{0}^{t} e^{-\delta x} i(x) \, dx$, and the APV of a t-year annuity of premium payments $\bar{a}_{\bar{t}|}^{s} \triangleq \int_{0}^{t} e^{-\delta x} s(x) \, dx$, the level net premium for a unit annuity claim payment plan is given by $\bar{P}\left(\bar{a}_{\overline{\eta}}^i\right) \triangleq \frac{\bar{a}_{\overline{\eta}}^i}{\bar{a}_{\overline{\eta}}^s}$. Here, δ represents the force of interest. As in life insurance, identities linking these APVs can be derived; see Feng and Garrido (2011). For instance, in the SIR model over an infinite term, we have

$$\left(1 + \frac{\alpha}{\delta}\right) \bar{a}_{\overline{\infty}}^i + \bar{a}_{\overline{\infty}}^s = \frac{1}{\delta}.$$

The intuitive interpretation is that, if each insured in the whole insured population is provided with a unit perpetual annuity, the APV of payments to class S is given by $\bar{a}_{\overline{\infty}}^s$ and the APV of payments to class I is given by $\bar{a}_{\overline{\infty}}^i$.

From this relation the net level premium for a policy of an infinite term with both premium and claim annuity is given by:

$$\bar{P}\left(\bar{a}_{\overline{\infty}}^i\right) = \frac{\bar{a}_{\overline{\infty}}^i}{\bar{a}_{\overline{\infty}}^s} = \frac{\delta\, \bar{a}_{\overline{\infty}}^i}{1 - (\delta + \alpha)\bar{a}_{\overline{\infty}}^i}.$$

If instead, the infectious disease insurance pays a lump sum compensation when an insured person is diagnosed infected, and immediately hospitalised, then medical expenses are to be paid immediately in a lump sum, terminating the insurance plan's obligation. The APV of benefit payments, denoted $\bar{A}_{\overline{\infty}}^i$, is thus given by

$$\bar{A}_{\overline{\infty}}^i \triangleq \beta \int_0^\infty e^{-\delta t} s(t)\, i(t)\, dt\,, \tag{2.11}$$

since the probability of being newly infected at time t is $\beta\, s(t)\, i(t)$. In the SIR model, this leads to additional useful identities:

$$\frac{1}{\delta} \bar{A}_{\overline{\infty}}^i + \bar{a}_{\overline{\infty}}^s = \frac{1}{\delta} s(0)\,, \tag{2.12}$$

and

$$\frac{1}{\delta} i(0) + \frac{1}{\delta} \bar{A}_{\overline{\infty}}^i = \frac{\alpha}{\delta} \bar{a}_{\overline{\infty}}^i + \bar{a}_{\overline{\infty}}^i\,. \tag{2.13}$$

Then net level premium $\bar{P}(\bar{A}_{\overline{\infty}}^i)$ for an infinite term insurance plan with lump sum compensation and annuity premium payments is given by the Equivalence Principle:

$$\bar{P}\left(\bar{A}_{\overline{\infty}}^i\right) \triangleq \frac{\bar{A}_{\overline{\infty}}^i}{\bar{a}_{\overline{\infty}}^s} = \frac{(\alpha + \delta)\bar{a}_{\overline{\infty}}^i - i(0)}{1 - (\alpha + \delta)\bar{a}_{\overline{\infty}}^i}\,. \tag{2.14}$$

Finally, if the coverage includes also a death benefit, say of one monetary unit, paid immediately at the moment of death, then its APV, denoted by $\bar{A}_{\overline{\infty}}^d$, is given by

$$\bar{A}^d_{\overline{\infty|}} \triangleq \alpha \int_0^\infty e^{-\delta t} i(t)\, dt = \alpha\, \bar{a}^i_{\overline{\infty|}}.$$

Therefore, the net level premium for an infinite term plan with both, a unit lump sum death benefit and health–care claim benefits, is:

$$\bar{P}\left(\bar{a}^i_{\overline{\infty|}} + \bar{A}^d_{\overline{\infty|}}\right) \triangleq \frac{\bar{a}^i_{\overline{\infty|}} + \bar{A}^d_{\overline{\infty|}}}{\bar{a}^s_{\overline{\infty|}}} = \frac{\delta(1+\alpha)\bar{a}^i_{\overline{\infty|}}}{1 - (\alpha + \delta)\bar{a}^i_{\overline{\infty|}}}.$$

Similarly, the net level premium for a plan with both coverages, a lump sum benefit for hospitalisation costs and a lump sum death benefit is given by:

$$\bar{P}\left(\bar{A}^i_{\overline{\infty|}} + \bar{A}^d_{\overline{\infty|}}\right) \triangleq \frac{\bar{A}^i_{\overline{\infty|}} + \bar{A}^d_{\overline{\infty|}}}{\bar{a}^s_{\overline{\infty|}}} = \frac{(\delta + \alpha + \delta\alpha)\bar{a}^i_{\overline{\infty|}} - i(0)}{1 - (\alpha + \delta)\bar{a}^i_{\overline{\infty|}}}.$$

The above net premiums are expressed in terms of $\bar{a}^i_{\overline{\infty|}}$, which is a Laplace transform of $i(t)$. Although an implicit integral solution is known in the SIR model, no general explicit solution is available for $s(t)$ and $i(t)$. Different numerical methods and approximations which have been proposed provide satisfactory solutions for insurance applications, even for finite term policies; see Feng and Garrido (2011).

2.3.2 Reserves

Reserves are made of assets set aside by an insurer in anticipation of claim payments in the future. It can be determined prospectively at any time t, as the accumulated value of future premiums less that of future benefits. Alternatively, reserve can also be defined in a retrospective manner (see, for example, Chap. 7 of Dickson et al. (2013)). Reserves are a critical tool for insurers to measure their liabilities towards policyholders. When the reserve is adequately set, the insurer should have sufficient funds to cover claims as they become due. In classical life insurance, reserves build up from the beginning of the policy term, as the insurer accumulates premiums, to ultimately run out at the end of the policy term, when all benefits have been paid out to the policyholders. In other words, the reserve as a function of time, typically exhibits a bell shape. However, for epidemic insurance, the reserve function may exhibit quite different patterns, due to the dynamics of an epidemic. This section presents different shapes of reserve functions, and the conditions under which they arise in an epidemic insurance using the SIR and SIS models.

2.3.2.1 SIR Model

Here assume that susceptible individuals pay premiums at a constant rate π, and once infected, the insurer pays hospitalisation benefits, say at a constant rate of 1. For this particular policy, the insurer's reserve level is given by:

Table 2.1 Possible shapes of the reserve function

Shape of $V(\pi, t)$	Interval for values of π
Increasing concave	$\left[\frac{1}{R_\infty} - 1, \infty\right)$
Increasing concave-then-convex	$\left[\frac{1}{R_{t_m}} - 1, \frac{1}{R_\infty} - 1\right)$
Nonmonotonic concave-then-convex	$\left[\frac{1}{R_0} - 1, \frac{1}{R_{t_m}} - 1\right)$
Nonmonotonic convex	$\left[0, \frac{1}{R_0} - 1\right)$

$$V(\pi, t) = \pi \int_0^t s(x)dx - \int_0^t i(x)dx, \qquad (2.15)$$

where, for simplicity, we take $\delta = 0$. Feng and Garrido (2011) shows that there are four possible shapes of $V(\pi, t)$, as a function of time. It turns out that the shape is dictated by the effective reproduction number, R_t, as summarised in Table 2.1.

In Table 2.1, the time t_m is defined as to satisfy the equation $s(t_m) = \exp\left(1 - \frac{\beta}{\alpha}c\right)$, where $c = 1 - \frac{\alpha}{\beta}\ln s(0)$. For further details, readers can consult Appendix 5 of Feng and Garrido (2011). It is worth noting that a related quantity from the table, namely $1 - \frac{1}{R_0}$, is called the *herd immunity* threshold; see Fine et al. (2011).

2.3.2.2   SIS Model

Now consider the insurer's reserve function (2.15) for a SIS model. The following results concern its first and second derivatives.

Proposition 2.1 *Assume that the infection rate exceeds the recovery rate, $\beta > \alpha$. Then,*

(1) $V(\pi, t)$ is non-decreasing on $\pi \in [\frac{1}{s(\infty)} - 1, \infty)$ and non-increasing on $\pi \in (-\infty, \frac{1}{s(\infty)} - 1)$ if $s(0) > \frac{\alpha}{\beta}$.

(2) $V(\pi, t)$ is non-decreasing on $\pi \in [\frac{1}{s(0)} - 1, \infty)$ and non-increasing on $\pi \in (-\infty, \frac{1}{s(0)} - 1)$ if $s(0) \leq \frac{\alpha}{\beta}$.

Convexity results for $V(\pi, t)$ are straightforward since:

$$V''(\pi, t) = \pi s'(t) - i'(t) = (\pi + 1)i(t)(\alpha - \beta s(t)).$$

Proposition 2.2 *If the infection rate exceeds the recovery rate, $\beta > \alpha$, then*

(1) $V(\pi, t)$ is concave on $\pi \in [-1, \infty)$ and convex on $\pi \in (-\infty, -1)$ if $s(0) > \frac{\alpha}{\beta}$.

(2) $V(\pi, t)$ is convex on $\pi \in [-1, \infty)$ and concave on $\pi \in (-\infty, -1)$ if $s(0) \leq \frac{\alpha}{\beta}$.

2.3.3   Further Extensions

Feng and Garrido (2011) has motivated additional work on deterministic insurance models. Perera (2017) considers control strategies in the simple SIR model as well as the variation of premiums with respect to the model parameters. Then Nkeki and Ekhaguere (2020) constructs the SIDRS model and studies its insurance applications. Billard and Dayananda (2014a) and Billard and Dayananda (2014b) develop a multi–stage HIV/AIDS model, where a waiting time distribution models the time one individual holds in one state. Premiums are defined for different insurance functions, and health–care cost adjustments are also included. Shemendyuk et al. (2019) investigates the deterministic and stochastic SIR models with multiple centres and migration fluxes. The optimal health–care premium is determined by considering different vaccine allocation strategies. Basic ideas in optimal resource allocation and contingency planning are explored in greater detail in Sect. 2.4.

Building on the work of Lefèvre et al. (2017), that explores the interplay between stochastic epidemiology and actuarial modelling, Lefèvre and Picard (2018a) generalised the SIR model to a controlled epidemic model, where the infectious are quarantined to ease the severity of the disease, and studies the epidemic outcomes and path integrals in terms of pseudo–polynomials. Then Lefèvre and Simon (2018) considers cross–infection between two linked populations. A general approach to study the Laplace transform of these integral functionals was developed by Lefèvre and Picard (2018b). More recently, Lefèvre and Simon (2020) proposes a general block–structured Markov processes for epidemic modelling.

2.3.4   Case Studies: COVID-19

This subsection illustrates a practical application of Sect. 2.3.2. Applying historical data, we use three epidemic models (SIR, SEIRD, and stochastic SIR) to describe the outbreak of COVID-19 and investigate their possible insurance applications. The data comes from Worldometer (2020), where as of Oct. 3, 2020, the U.S. population stood at $N = 328,300,000$ people, and COVID-19 had resulted in $I_1(0) = 2,623,708$ infected individuals, $E(0) = 335,272$ exposed individuals, $D(0) = 214,637$ deceased individuals and $R(0) = 4,827,450$ recovered individuals.

2.3.4.1   SIR Model

In the case of COVID-19, the Removed compartment in the SIR model can be further divided into two sub–compartments, namely Recovered ($\tilde{R}$) and Death (D). According to Bastos and Cajueiro (2020), the underlying differential equations are:

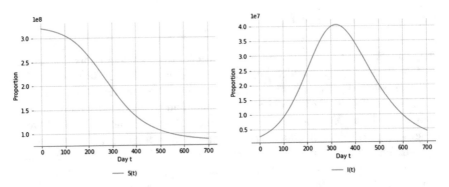

Fig. 2.2 Percentage of population in compartments S and I

$$S' = -\beta S \frac{I}{N},$$

$$I' = \beta S \frac{I}{N} - \frac{\tilde{\alpha}}{1-d} I,$$

$$\tilde{R}' = \tilde{\alpha} I,$$

$$D' = \frac{d}{1-d} \tilde{\alpha} I,$$

$$(2.16)$$

where β is the infection rate, α is the recovery rate and d is the death rate due to COVID-19. From (2.16), one can easily reduce the model to the baseline SIR model by aggregating the $\tilde{R}$ and D compartments into a single compartment R, resulting in the system of differential equations in (2.1) and (2.2) with $\alpha = \tilde{\alpha}/(1-d)$.

For parameter estimation, Bastos and Cajueiro (2020) solve the following minimisation problem:

$$\min_{\beta,\alpha,d} \sum_{t=1}^{T_0} \left[\left[I(t) - \hat{I}(t) \right]^2 + \left[R(t) - \hat{R}(t) \right]^2 \right],$$

where T_0 is the end of estimation period.

Using COVID-19 data for the U.S., from Worldometer (2020), the estimated parameters are $\beta = 0.03014$ and $\alpha = 0.01635$, where we consider $T_0 = 92$ days from Oct. 3, 2020 to Jan. 2, 2021; interested readers can refer to the data analysis section in Bastos and Cajueiro (2020). Assuming that the initial conditions are $N = 328{,}200{,}000$, $I(0) = 2{,}623{,}708$ and $R(0) = 4{,}827{,}450$, Fig. 2.2 shows the dynamics of the population in compartment S and I for the next 700 days.

Similar to the setup in Feng and Garrido (2011), assume that the whole population is enrolled in an insurance plan at the beginning of the COVID-19 outbreak. Susceptible individuals pay a premium of π per day to the fund, and in return, infected individuals receive $1{,}000$ per day, until removal from the infectious state, to cover medical costs.

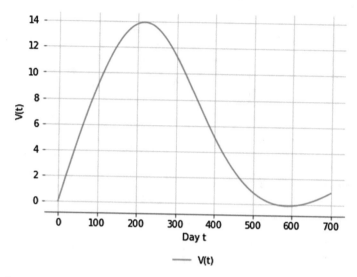

Fig. 2.3 Reserve of the epidemic insurance plan

Now, to determine the daily premium rate π of a 700-day insurance policy, it should be set so that the reserve function is non–negative over the entire policy term. That is, we set $V(\pi, t) \geq 0$ in Eq. (2.15), which gives

$$\pi \geq \max_t \frac{\int_0^t i(x)dx}{\int_0^t s(x)dx}.$$

Numerically, the premium level can be solved to be $\pi^* = 111.80$ and the cash value of the insurance fund at the end of 700 d is $V(111.80, 700) = 910.98$. In other words, with a daily premium payment of π^*, $910.98 is paid to the survivors at the end of the policy period. Figure 2.3 shows the change in reserves with respect to time during the pandemic, where the reserve function is displayed in thousands.

We see that a daily premium paid by the susceptible class of about 10% of the daily benefit paid to the infected class, leaving a positive final cash value left of the order of one day of benefit.

2.3.4.2   SEIRD Model

The baseline SIR model cannot capture some important features of the COVID-19 outbreak. For example, although it is a highly contagious disease, it takes some time for infected individuals to show symptoms and spread the virus. This time is often referred to as the incubation period. Therefore, infected but not infectious individuals should be classified as exposed. It is only after the incubation period that these exposed individuals become infectious. Furthermore, infected individuals

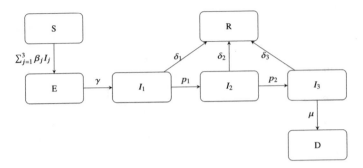

Fig. 2.4 Population flow of SEIRD model

Table 2.2 SEIRD parameters, U.S. data

Parameter	Definition	Estimation
β_i	Rate of infection in compartment I_i for $i = 1, 2, 3$	$\beta_1 = 0.8$
		$\beta_2 = 0.15$
		$\beta_3 = 0.05$
γ	Rate of transmission from compartment S to E	$\gamma = 0.2$
δ_i	Rate of recovery in compartment I_i for $i = 1, 2, 3$	$\delta_1 = 0.133$
		$\delta_2 = 0.125$
		$\delta_3 = 0.075$
p_i	Rate of transmission from compartment I_i to I_{i+1} for $i = 1, 2$	$p_1 = 0.033$
		$p_2 = 0.042$
μ	Rate of death in compartment I_3	$\mu = 0.05$

should be divided into different stages according to their clinical record: infected with mild symptoms, severe symptoms, or critical symptoms. Infected individuals with critical symptoms need to seek professional medical treatment.

To that end, consider a 7–compartment SEIRD model, as presented in Sect. 2.2.2.2. A flow chart of the model is presented in Fig. 2.4.

The interpretation and estimation of the parameters are summarised in Table 2.2. All parameters come from clinical research findings; interested readers can refer to Table 1 in Hill (2020). For instance, Linton et al. (2020) shows that the average incubation period is $\frac{1}{\gamma} = 5$ days, and so the rate of transmission from compartments E to I_1 is $\gamma = 0.2$.

The evolution of COVID-19 in the U.S. can now be simulated for the SEIRD model using the parameters in Table 2.2. Given the initial values for each compart-

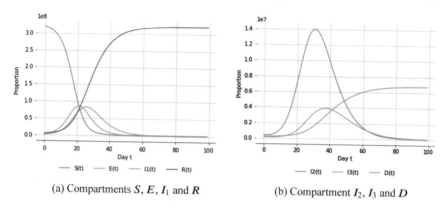

(a) Compartments S, E, I_1 and R (b) Compartment I_2, I_3 and D

Fig. 2.5 Evolution in each compartment w.r.t. time t

Table 2.3 Premium rating for the AH and SH insurance plans

Plan	P.V. benefits	P.V. premiums	Premium level
AH	2.456×10^{12}	5.772×10^9	425.50
SH	2.339×10^{12}	5.772×10^9	405.23

ment, $E(0) = 335{,}272$, $I_1(0) = 2{,}098{,}966.4$, $I_2(0) = 393{,}556.2$, $I_3(0) = 131{,}185.4$ $R(0) = 4{,}827{,}450$, $D(0) = 214{,}637$ and $N = 328{,}300{,}000$, the evolution of the epidemic is shown in Fig. 2.5. Since the parameters are not estimated by the same optimisation problem as in the SIR model, a different time horizon of $t = 100$ days, instead of 700 days, is used here for the SEIRD model.

In terms of actuarial modelling, assume again that U.S. residents are enrolled in one of the insurance plans proposed in Feng and Garrido (2011); the Annuity for Hospitalisation (AH) Plan or the other the Lump–Sum for Hospitalisation (SH) Plan. The insurance company provides $1,000 per day to the individuals in compartments I_1 and I_2, to cover medical, examination and consultation fees. There is also compensation for individuals in compartment I_3 to cover treatment fees. Hospitalised individuals benefit payments of $1,000 per day from the AH plan, while those enrolled in the SH plan, receive a lump–sum payment of $10,000 at the time of hospitalisation. The discounted total benefit payments, and the fair premium level for both plans are summarised in Table 2.3.

Recall that the parameters in SIR model are estimated by the minimisation problem, while the parameters in SEIRD model come from clinical research findings. Consequently, the evaluation periods are different (100 days in SEIRD versus 700 days in SIR) and the daily premium rates are seen to be about 40% of daily benefit rates in SEIRD, compared to about 10% in the SIR model.

2.3.4.3 Stochastic SIR Model

Lefèvre et al. (2017) considers an epidemic insurance model based on the stochastic SIR model, in Sect. 2.2.2.3. It is assumed that the policyholders pay premiums at a constant rate π per unit time, while they remain in the susceptible class. The expected value of all premiums received is

$$\mathbb{E}\left[\int_0^T \pi S(u)\, du\right]. \tag{2.17}$$

The insurer reimburses the medical expenses of infected policyholders, continuously, at a rate of c_1 per unit time. Furthermore, immediately upon removal, policyholders receive a lump sum of amount c_2. So the insurer's expected liability is

$$\mathbb{E}\left[\int_0^T c_1 I(u)\, du + c_2 R(T)\right] \tag{2.18}$$

By the Equivalence Principle (2.10) and the fact that $I(T) = 0$, the premium level is

$$\pi = \frac{c_1 \mathbb{E}\left(\int_0^T I(u)\, du\right) + c_2[N - \mathbb{E}(S(T))]}{\mathbb{E}\left(\int_0^T S(u)\, du\right)}. \tag{2.19}$$

The three expectations above are calculated using martingale arguments and recursive methods. Interested readers are referred to Lefèvre et al. (2017) for the detailed recursive formulas and proofs.

What follows is a numerical illustration of the stochastic SIR model. For simplicity, assume constant removal and infection rates, where the infection rate is $\beta = 1.5$ and the recovery rate is $\alpha = 1$. At time 0, the population counts in the three compartments are assumed to be $S(0) = 30$, $I(0) = 3$ and $R(0) = 0$. Figure 2.6(a) shows the probability mass function of $S(T)$; it suggests a relatively large final number of susceptibles at the end of the pandemic, with a high probability. Figure 2.6(b) shows the expectation of $S(T)$, as a function of different infection rates; if β is small (resp. large), more (resp. less) individuals remain uninfected at the end of the

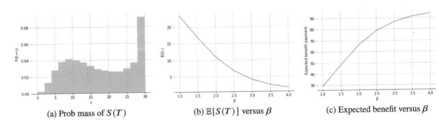

(a) Prob mass of $S(T)$ (b) $\mathbb{E}[S(T)]$ versus β (c) Expected benefit versus β

Fig. 2.6 Key quantities of the stochastic epidemic insurance model

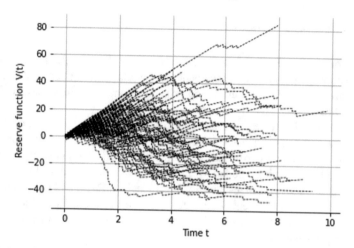

Fig. 2.7   Simulations of the reserve function $V(t)$, for $t \in [0, 10]$, 100 scenarios

pandemic and hence, leading to larger (resp. smaller) expectations of $S(T)$. Finally, Fig. 2.6(c) illustrates the fact that the expected benefit (the numerator in (2.19)) is an increasing function of the infection rate β; as the epidemic worsens, more payments are made to policyholders in compensation of their medical expenses. Applying the recursive formulas outlined in Lefèvre et al. (2017) and assuming that $c_1 = 1, c_2 = 2$, gives a fair premium level of $\pi = 0.576$. In other words, for the insurer to break even on average, the susceptible policyholders need to pay premiums continuously of 0.576 per unit time for each unit $c_1 = 1$ of continuous benefits and $c_2 = 2$ of lump sum units.

Understanding the policy value at time t of an insurance plan is important to ensure solvency of the company. To this end, define the reserve process of the insurer as

$$V(t) = \pi \int_0^t S(u)\, du - c_1 \int_0^t I(u)\, du - c_2 [N - S(t) - I(t)].$$

Figure 2.7 graphs the reserve function $V(t)$ of the stochastic SIR model for $t \in [0, 10]$. One hundred simulated scenarios of the stochastic reserve function are shown. Some produce a positive reserve, when most individuals remain in compartment S and fewer benefit payments go to policyholders. Other scenarios show many infected and recovered cases, producing very negative reserves when benefit payments out-pace collected premiums. However, from (2.19), we conclude that the average reserve function at time T should be zero, given the fair level of premium π and a sufficiently large number of scenarios.

2.4 Resource Management

This section discusses another actuarial application of epidemic models. Specifically, we propose a resource management framework for contingency planning and resource allocation, to help prevent and respond to public health crises such as the COVID-19 pandemic.

The outbreak of an infectious disease usually leads to a surge in the demand for medical care and resources, such as ventilators and personal protective equipment. Healthcare systems can experience a shortage of medical supplies that can have devastating effects, such as the loss of lives or the inability to control the spread of disease. A contingent plan in resource management is a critical tool for governments, healthcare systems, and essential businesses to mitigate these inevitable pandemic risks. Stocking resources is one possible mitigation strategy to help meet the demand surge during a pandemic.

Resource management involves demand predictions, ex–ante planning prior to a pandemic and allocation of limited resources during the pandemic. Here, we introduce an overarching framework for an alliance of different regions to optimise stockpiling and resources allocation at different pandemic stages in order to best utilise limited resources. This framework was originally proposed in Chen et al. (2020). Here we use inter–state resources pooling, as an illustrative example, but applications can also include international collaboration for the production, procurement, distribution and pooling of critical medical resources, such as ventilators, pharmaceuticals or vaccines.

This resource management framework can be summarised in three pillars:

Pillar I: Regional and Aggregate Resources Supply and Demand Forecast. Any pre–pandemic preparation plan should consist of supply and demand assessment and forecast. The supply side should include inventory assessments of critical resources and supplies, the maximum capacity of services, the capability of emergency acquisition and production. The demand side requires an understanding of the dynamics of a potential pandemic across regions and across borders. Historical data and predictive models can be used to project the evolution of a pandemic and the resulting surge demand on the healthcare system.

Pillar II: Centralised Stockpiling and Distribution. A central authority coordinates the efforts to develop a national preparedness strategy and to set up reserves of critical resources including preventive, diagnostic and therapeutic resources. A response plan is also necessary to understand how the central authority can deliver resources to different regions quickly to meet surge demands and to balance competing interests and priorities.

Pillar III: Central-Regional Resources Allocation. A pandemic response plan is critical for a central authority to contain and control the spread of a pandemic in all regions under its jurisdiction. As demand may exceed any best–effort pre–pandemic projections, the authority needs to devise optimal strategies that best

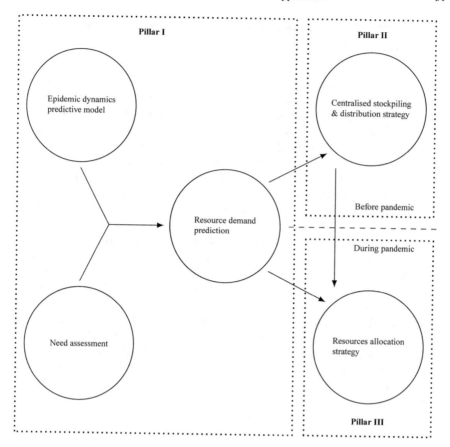

Fig. 2.8 Three-pillar pandemic risk management framework

utilise limited existing resources and minimise the economic cost of supply–demand imbalances. A coordination strategy needs to be in place to ensure smooth communications with regional authorities. The allocation strategy should be based on scientifically sound methods taking into account spatio–temporal differences across regions to ensure fairness and impartiality.

Figure 2.8 illustrates this framework and its underlying workflow. The rest of this section takes a closer look at each pillar and uses numerical examples to illustrate how this global framework can be implemented in practice.

2.4.1 Pillar I: Regional and Aggregate Resources Demand Forecast

Compartmental epidemiological models can be well integrated in this framework, where the estimation and prediction of different medical resources are derived directly from the evolution of the compartments. To demonstrate how contingency planning and allocation can be optimised for scarce resources we use ventilators as an example. Ventilators are typically durable resources that can perform their required functions for a long period of time without significant expenditures of maintenance or repair. Chen et al. (2020) provides a detailed account of single–use/disposable resources such as personal protective equipment, using similar arguments.

To assess the needs for those resources, which is a key component in Pillar I, consider the number of resources needed as a function of different compartments. Ventilators are not necessarily needed unless patients are in critical conditions and require additional respiratory aid for survival. So, naturally the demand for ventilators can be seen as a function of the number of patients who require intensive care. To predict the evolution of critical cases, we adopt the SEIRD model as presented in Sect. 2.2.2.2. An important feature of this model is that it distinguishes between patients that do not need intensive care and those who need it, which enables us to make estimations and predictions on ventilator demand more accurately.

After the SEIRD model outputs the evolution of the number of cases requiring intensive care, then predictions on ventilator demand can be made. Based on the findings in the medical literature (c.f. Yang et al. (2020), Grasselli et al. (2020)), there exist estimates of the percentage α of the infectives with intensive care that require the use of mechanical ventilators. Regional differences can be addressed in separate regional compartment models. These estimates can be used to project the ventilator demand as $X_j^{\text{VEN}(i)} = \alpha I_{3,j}^{(i)}$, where i indicates the ith region in the alliance and j indicates the jth day of the pandemic.

To illustrate how this can be done in practice, consider a hypothetical three–state alliance that includes New York (NY), Florida (FL) and California (CA) as participating states, regardless of any barriers that might prevent them from a full collaboration in practice. Further assume that 90% of ICU cases require ventilators. Once the parameters in the SEIRD model are estimated using real data from those states, we can create projections of the demand for ventilators as shown in Fig. 2.9, where Fig. 2.9(a) represents regional ventilators demand predictions and Fig. 2.9(b) represents the aggregate demand prediction in all three states.

In summary, the first pillar of resource management can be set–up with the help of compartmental models to obtain predictions on the demand for various resources over time.

Next, shift focus on how contingency planning and resources allocation can be carried out to minimise the impact of epidemics and pandemics on the economy, given these demand projections.

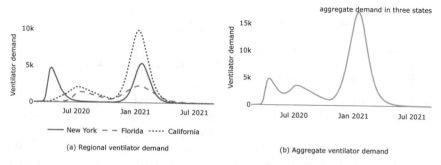

Fig. 2.9 Ventilator regional and aggregate demand prediction in NY, FL, and CA

2.4.2 Pillar II: Centralised Stockpiling and Distribution

As the pandemic unfolds, many hospitals and healthcare facilities can run out of pharmaceuticals and other essential resources, before emergency production ramps up and additional supplies become available. A *centralised stockpiling strategy* is intended to provide a stop–gap measure to meet the surge in resources demand at the early stage of the pandemic.

One should keep in mind that a practical stockpiling strategy is often an act of *balance between adequate supply and economic cost*. On the one hand, under–stocking is a common choice as resources and their storage can pose a heavy cost, while the actual demand during the pandemic outbreak could deviate from its projections. On the other hand, excessive stockpiling for a long term can lead to unnecessary waste.

In the second pillar of this proposed framework, using the estimated aggregate resources demand, the central authority could develop stockpiling and distribution strategies in normal periods, before any pandemic. In the case of ventilators, the central authority would have to determine an optimal initial stockpile size K_0 of resources to maintain in some centralised location. In addition, to meet surges in demand, the authority may need to reach contractual agreements with suppliers for emergency orders, which may be limited by the maximum production rate, say of a units per day during a pandemic. Since ventilators are durable, the stock of ventilators does not decrease over time due to usage. Assume that they can be deployed to different regions at negligible cost. Therefore, the total number of available ventilators in the entire alliance is given by $K_j = K_0 + aj$, on the j-th day after the onset of the pandemic. Hence, the only decision variable of the central authority in the case of ventilators is the initial stockpile size K_0.

Consider the following optimisation model for an initial stockpiling size.

$$
\min_{K_0 \geq 0} \sum_{j=1}^{m} \omega_j \left(\frac{\theta_j^+}{2} \left(X_j - (K_0 + aj) \right)_+^2 + \right.
$$

$$
\left. \frac{\theta_j^-}{2} \left(X_j - (K_0 + aj) \right)_-^2 + c_j \left(K_0 + aj \right) \right) + c_0 K_0, \tag{2.20}
$$

where m is the number of days of the pandemic, ω_j is a weight for significance of precision for the costs on the j-th day of the pandemic, θ_j^+ is an economic cost per square unit of shortage, θ_j^- is an opportunity cost per square unit of oversupply, c_j is the aggregate cost of possession per unit of ventilators per day, c_0 is the initial stockpile cost, which may include both the acquisition cost and the expected cost of possession (storage, maintenance, inventory logistics, opportunity cost). The quadratic form above can be interpreted as follows. While one copy of the quantity $X_j - (K_0 + aj)$ represents the amount of resource imbalance (shortage or surplus), the other copy $(\theta_j^\pm / 2)[X_j - (K_0 + aj)]_\pm$ can be viewed as the (linear) variable cost of the imbalance. The quadratic form is the product of cost per unit and the unit of imbalance, which yields the overall economic cost of imbalance. The weight w_j can be used for different purposes. For example, it may be reasonable to make the weight proportional to the daily demand X_j as the demand–supply imbalance can have a greater impact on population dense areas than otherwise, which is the weighting scheme adopted here for numerical examples.

To find the solution to this problem, first calculate the projected shortage without any initial stockpile, $Y_j := X_j - aj$, for $j = 1, \cdots, m$, which is the accumulated demand less the accumulated supply apart from the initial stockpile. Then, sort them in ascending order and denote the sorted sequence by $\{Y_{[j]}, j = 1, \cdots, m\}$, where $Y_{[j]}$ represents the j-th smallest projected shortage. The purpose of sorting the sequence is to find a $J \in \{1, 2, \cdots, m\}$, such that the following inequality holds,

$$
Y_{[J-1]} \leq \frac{\sum_{j=1}^{J-1} \omega_{[j]} \theta_{[j]}^- \left(Y_{[j]} - \frac{c_{[j]}}{\theta_{[j]}^-} \right) + \sum_{j=J}^{m} \omega_{[j]} \theta_{[j]}^+ \left(Y_{[j]} - \frac{c_{[j]}}{\theta_{[j]}^+} \right) - c_0}{\sum_{j=1}^{J-1} \omega_{[j]} \theta_{[j]}^- + \sum_{j=J}^{m} \omega_{[j]} \theta_{[j]}^+} \leq Y_{[J]}.
$$

Once J is identified, the optimal stockpile K_0^* is given by

$$
K_0^* = \max \left\{ \frac{\sum_{j=1}^{J-1} \omega_{[j]} \theta_{[j]}^- \left(Y_{[j]} - \frac{c_{[j]}}{\theta_{[j]}^-} \right) + \sum_{j=J}^{m} \omega_{[j]} \theta_{[j]}^+ \left(Y_{[j]} - \frac{c_{[j]}}{\theta_{[j]}^+} \right) - c_0}{\sum_{j=1}^{J-1} \omega_{[j]} \theta_{[j]}^- + \sum_{j=J}^{m} \omega_{[j]} \theta_{[j]}^+}, \ 0 \right\}.
$$

The proof can be found in Chen et al. (2020). This result shows that the optimal initial stockpile K_0 is the weighted average of all projected shortages, discounted by the cost of possession, relative to the economic cost of shortage, $Y_{[j]} - c_{[j]} / \theta_{[j]}^\pm$. The adjustment term $c_{[j]} / \theta_{[j]}^\pm$ indicates that, the higher the cost of possession relative to the economic cost of imbalance, the fewer ventilators should be acquired.

Figure 2.10 depicts optimal initial stockpile size in the case study. When the resource shortage costs are the same or less than the resource surplus, Fig. 2.10a

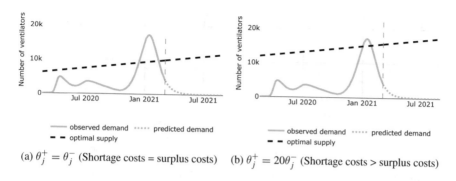

(a) $\theta_j^+ = \theta_j^-$ (Shortage costs = surplus costs)   (b) $\theta_j^+ = 200\theta_j^-$ (Shortage costs > surplus costs)

Fig. 2.10 Optimal initial stockpile size K_0 under different weights of economic cost

shows that the strategy requires less initial stockpile due to the excessive amount of supply after the pandemic dies out. By contrast, if the shortage costs weigh more than those of a surplus, the strategy is to reduce shortage in the early stages at the expense of increasing oversupply in the late stages; see Fig. 2.10b.

2.4.3   Pillar III: Centralised Resources Allocation

At the time of severe resource shortages, a coordinated effort becomes necessary to obtain additional supplies and to ration limited existing resources. There are two common types of resources allocation problems in the course of a pandemic, both of which can be cast in Pillar III of the proposed framework.

1. *Macro level resources pooling.* A central authority acts in the best interest of a union of many regions to increase supply, as well as to coordinate the distribution of existing and additional resources among different regional healthcare providers.
2. *Micro level rationing.* Facing an imbalance of demands and supplies in medical equipment and resources, hospitals often have to make difficult but necessary decisions to ration limited existing resources, as well as new supplies.

While in both cases the aim of the allocation exercise is to deliver limited resources where they are most needed, the macro level pooling addresses spatio–temporal differences and the micro level rationing focuses on healthcare effectiveness and fairness. The third pillar of the proposed pandemic risk management framework is about the allocation of limited resources to different regions, based on the proposed optimal centralised stockpiling and distribution strategies.

Figure 2.11 puts the regional resources demand and optimal aggregate supply together for the ease of exposition.

Following the previously mentioned principles, consider the allocation of existing resources in a healthcare system with n regions during a pandemic that lasts for m days. We always use the superscript (i) to indicate quantities for the i-th region. Bear

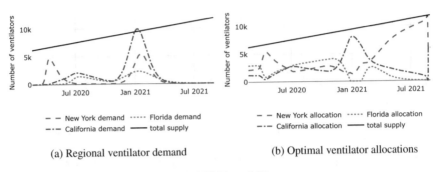

Fig. 2.11 Optimal ventilator allocations in NY, FL, and CA

in mind that there could still be aggregate shortage of supply for ventilators in all the alliance regions. The central authority would have to take a holistic view of competing interests among participating regions. On each day, during the pandemic, when the aggregate demand exceeds the aggregate supply, the central authority should choose to allocate resources taking into account spatial differences in demand and supply. This motivates the following optimisation model for ventilator allocations:

$$\min_{\substack{K_j^{(i)} \geq 0; i=1,2,\ldots,n; \\ j=1,2,\ldots,m}} \sum_{j=1}^{m} \sum_{i=1}^{n} \omega_j^{(i)} \left(\frac{\theta_j^{+(i)}}{2} \left(X_j^{(i)} - K_j^{(i)} \right)_+^2 + \frac{\theta_j^{-(i)}}{2} \left(X_j^{(i)} - K_j^{(i)} \right)_-^2 \right)$$

such that $\sum_{i=1}^{n} K_j^{(i)} = K_j,$ for $j = 1, 2, \ldots, m,$

where $\omega_j^{(i)}$ is a weight assigned the j-th day of the pandemic in the i-th allied region, $\theta_j^{+(i)}$ is an economic cost per squared unit of shortage, $\theta_j^{-(i)}$ is an opportunity cost per square unit of oversupply. The quadratic terms $\frac{\theta_j^{\pm(i)}}{2} \left(X_j^{(i)} - K_j^{(i)} \right)^2$ represent the economic cost for demand–supply imbalance. Note that $\frac{\theta_j^{\pm(i)}}{2}$ measures the rate of increase in cost per unit, and hence $\frac{\theta_j^{\pm(i)}}{2} \left(X_j^{(i)} - K_j^{(i)} \right)$ represents the linear variable cost per unit. The variable cost in principle reflects the law of demand, under which the price increases with the quantity demanded. Therefore, the total cost is the product of the variable cost per unit $\frac{\theta_j^{\pm(i)}}{2} \left(X_j^{(i)} - K_j^{(i)} \right)$ and the total unit of imbalance $\left(X_j^{(i)} - K_j^{(i)} \right)$. The economic cost is used to account for both potential loss of lives due to the lack of resources and the opportunity cost of idle medical resources due to oversupply. This structure of economic cost is used not only for its mathematical tractability, but also to penalise large imbalances between demand and supply. The weight $\omega_j^{(i)}$ can be used to measure the relative importance of the resource allocation for region i at time t_j to other regions and time points.

The constraint $\sum_{i=1}^{n} K_j^{(i)} = K_j$ indicates that resources allocated to different regions must add up to the total amount of supply available to the central authority. The evolution of supply $\{K_j, j = 1, 2, \ldots, m\}$ is based on the centralised stockpiling strategy discussed in previous sections. The evolution of demand $\{X_j^{(i)}, i = 1, \ldots, n, j = 1, 2, \cdots, m\}$ can be based on forecasts from epidemiological models fitted to most recent local data.

The analytical solution to this allocation problem is summarised as a holistic allocation algorithm and explained in full detail in Chong et al. (2021). Here we only explain one particular case of the solution. As the allocation is carried out from period to period, we shall suppress the subscript j for brevity, and use $X^{(i)}$ for ventilator demand in region i and $K^{(i)}$ for the quantity of allocated resources in the same region.

If there is an overall surplus in the healthcare system at time j, i.e. $K > \sum_{r=1}^{n} X^{(r)} = X$, then only the economic cost for oversupply $\theta^{-(i)}$ applies and the optimal allocation of existing supply to the i-th region is given, for all $i = 1, 2, \ldots, n$, by

$$
K^{(i)} = \left(1 - \frac{\frac{1}{\omega^{(i)}\theta^{-(i)}}}{\sum_{r=1}^{n} \frac{1}{\omega^{(r)}\theta^{-(r)}}} \right) X^{(i)} + \frac{\frac{1}{\omega^{(i)}\theta^{-(i)}}}{\sum_{r=1}^{n} \frac{1}{\omega^{(r)}\theta^{-(r)}}} \left(K - \sum_{r \neq i} X^{(r)} \right). \tag{2.21}
$$

The economic interpretation of the allocation formula (2.21) is that the optimal supply for region i results from a balance of two competing optimal solutions.

- **Self–concerned optimal supply**: $X^{(i)}$
  If region i can ask for as much as it needs, then this amount shows the ideal supply in the best interest of the region alone. The demand and supply for all other regions are ignored in its consideration.
- **Altruistic optimal supply**: $K - \sum_{r=1;r \neq i}^{n} X^{(r)}$
  If the region i places the interests of all other regions above its own, then the medical supply goes to other regions and region i ends up with the leftover amount.

The central authority has the responsibility to mediate among regions competing for resources. Formula (2.21) indicates that optimality for region i, in consideration of the entire system, is reached by a weighted average of two extremes, namely the self–concerned optimal and the altruistic optimal supplies. The average of two optimal supplies is determined by the harmonic weighting $\frac{1}{\omega^{(i)}\theta^{-(i)}} \Big/ \sum_{r=1}^{n} \frac{1}{\omega^{(r)}\theta^{-(r)}}$ as opposed to arithmetic weight $\omega^{(i)}\theta^{-(i)} \Big/ \sum_{r=1}^{n} \omega^{(r)}\theta^{-(r)}$. It is shown in Chong et al. (2021) that in multi–objective Pareto optimality the harmonic weighting is always used for balancing competing interests of participants whereas the arithmetic weighting serves the purpose of balancing competing objectives of the same participant.

2.5 Conclusion

In summary, this chapter explores several well–known compartmental models in epidemiology (Sect. 2.2) and studies their actuarial applications, in two different compartmental models. Key actuarial quantities are studied for an epidemic insurance plan in Sect. 2.3, such as annuities and benefits. In connection to the COVID-19 pandemic, three case studies are presented in Sect. 2.3.4, modelling the reserve level of an epidemic insurance provider. A second actuarial application of compartmental models in resource management is presented in Sect. 2.4 and contingency planning during a pandemic. To optimise stockpiling and medical resource allocation, we propose an overarching framework that includes supply and demand forecasts, centralised distribution, and regional allocations.

The first results obtained for the COVID-19 case study are that, as expected, premium levels for possible epidemic insurance coverages vary greatly, depending on the compartmental model calibrated to the available epidemic data. It highlights the importance of developing compartmental models as well suited as possible to the particular characteristics of the epidemic under study. Section 2.3 also shows that it is possible to set viable reserves, for epidemic insurance coverages, for different compartmental models, not just for the SIR. Similarly, the results in Sect. 2.4 illustrate possible ways to quantify resources and manage them optimally during a pandemic.

As both actuaries and epidemiologists specialise in quantifying risks, each in their own ways, we hope that this chapter motivates future research in bridging the two fields, to gain further insight into how to quantify the impact of pandemics.

Acknowledgements This work is supported by a research grant by the Canadian Institute of Actuaries in response to COVID-19, by the Natural Sciences and Engineering Research Council (NSERC) of Canada grant RGPIN–2017–06643 and also by an endowment from the State Farm Companies Foundation.
The authors are grateful to Prof. Wilkie and the Editors for their constructive comments on an earlier version of this chapter.

References

L. Allen, A primer on stochastic epidemic models: formulation, numerical simulation, and analysis. Infect. Dis. Model. **2**(2), 128–142 (2017)

H. Andersson, T. Britton, *Stochastic Epidemic Models and Their Statistical Analysis*, vol. 151 (Springer Science & Business Media, 2012)

S. Bastos, D. Cajueiro, Modeling and forecasting the early evolution of the COVID-19 pandemic in Brazil (2020). arXiv:2003.14288

L. Billard, P. Dayananda, A multi-stage compartmental model for HIV-infected individuals: I-waiting time approach. Math. Biosci. **249**, 92–101 (2014a)

L. Billard, P. Dayananda, A multi-stage compartmental model for HIV-infected individuals: II-application to insurance functions and health-care costs. Math. Biosci. **249**, 102–109 (2014b)

F. Brauer, C. Castillo-Chavez, *Mathematical Models in Population Biology and Epidemiology*, vol. 2 (Springer, 2012)

X. Chen, W. Chong, R. Feng, L. Zhang, Pandemic risk management: resources contingency planning and allocation (2020). arXiv:2012.03200

W.F. Chong, R. Feng, L. Jin, Holistic principle for risk aggregation and capital allocation. Ann. Oper. Res. 1–34 (2021). https://doi.org/10.1007/s10479-021-03987-4

D. Dickson, M. Hardy, H. Waters, *Actuarial Mathematics for Life Contingent Risks* (Cambridge University Press, 2013)

R. Feng, J. Garrido, Actuarial applications of epidemiological models. N. Am. Actuar. J. **15**(1), 112–136 (2011)

R. Feng, L. Jin, S.-H. Loke, Interplay between epidemiology and actuarial modeling. Submitted to the Casualty Actuarial Society E-Forum (2020)

P. Fine, K. Eames, D. Heymann, "Herd immunity": a rough guide. Clin. Infect. Dis. **52**(7), 911–916 (2011)

G. Grasselli, A. Zangrillo, A. Zanella, M. Antonelli, L. Cabrini, A. Castelli, D. Cereda, A. Coluccello, G. Foti, R. Fumagalli, G. Iotti, N. Latronico, L. Lorini, S. Merler, G. Natalini, A. Piatti, M. Ranieri, A. Scandroglio, E. Storti, M. Cecconi, A. Pesenti, et al. Baseline characteristics and outcomes of 1591 patients infected with SARS-CoV-2 admitted to ICUs of the Lombardy region, Italy. JAMA **323**(16), 1574 (2020)

H. Hethcote, An immunization model for a heterogeneous population. Theor. Popul. Biol. **14**(3), 338–349 (1978)

A. Hill, Modeling COVID-19 spread versus healthcare capacity (2020). https://alhill.shinyapps.io/COVID19seir/

W. Kermack, A. McKendrick, A contribution to the mathematical theory of epidemics. Proceedings of the royal society of London. Ser. A, Contain. Pap. Math. Phys. Character **115**(772), 700–721 (1927)

W. Kermack, A. McKendrick, Contributions to the mathematical theory of epidemics. II. -the problem of endemicity. Proc. R. Soc. London. Ser. A, Contain. Pap. Math. Phys. Character **138**(834), 55–83 (1932)

C. Lefèvre, P. Picard, Final outcomes and disease insurance for a controlled epidemic model. Appl. Stoch. Model. Bus. Ind. **34**(6), 803–815 (2018a)

C. Lefèvre, P. Picard, A general approach to the integral functionals of epidemic processes. J. Appl. Probab. **55**(2), 593–609 (2018b)

C. Lefèvre, M. Simon, Cross-infection in epidemics spread by carriers. Stoch. Model. **34**(2), 166–185 (2018)

C. Lefèvre, M. Simon, SIR-type epidemic models as block-structured Markov processes. Methodol. Comput. Appl. Probab. **22**(2), 433–453 (2020)

C. Lefèvre, P. Picard, M. Simon, Epidemic risk and insurance coverage. J. Appl. Probab. **54**(1), 286–303 (2017)

N. Linton, T. Kobayashi, Y. Yang, K. Hayashi, A. Akhmetzhanov, S.-M. Jung, B. Yuan, R. Kinoshita, H. Nishiura, Incubation period and other epidemiological characteristics of 2019 novel coronavirus infections with right truncation: a statistical analysis of publicly available case data. J. Clin. Med. **9**(2), 538 (2020)

C. Nkeki, G. Ekhaguere, Some actuarial mathematical models for insuring the susceptibles of a communicable disease. Int. J. Financ. Eng. 2050014 (2020)

S. Perera, An insurance based model to estimate the direct cost of general epidemic outbreaks. Int. J. Pure Appl. Math **117**(14), 183–189 (2017)

A. Shemendyuk, A. Chernov, M. Kelbert, Fair insurance premium level in connected SIR model under epidemic outbreak (2019). arXiv:1910.04839

Worldometer, United states coronavirus data (2020). https://www.worldometers.info/coronavirus/country/us/

X. Yang, Y. Yu, J. Xu, H. Shu, J. Xia, H. Liu, Y. Wu, L. Zhang, Z. Yu, M. Fang, T. Yu, Y. Wang, S. Pan, X. Zou, S. Yuan, Y. Shang, Clinical course and outcomes of critically ill patients with SARS-CoV-2 pneumonia in Wuhan, China: a single-centered, retrospective, observational study. Lancet Respir. Med. **8**(5), 475–481 (2020)

S. Zhao, Q. Lin, J. Ran, S. Musa, G. Yang, W. Wang, Y. Lou, D. Gao, L. Yang, D. He et al., Preliminary estimation of the basic reproduction number of novel coronavirus (2019-nCoV) in China, from 2019 to 2020: a data-driven analysis in the early phase of the outbreak. Int. J. Infect. Dis. **92**, 214–217 (2020)

Open Access This chapter is licensed under the terms of the Creative Commons Attribution 4.0 International License (http://creativecommons.org/licenses/by/4.0/), which permits use, sharing, adaptation, distribution and reproduction in any medium or format, as long as you give appropriate credit to the original author(s) and the source, provide a link to the Creative Commons license and indicate if changes were made.

The images or other third party material in this chapter are included in the chapter's Creative Commons license, unless indicated otherwise in a credit line to the material. If material is not included in the chapter's Creative Commons license and your intended use is not permitted by statutory regulation or exceeds the permitted use, you will need to obtain permission directly from the copyright holder.

Chapter 3
Some Investigations with a Simple Actuarial Model for Infections Such as COVID-19

A. D. Wilkie

Abstract In this chapter the author adds an infection feature to an actuarial multiple state model to give a simple model for an infection such as COVID-19. The model is simple enough to be replicated in an Excel worksheet, with one row per day of calculations. The whole population is treated as homogenous, with no distinction by age, sex or anything else; to that extent it is unrealistic, but to include these features would complicate it considerably. To fit it to observed data requires successive optimisation by programme, and this is described. Different variations of the model allow it to fit better and take account of, for example, immunisation by vaccine. It is shown to fit the past events in the United Kingdom (U.K.) quite well, and it has also been fitted to other countries, but this is not shown in this chapter. It is also observed that this, or any other model, is of less use for forecasting the future, because it cannot predict the behaviours of governments or of populations. But various assumptions can be made about the future, as at the latest date of calculation (1 March 2021), and interesting consequences are shown.

3.1 Introduction

Actuarial multiple state models have great similarities with epidemiological models for infection. In this Chapter the author combines the two to give a simple model for an infection such as COVID-19. In Sect. 3.2 multiple state actuarial models are described. In Sect. 3.3 the elements of the simple model are explained, and comparisons with SIR models are made in Sect. 3.4. The necessary initial assumptions for the model are stated in Sect. 3.5. How the variable daily parameters are estimated is explained in Sect. 3.6.

An alternative, Model 2, is introduced in Sect. 3.7, and results for the U.K. (up to the latest date of calculation) for the two models are shown in Figs. 3.1, 3.2, 3.3, 3.4 with comments in Sect. 3.8. Two new Models, 3 and 4, are introduced in Sect. 3.9,

A. D. Wilkie (✉)

InQA Limited and Heriot-Watt University, Dennington, Ridgeway, Horsell, Woking GU21 4QR, UK

e-mail: david.wilkie@inqa.com

© The Author(s) 2022

M. C. Boado-Penas et al. (eds.), *Pandemics: Insurance and Social Protection*, Springer Actuarial, https://doi.org/10.1007/978-3-030-78334-1_3

with results shown in Figs. 3.5, 3.6, 3.7, 3.8 and further comments on the results in Sect. 3.10.

Some hypothetical projections from the latest date of calculation are described in Sect. 3.11 and their consequences are shown.

Some remaining problems are discussed in Sect. 3.12, reference is made in Sect. 3.13 to other countries, and Sect. 3.14 concludes.

3.2 Multiple State Actuarial Models

Actuaries are familiar with several multiple state models. The simplest is the life table. A person can be in one of two states, Living or Dead and can move from one to the other, but in this case only in one direction. We assume that at age, or time, x there are $L(x)$ persons in state Living, and $D(x)$ persons in state Dead. The total $T = L(x) + D(x)$ is necessarily constant.

To describe the rate of transfer between states we can use either a continuous or a discrete model. The continuous model uses the derivate:

$$\mathrm{d}L(x)/\mathrm{d}(x) = -\mu(x) \cdot L(x) = + \,\mathrm{d}D(x)/\mathrm{d}(x)$$

where $\mu(x)$ is the "force of mortality" or "continuous transition intensity" from living to dead. The discrete model uses a unit time step, often, but not necessarily, a year:

$$L(x+1) = L(x) - q(x) \cdot L(x) \qquad D(x+1) = D(x) + q(x) \cdot L(x)$$

where $q(x)$ is the one-year probability of death of a life aged x.

A more complicated actuarial model is that for income protection (IP) insurance for sickness, where the states are: Healthy, Sick and Dead, with exit from Sick either by recovery back to Healthy or by death to Dead, and these rates depend both on age at start of sickness, x, and duration in the Sick state, z, see Continuous Mortality Investigation (1991). For this a continuous model is best, with the numbers in different states as: $H(x)$, $D(x)$ and for sickness a continuum with density $S(x, z)$. Transfers are then:

new Sicknesses from Healthy at rate $\sigma(x) \cdot H(x)$
new Deaths from Healthy at rate $v(x) \cdot H(x)$
recoveries back to Healthy from $Sick(x, z)$ at rate $\rho(x, z) \cdot S(x, z)$
new Deaths from $Sick(x, z)$ at rate $\mu(x, z) \cdot S(x, z)$.

A discrete model is harder to define in this case, because of the possible circular movements from Healthy to Sick and back, and the multiple reasons for exit from Sick. In practice a daily step has to be assumed in the data for the estimation of the (assumed) continuous rates.

3.3 A Simple Daily Model for Infection

An infection model is similar to the IP model for sickness, with a fundamental difference in one respect. Whereas in the IP model the transitions from Healthy to Sick depend only on the numbers in the Healthy state, in an infection model they depend also on the numbers in the Sick (or Infected) state. In the simple model we ignore age and replace x by time t, measured now in days. Initially we also ignore duration of infection, z, and denote those Infected at time t as $I(t)$. We then assume that an infected person can potentially pass the infection to others at a rate of $r(t)$ per day, but that only those who are not infected already can move to the Infected state.

A continuous model would thus give us:

$$\mathrm{d}H(t)/\mathrm{d}(t) = -r(t) \cdot I(t) \cdot H(t)/\big(H(t) + I(t)\big).$$

We omit $D(t)$ from the denominator, because we assume that they cannot become infected at all. This is essentially the SIR infection model described in Chap. 2 of this book, with different names for the states.

For simplicity in practice I have used a discrete model, with daily steps, and with the Infected at time t subdivided by days infected, d, giving us $I(t, d)$. For this model I ignore deaths and sicknesses other than from the relevant infection, and I ignore age, sex and any other variability in the population.

Time is measured in discrete days, t, and changes happen at the end of each day. On day 0 there are $H(0)$ in the population, none of whom are infected. On each day t, there are $HI(t)$ new infections (I explain below how these are calculated) moving from Healthy to Infected; thus on day $t + 1$ there are $I(t + 1, 1)$ infected. On each day these may die, with mortality rate $m(d)$, so:

$$I(t + 1, d + 1) = \big(1 - m(d)\big) \cdot I(t, d)$$

and new deaths on day t are

$$ID(t) = \sum m(d) \cdot I(t, d).$$

The total number Infected on day t are

$$I(t) = \sum I(t, d)$$

and the total Living on day t are

$$L(t) = H(t) + I(t).$$

The proportion healthy among the living is

$$g(t) = H(t)/L(t).$$

The total number of deaths by day $t + 1$ is

$$D(t + 1) = D(t) + ID(t),$$

and the grand total of the population on day t is

$$T(t) = L(t) + D(t).$$

This is constant, but it is useful to calculate it to check that the calculations are accurate.

On day t there is a basic infection rate $r(t)$, which can vary by t. Those infected with duration d have a relative infection rate of $pr(d)$ so their infection rate is $pr(d).r(t)$. The number of new infections generated by those infected on day t with duration d is

$$HI(t, d) = pr(d) \cdot r(t) \cdot g(t) \cdot I(t, d),$$

The total number of new infections on day t is

$$\sum HI(t, d) = HI(t),$$

as noted above.

To start the epidemic we need one or more initial infections. We put these in as so many at the end of day t_0, so that

$$HI(t_0) = I(t_0 + 1, 1) = 1,$$

or such other number as we wish.

Once infected, persons remain in the count of infected, but their relative infectiousness $pr(d)$ may reduce to zero by some duration, say M, so that all of those infected on or after duration M can be added together in $I(t, M)$, with zero infectiousness. We think of them as recovered, but do not, as in the SIR model remove them altogether. They cannot become infected again, so are correctly in the denominator but not the numerator of $g(t)$. All this is true for many infectious diseases and I assume that it is true for COVID-19, but it may not be. For other diseases it may also not be true, and once infected, always infectious. This can be represented in this model by keeping $pr(M)$ non zero.

3.4 Comparisons with the SIR Model

In the simplest SIR model there is no allowance for duration of infection, so no variation in infectiousness by duration. Nor is there any explicit mortality, since those in the R state are recovered. But this simple model has many variations that can include some of these features.

A value in the SIR model that is often publicly quoted is denoted R_0 or just "the R number". This is the number of infections caused by the (hypothetical) first patient. If $R = 3$, then one infected passes the disease on to three others, each of them to three more, etc. It is given as an absolute number, but it needs also a time scale: infecting three in a day is very different from three in a month. It seems that the publicly quoted number in the U.K. is about "per week".

In the simple model described above, a value of R_0 can be calculated, along with a time scale. A new infected is able to infect others depending on his/her duration of infectiousness. The number of new infections on day d depends on the proportion who survive to day d, and their rate of infectiousness. So we first calculate the proportion of survivors, allowing for mortality, by putting

$$p(d = 1) = 1,$$

and then

$$p(d + 1) = p(d) \cdot (1 - m(d)),$$

for as far as we need, and assuming that those of duration M and greater are no longer infectious.

We then calculate

$$S = \sum \{p(d) \cdot pr(d)\}$$

and our estimate of R_0 on day t is

$$R_0(t) = S \cdot r(t).$$

The average time till infection is given by first calculating

$$U = \sum \{d \cdot p(d) \cdot pr(d)\}$$

and then the average time

$$T = U/S.$$

In this simplest model the values of, S, T and U do not vary with time t.

However, the effectiveness of infection at time t is diminished by the factor $g(t)$, the proportion of the population available to infect. Initially this is close to unity, but as the disease progresses, more of the population may have become infected, and cannot be reinfected, so that a level of population immunity may be reached ("herd" immunity is a term appropriate for cattle). The effective R value at time t is therefore

$$Rg(t) = R_0(t) \cdot g(t).$$

This damping by $g(t)$ increases with time t.

3.5   Enhancements for COVID-19 and Initial Assumptions

The model so far described suits situation where all infections are known about as soon as they occur, as with a small closed population and a disease that is immediately apparent. However, this is not the case far COVID-19 and we describe the features as they became apparent in the U.K. In other countries it may not be quite the same.

In the first few days an infected person may show no symptoms, so the fact of infection and infectiousness is not known. Then when symptoms do appear, the infected may not be tested and counted as infected unless the symptoms are bad enough for him/her to go to hospital. We represent this by postulating that, of all the new infections on day t, $I(t, 1)$, the symptoms do not appear until duration day K, and then only a fraction, $f(t)$, are recorded as having COVID-19. Then we assume that deaths only occur among those whose infection is bad enough to have been recorded.

In the simplest model effected so far for the U.K., I have assumed that day K is day 6, that day M is 36, that $f(t) = 0.1$ for all t, i.e. only one tenth of all cases are reported (but this changed later). I then assume that $pr(d) = 1$ for $d = 1$–5, $= 0.5$ for $d = 6$–15, and $= 0$ thereafter. This implies that in the first few days before symptoms have appeared, the infected is wholly infectious, but that when symptoms appear he/she stays isolated to some extent, so the relative level of infectiousness is halved. I then assume that $m(d) = 0$ for $d = 1$–5, equals some constant level m for days 6–15, and is zero thereafter.

All these assumptions are rather arbitrary. If one had access to full hospital data, one could estimate the values of the mortality rates, $m(d)$, by day of infection, but if these have been calculated by anyone, they do not seem to have been made public yet. With more testing of individuals, the value of $f(t)$ may well increase with time, but it is always difficult to estimate what faction of any large population is not in some category without large sampling, or careful random sampling.

With this model one can insert arbitrary parameters and see what effect different assumptions make. For example, if the infected continue to be infectious beyond day M, then ultimately the whole population becomes infected, but with anything short of this, there may be some value of $r(t)$ that is low enough to leave an uninfected residual. If the mortality of the infected is very high, then with a low enough level of infectiousness the epidemic dies out, perhaps quite quickly, but with a higher level then ultimately everyone dies. These are all extreme cases, but it helps to understand what can produce them, although COVID-19 does not seem to be so extreme.

3.6   Estimating Parameters Model 1

In order to compare this model with the available facts we need to use some available data. I have used the data collected by the European Centre for Disease Prevention and Control (2021), ECDC, which gives data, for a very large number of countries, of the number of New Cases and New Deaths for each day up to 14 December 2020. After that date ECDC gives the numbers in each week, from the start of 2020, but not for individual days. To get continuing daily data one has to look for the records of each country separately, which is much less convenient.

There are many ways in which one could choose parameters that would in some way match the actual numbers of cases and of deaths with those expected by the model. I follow an idea of Wüthrich (2020) and I have chosen one way to do it, but many others might be as good or better. After an initial starting period, as described below, I allow the rate of infection per day, $r(t)$, to vary for each time day t, and I estimate it for each t. I assume that the rates of morality, $m(d)$ are equal for day of infection from day 6 to day 15, and zero outside those days, but that at each estimation step they are the same for all times, t.

To estimate the rates for time t, I choose a 14-day period, with day t as day 7, so from day $t - 6$ to day $t + 7$. For days up to $t - 7$, I assume the values of $r(t)$ which have been estimated already, but the value of the mortality rates, for this estimation, is constant for all t and d and equals $m(t)$. I then choose the values of $r(t)$ and $m(t)$ by equating the expected (E) and actual (A) cumulative number of Cases (C) and the cumulative number of Deaths (D) as at day $t + 7$, the end of the 14-day period. This can be done by minimising the function $F(r, m)$ calculated as:

$$F(r, m) = \big(EC(t + 7) - AC(t + 7)\big)^2 + \big(ED(t + 7) - AD(t + 7)\big)^2$$

and at the optimum values of r and m, we get

$$F(r, m) = 0.$$

To start the estimation I choose a day when the numbers of cases and deaths are non-zero for most days, and estimate single rates, $r(t)$ and $m(t)$ up to that day. In the very early stages the numbers of new cases are very erratic, and the numbers of new deaths are often zero for a week or two after the numbers of new cases cease to be zero, and a comparison of actual and expected is poor.

Apart from the initial period, this process produces quite good values for $r(t)$ for the whole period, and the expected numbers of cases for each individual day are fairly close to the actual number. In the initial period of the epidemic, during the "first wave" this method also gave quite good comparisons for deaths, but as time went by the comparison of deaths became poor, so I developed Model 2 described below. Results for Model 1 are shown later along with those for Model 2.

3.7 Estimating Parameters Model 2

Since my Model 1 did not give a good correspondence between actual and expected daily deaths, I changed the model a little. Instead of keeping the mortality varying by duration, but not by time, I allow the rates to vary by time, so that each day, t, has a basic mortality rate, $m(t)$, and there are fixed ratios that vary by duration, d, denoted $pm(d)$ so that the mortality rate at duration d at time t is $m(t, d) = m(t).pm(d)$. This is a very similar pattern to the rate of infection, with daily rate $r(t)$, and intensity on day d of $pr(d)$.

I fix the values of $pm(d)$, to be zero except for days 6–15 inclusive, when they have value 1.0. This is the same pattern as described for Model 1. This can be accommodated in an Excel spreadsheet, with more columns, to show the values of $m(t, d)$, and of $p(d)$, the surviving infected, which now also vary with time t, as do T, U and S.

I then use almost the same estimation procedure for fitting the values of $r(t)$ and $m(t)$, except that instead of making m the same for all past days in one calculation, it varies like $m(t)$. I show the results in Figs. 3.1, 3.2, 3.3, 3.4, using the ECDC data for the U.K. data up to 14 December 2020, and adding data from Office of National Statistics (2021) up to 28 February 2021.

The inception rates for Models 1 and 2 are almost identical, so only one set is shown.

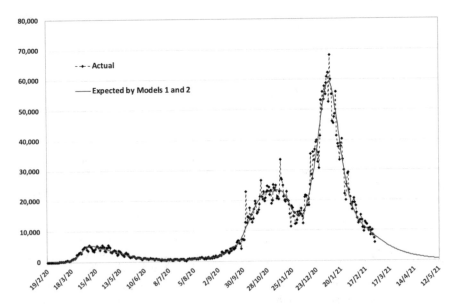

Fig. 3.1 Actual and expected daily reported new cases of COVID-19, U.K., Models 1 and 2

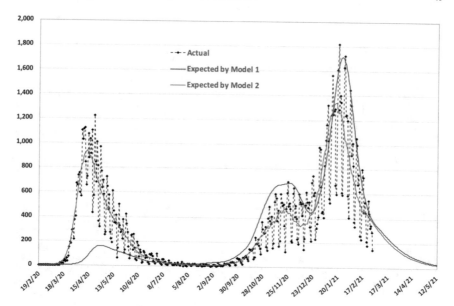

Fig. 3.2 Actual and expected daily reported deaths of COVID-19, U.K., Models 1 and 2

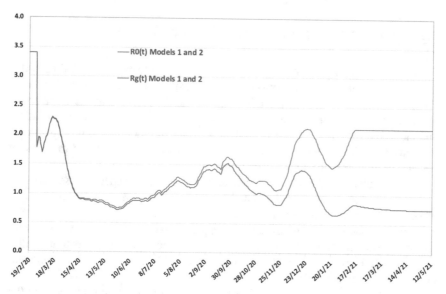

Fig. 3.3 Estimated values of infection rates, $R_0(t)$ and $R_g(t)$, U.K., Models 1 and 2

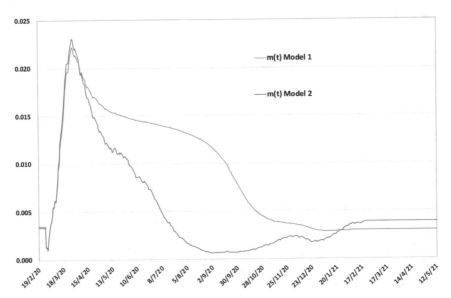

Fig. 3.4 Estimated values of mortality rates, $m(t)$, U.K., Models 1 and 2

3.8 Comments on Results of Models 1 and 2

We can see from Fig. 3.1 that the numbers of actual new cases reported each day are very erratic, nothing like consistent with the expected numbers with a binomial, Poisson or normal probability model. From Fig. 3.2 we see that the numbers of reported deaths are even more erratic. We have assumed uniformity in our population, and the true position is very far from this, so some variation would be expected as with any actuarial model. However, the gross irregularities shown are more likely because of great irregularities in reporting.

Both models show almost the same quite smooth curve for expected cases reported and fit the centre of the irregular actual events quite well. We see the smaller first wave of March and April 2020, reduced by the first lockdown. After a comparatively mild summer the second and third waves appear in late 2020 and early 2021, with far more reported cases than in the first wave, and more deaths, but with a smaller proportionate increase. We continue the model to mid-May 2021 and we comment on these results below.

In Fig. 3.2 we see that the actual numbers of deaths reported each day and the numbers expected by both models are fairly different from each other. The reported numbers are also very irregular, but the expected numbers with Model 2 are reasonably good, whereas Model 1 fits rather badly.

In Fig. 3.3 we see the estimated values of the infection rates, both the "gross" $R_0(t)$ described earlier, and the "net" $Rg(t)$. In the early stages of the epidemic, these are quite close because very few of those that are not currently infectious have been infected already. But as more of the population becomes infected, the more those who are infectious are in contact with those who have already been infected, and population immunity grows.

I explained above that I had initially assumed, on the basis of casual press comments, that only 10% of those infected were sufficiently ill to report their infection, and so have their infection recorded. I assume that $f(t) = 0.1$ for all t. By the middle of January 2021 some 3.4 million people had been reported as infected, which would imply that some 34 million people in the U.K. had been infected, or about half the population. So my estimate of $g(t)$ is about 0.5, and the net infection rate is about halved, going down from a gross rate of about 1.8 to about 0.9 and progressively lower as the rest of the population becomes infected. At these levels the epidemic quite rapidly dies out, and my "projections" of the numbers of new cases and deaths reduce rapidly, as implied in Figs. 3.1 and 3.2.

This is a typical result for any infection that continues with a constant value of $r(t)$ such that $R_0(t)$ is greater than 1. The curves of expected cases and of deaths rises exponentially initially, curve over at peak, and fall symmetrically on the way down. On a vertical logarithmic scale the curve closely resembles a hyperbola, but is not exactly equal to one. The maximum is when $g(t) = 0.5$, and the number of cases so far is at half its total number, as are the numbers of deaths with perhaps some time-lag. Figures 3.1 and 3.2 show that the peaks of new cases and new deaths were reached during January 2021, and by mid-April would be quite small. It assumes that the present level of lockdown continues indefinitely so that the value of $r(t)$ is held constant. This seems at the time of writing a very optimistic scenario. My assumption of half the population being infected seems wrong.

The increase in testing suggests that far more than 10% of cases of COVID-19 are being reported, but I see no indications of how many this would be. Indeed, if many people experience the infection, but show no symptoms at all, it seems difficult for anyone to estimate their numbers without either very extensive testing of a population, or testing of a very carefully selected sample. I see no way of estimating a variable $f(t)$ from the published data, but it would be possible, by extending the model, to insert an arbitrarily changing value of $f(t)$, going up from say 10% at the start to some much higher value, perhaps 70% at some later date.

Figure 3.4 shows my estimates of the aggregate mortality rate $m(t)$ on the two models. It seems to have been quite high, about 2% in the first wave, but much lower, below 1/2% by the end of the period. It is reported that treatment has greatly improved, so this is not implausible. But the estimates are greatly affected by the estimate of the proportion reported. If the observed deaths (all of which are included in the reported cases) come from only 10% of those actually sick, then the population "case fatality rate" is only one tenth of that observed among the reported cases, but if the reported cases are half the total actual cases, then the observed deaths come from a smaller total number of cases and the cases and the rates should just be halved. It is possible that my observed reductions in mortality are in fact caused by the increase in testing, and consequent changes in $f(t)$.

3.9   Further Extensions: Models 3 and 4

Taking account of these various considerations I add two more models. In Model 3 I vary $f(t)$, starting as before at 10%, then increasing by 1/2% per day from 16 July 2020 to reach 70% by 12 November 2020 and leaving it at 70% thereafter. This is rather arbitrary, but it does produce the result that by 31 December 2020, when the number of reported cases exceeded 2.4 million, my estimate of the number of actual cases is about 6.9 million or just over 10% of the total population of the U.K. (I assume that this is 67.8 million). This accords with press reports of other estimates made at that time.

In early 2021 vaccination on a large scale started in the U.K., and elsewhere. This can easily be inserted into the model. One has an extra status, *Vaccinated*, with $V(t)$ persons in it. People are transferred from Healthy at so many per day depending on available and estimated figures. They come into the denominator of $g(t)$, but are excluded from the numerator; it is assumed that the infectious can meet the vaccinated, but not infect them. So with more vaccinations the value of $g(t)$ is reduced, and consequently the net infection rate reduces and the epidemic diminishes. This can be fitted into my simple model.

In Model 4 I introduce vaccination. I simplify this greatly by assuming that only one vaccination is necessary to give full immunity, with no delays. I use the published figures for first vaccinations given, with a little estimation initially, starting on 26 December 2020 with about 140,000 vaccinations per day, increasing to about 350,000 per day during February 2021, and I then assume 350,000 per day thereafter, until all Healthy are vaccinated by about the middle of June. This is unrealistic in that I assume that the first vaccination is wholly effective, and that all persons of all ages, except for those who have been infected with COVID, are vaccinated, including "antivaxers" and infants and children.

The results are shown in Figs. 3.5, 3.6, 3.7, 3.8 along with those for Model 2 (I now discard Model 1).

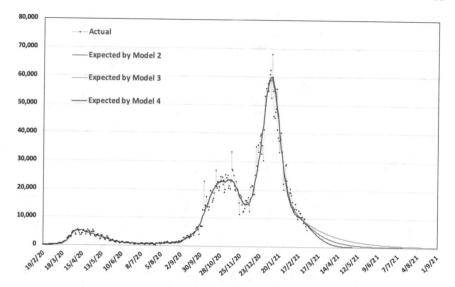

Fig. 3.5 Actual and expected daily reported new cases of COVID-19, U.K., Models 2–4

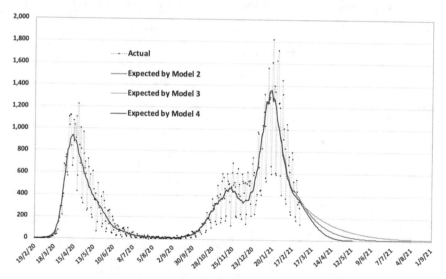

Fig. 3.6 Actual and expected daily reported deaths of COVID-19, U.K., Models 2–4

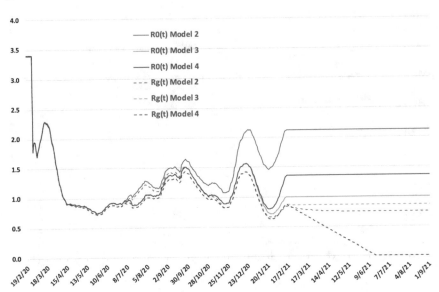

Fig. 3.7 Estimated values of infection rates, $R_0(t)$ and $R_g(t)$, U.K., Models 2–4

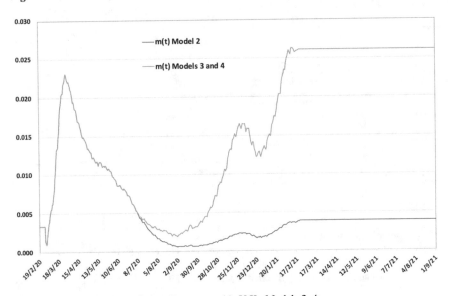

Fig. 3.8 Estimated values of mortality rates, $m(t)$, U.K., Models 2–4

3.10 Comments on Results of Models 3 and 4

We can see from Fig. 3.5, as in Fig. 3.1 that the numbers of actual daily new cases have dropped during the latter part of January 2021. Presumably this is the result of the most recent lockdown at that time. My estimated $R_0(t)$ drops to just below 1 for Model 3, but is higher for Model 4. But the net rate, $Rg(t)$, for Model 3 is below 1, at about 0.8, and for Model 4 is much lower. All three curves of expected new cases in Fig. 3.5 show them declining quite rapidly, with the too optimistic Model 2 being lowest, Model 3, with variable $f(t)$, the highest and Model 4, with vaccines, intermediate. By September, the latest date shown here, the expected number of cases in Model 3 is about 100 per day, with a handful of deaths. This Model assumes that the latest estimated value of $r(t)$ continues indefinitely, which is unlikely to be the case if lockdown measures are relaxed as the cases reduce.

Fig. 3.6 shows the same patterns of results for expected deaths.

It is interesting, however, that with the assumptions of Model 3 the epidemic dies out long before everyone has been infected. Continuing the model projections gives ultimately about 10 million infections, of which about half, or 5 million, are reported, and there are about 140,000 deaths in total. On these assumptions, population immunity is not achieved, nor required. But perhaps a permanent lockdown is required for this model to be valid, which would have other severe consequences, social and economic, which I do not go into here.

With Model 4 the estimated $r(t)$ rate when vaccinations start rises above that estimated with Model 3. As many become vaccinated, one needs a higher transmission rate to get the same number of expected cases, to match the actual numbers. But the net rate $Rg(t)$ falls well below that for Model 3 because of the vaccinations. On my assumptions the entire uninfected population is vaccinated by later in June 2021, and a few weeks before that the expected numbers of reported new cases and of deaths have dropped to a handful. Within my homogeneous population I do not separate out children, but very young children might well not be vaccinated at all; and also there will be some people that decline to be vaccinated or have some medical reason for not being vaccinated. I also assume no further imports of infection, so Model 4 is likely to prove too optimistic.

3.11 Projection Models

Using my estimation results as at the end of February 2021 it is possible to make experimental projections of the numbers, on different, but arbitrary, assumptions. I make four of these, called 5A, 5B, 5C and 5D and the projected numbers of new reported cases and of deaths are shown in Figs. 3.9 and 3.10.

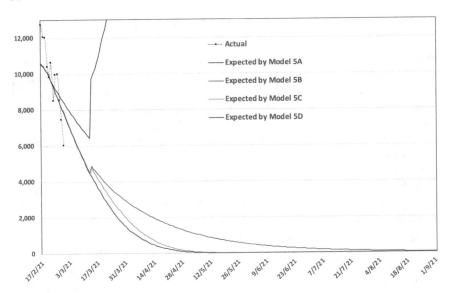

Fig. 3.9 Expected daily reported new cases of COVID-19, U.K., Models 5A–5D

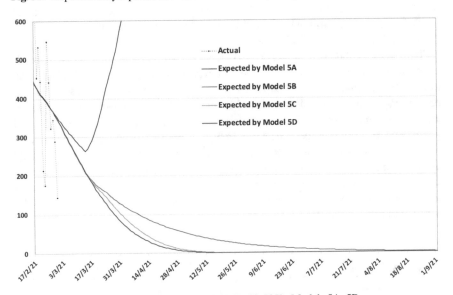

Fig. 3.10 Expected daily reported deaths of COVID-19, U.K., Models 5A–5D

In Model 5A I continue Model 3, assuming that no vaccinations have happened (or alternatively that vaccinations have no effect whatever), and then that on 8 March 2021 most lockdown measures are relaxed, and the $r(t)$ rate goes up immediately to 0.155, about the same as my highest estimated rate in the middle of December 2020, but still well below my highest estimated rate in March 2020 of almost 0.23. The result of this would be a rapid increase in both cases and deaths, with reported cases in June and July exceeding 200,000 a day, and deaths exceeding 8,000 a day, both far higher than in previous waves. These are a long way off the scale of the charts.

By the end of July this projection has $g(t)$ falling to below 0.5, so half the population would have been infected. The numbers of new cases and new deaths would decline to a low level by the end of 2021, by which time total deaths would have reached about 850,000 (compared with about 116,000 at the end of February 2021), but $g(t)$ would still be about 0.45, so not much more than half the population would have been infected. I do not suggest this as a likely projection, but as a warning "what if" one.

In Model 5B I continue Model 4, allowing for the vaccinations that have taken place, but assuming that they stop entirely on 8 March 2021, at the same time as the $r(t)$ rate goes up to 0.155, as in Model 5A. The effect is quite different from that of Model 5A. There is a small jump up in the expected numbers of cases with a continuing fall thereafter, and a continuing fall in deaths. The number of vaccinations is already enough to reduce $Rg(t)$ to well below unity.

In Model 5C I modify Model 5B by assuming that vaccinations continue as in my Model 4, until the whole population is vaccinated by later in June, but with the same relaxation in lockdown as in Models 5A and 5B, rising to 0.155 on 8 March 2021. The projected numbers of aces and of deaths is a bit lower than in Model 5B. and they fall to zero once the whole population is vaccinated.

In Model 5D I follow the dates of what has already been announced by the U.K. government (actually only in respect of England, but I make no distinction in my model between the different nations in the U.K., and England accounts for much the largest part of it). I assume roughly equal steps in increasing $r(t)$, from an estimated 0.137 at the end of February to 0.140 on 8 March 2021, to 0.144 on 29 March, to 0.148 on 12 April, to 0.151 on 17 May and finally to 0.155 on 21 June. These dates have been announced as the earliest dates of steps for unlocking, but I assume that they are in fact realised. The results show projected new cases and deaths even lower than in either of the two previous models.

These experiments suggest that vaccination is likely to do the required job, and to allow certain amount of relaxation of lockdown. But my assumptions that the first vaccine is 100% effective, that the whole population is vaccinated, that there are no imported cases, and no nasty new variants of the virus, are all on the optimistic side.

3.12   Problems and Unknowns

I have mentioned above some of the difficulties with this simple model. I make many assumptions, and I do not know whether these are correct. I assume that those infected show symptoms on day 6; this should perhaps be a distribution of different days. I assume initially that only 10% of cases show bad enough symptoms to go to hospital or even be tested, with this the proportion increasing with Models 3 and 4, but even those assumptions may be wrong.

I assume that those infected are fully infections for five days, but only half as infectious for the next ten days, and not at all thereafter. But many cases are reported of people being sick for much longer than this, and perhaps for much less, so their infectiousness may also vary. I assume equal mortality rates for days 6–15, but this is arbitrary, and cases are reported of much longer periods before death has occurred.

I assume that after some days (15 so far), those infected cannot pass on the disease and cannot themselves get it again. This latter seems in a few cases to have been incorrect, and we cannot know how long any immunity might last, until a longer period has passed. Nor for certain do we know what immunity vaccines will provide.

I assume a single homogenous population, but it is well known that mortality rates vary greatly by the age of the individual and to a lesser extent the sex, and also to some extent the ethnic status. The extent to which different individuals are able to isolate themselves may vary very much by age, and perhaps by other factors. The fit elderly may find it easy to be very isolated at home, and so run very little risk of infection. The seriously ill elderly may be in a care home, and at high risk. So the population available for infection may vary with time quite a lot, depending on what people do and can do; go to work or work from home; live alone or with an extended family; have children at school or not at school. In different stages of lockdown these may have had very varying influence.

Elaborations that would be needed, or at least desirable, in a more comprehensive model would be variation by age and sex, perhaps stratification into a smallish number of discrete classes. This would require the attribution of a different mortality rate to each class, perhaps calculated as an overall $m(t)$ multiplied by a class-specific ratio, to give a value for each class. One might be able to get such relative values from published data. Then one would wish to have a square table of contacts between members of each class with each other class, to vary the relative level of infectiousness across classes; it might be very difficult to get estimates for this.

A further elaboration would be to model region or localities separately. It has been clear that the disease has affected different areas differentially, some having high rates and others low at the same time, but then perhaps then reversing; some having increasing rates and others decreasing rates. Further, different levels of lockdown have been imposed in the different parts of the U.K., both in the separate countries and in separate areas within those countries. Allowing all the variability described above for age classes, and the further complications of contacts between areas might well magnify the model out of the realm of practicability.

I have also assumed that, apart from a single imported infection at some starting date, there has been no contact with other countries. This is manifestly untrue. Almost certainly there were several early imports from different places, and these have probably continued. It would also be the case that other countries have had imports from the U.K. and elsewhere. To identify these cases would be very difficult, and it is not obvious how one would model this feature, except by arbitrarily introducing a new imported infection every so often.

3.13 Other Countries

It is possible to fit these models to the daily data from other countries, but it is difficult to interpret any results without local knowledge of the conditions in those countries. The data may be collected by different regions within another country, and may be defined differently; the reliability of reporting may differ; the social and economic conditions of countries differ, and the various lockdown measures taken or not taken by the relevant authorities may also differ very much. I do not show any results for other countries here.

3.14 Conclusions

Experimenting with models such as I have described may give some insight into features of an epidemic, and they can be made to fit past data tolerably well. But they seem to be of less use in prediction. They can give "what if?" results, showing what would happen if certain assumptions about a stationary, or a changing, future were to occur, but less use in knowing what will actually happen. That depends very much on the actions of governments, the responses of individuals to governments and to the disease, on medical improvements in treating those affected, on whether vaccines will be available and on their possible efficacy, and also on what the virus itself does, with possible new mutations with different characteristics.

However, I hope that the experiments I have done shed some light on infection models for those not previously familiar with them, and also show how experimental projections might allow a better understanding of the effects of different government actions. But I hope that the models used by those who advise governments are considerably better than this very simple one.[1]

---

[1] The author has placed specimen Excel worksheets, programmes, results, updates and other material on his website at: https://davidwilkieworks.wordpress.com/.

References

Continuous Mortality Investigation The Analysis of Permanent Health Insurance Data. *CMI Reports* **12**, 1–263 (1991). https://www.actuaries.org.uk/system/files/documents/pdf/cmir12all.pdf

European Centre for Disease Prevention and Control (2021). https://www.ecdc.europa.eu/en/publications-data

Office of National Statistics (2021). https://coronavirus.data.gov.uk/details/download

M.V. Wüthrich, Corona COVID-19 analysis: Switzerland and Europe, 18 April 2020. SSRN: https://ssrn.com/abstract=3565765

Open Access This chapter is licensed under the terms of the Creative Commons Attribution 4.0 International License (http://creativecommons.org/licenses/by/4.0/), which permits use, sharing, adaptation, distribution and reproduction in any medium or format, as long as you give appropriate credit to the original author(s) and the source, provide a link to the Creative Commons license and indicate if changes were made.

The images or other third party material in this chapter are included in the chapter's Creative Commons license, unless indicated otherwise in a credit line to the material. If material is not included in the chapter's Creative Commons license and your intended use is not permitted by statutory regulation or exceeds the permitted use, you will need to obtain permission directly from the copyright holder.

Chapter 4
Stochastic Mortality Models and Pandemic Shocks

Luca Regis and Petar Jevtić

Abstract After decades of worldwide steady improvements in life expectancy, the COVID-19 pandemic produced a shock that had an extraordinary immediate impact on mortality rates globally. This shock had largely heterogeneous effects across cohorts, socio-economic groups, and nations. It represents a remarkable departure from the secular trends that most of the mortality models have been constructed to capture. Thus, this chapter aims to review the existing literature on stochastic mortality, discussing the features that these models should have in order to be able to incorporate the behaviour of mortality rates following shocks such as the one produced by the COVID-19 pandemic. Multi-population models are needed to describe the heterogeneous impact of pandemic shocks across cohorts of individuals. However, very few of them so far have included jumps. We contribute to the literature by describing a general framework for multi-population models with jumps in continuous-time, using affine jump-diffusive processes.

4.1 Stochastic Mortality Models and the COVID-19 Shock

The life expectancy of human individuals worldwide has been steadily increasing since World War II. The Organisation for Economic Co-operation and Development (OECD) estimates that life expectancy worldwide increased by more than 25 years in the last 70 years, moving from 45.7 years in 1950 to 72.6 in 2019. Mortality rates have constantly been declining at all ages, spanning from infants to the elderly. This phenomenon relies mainly on the economic progress of nations, which improved people's well-being, habits, nutrition, and healthcare consumption.

L. Regis (✉)
ESOMAS Department, University of Torino and Collegio Carlo Alberto, Corso Unione Sovietica 218/bis 10134, Torino, Italy
e-mail: luca.regis@unito.it

P. Jevtić
School of Mathematical and Statistical Sciences, Arizona State University, Wexler Hall, Rm 341 |, 871804, Tempe, AZ 85287-1804, USA
e-mail: petar.jevtic@asu.edu

© The Author(s) 2022
M. C. Boado-Penas et al. (eds.), *Pandemics: Insurance and Social Protection*,
Springer Actuarial, https://doi.org/10.1007/978-3-030-78334-1_4

Advances in medicine have allowed us to prevent and cure common and less common diseases. This progress, with different intensities, was shared by all regions and countries in the world. Importantly, improvements in life expectancy and mortality rates constantly exceeded the expectations. While being good news for humanity, higher-than-expected mortality improvements generated unexpected increases in the value of the liabilities of life insurance companies engaged in the annuity business, of pension funds and public pension schemes. This fact led actuaries to focus on the so-called longevity risk in the last thirty years, i.e., the risk of unexpected improvements in the mortality of individuals. Modelling aims at capturing the uncertainty in the changes of future mortality rates. Modelling longevity risk became crucial for two reasons in particular: first, to assess the likelihood and impact of deviations from expectations; second, to price and hedge longevity risk via risk mitigation techniques such as the use of derivative contracts.

The seminal contribution by Lee and Carter (1992), who first described mortality rates via stochastic processes, paved the way for extensive literature which tried to capture the essential features observed in the mortality dynamics and project them in future. Progressively, the literature moved from the modelling of mortality rates of single populations to the joint modelling of the mortality of multiple socio-economic groups and populations, which display interconnected and specific features at the same time. Because the mortality improvements in the last 100 years in the vast majority of countries (with some notable exceptions due to the effects of wars, as for Italy and Germany during World War II) were following a substantially stable trend, only a few of the proposed models considered adding jumps in the mortality dynamics, and the application of such models has been limited. Nonetheless, the COVID-19 pandemics in 2020 reminded us that sudden shocks to mortality rates might occur, although with (hopefully) relatively low frequency. COVID-19 showed us how epidemics in a highly interconnected world could spread rapidly across continents, affecting people's lives and health conditions. This chapter reviews the literature on stochastic mortality models with jumps whose interest will likely surge in the coming years. To ground our analysis, we first describe briefly in Sect. 4.2 the immediate impact that COVID-19 had on the mortality rates in 2020. This effort helps us highlight that the 2020 shock had largely different effects:

- across countries, as some were better able to contain the spread of the virus than others and/or because their health system was better prepared to respond to the emergency;
- across ages and sexes, because the observed lethality of COVID-19 was higher for males and older people;
- across socio-economic groups, especially in countries with mostly private health systems, due to unequal wealth levels and health-care quality and consumption.

We also point out that it is, at present, unclear whether the shock to mortality rates will have persistent effects or transitory effects only. Direct and lasting effects of "long-COVID" and the impact that the strict lock-down measures adopted in many countries are having on people's habits, health, and economic situation will become apparent in the future. Also, there were indirect effects due to the stress placed by COVID-19 on the health systems, which had to limit to some extent non-COVID-related treatments of other diseases and screenings. Individuals who reduced health-related consumption may have consequences that will become apparent after some time.

With these aspects in mind, Sects. 4.3 and 4.4 review the stochastic mortal-ity models that account for jumps in the dynamics of mortality rates, developed either in discrete-time or in continuous-time set-ups. Section 4.3 focuses on single-population models, while Sect. 4.4 considers multi-population ones. We believe this last stream of literature is particularly relevant in light of the heterogeneous effects of the COVID-19 pandemic shock on mortality rates because it can account for common sudden shocks, which may have different impacts across countries and/or socio-economic groups. Recognizing a gap in the literature because, to our knowl-edge, none of the continuous-time multi-population models proposed so far have considered the presence of jumps, in Sect. 4.5 we generalise the continuous-time multi-population framework in Jevtić and Regis (2019) to the case of jump-diffusive affine processes. Section 4.6 provides some concluding remarks.

4.2 The Impact of COVID-19 on Mortality Rates

Up to 2020 almost uninterrupted improvements in mortality rates were observed in all countries since the end of World War II. Figure 4.1 supports this statement, by

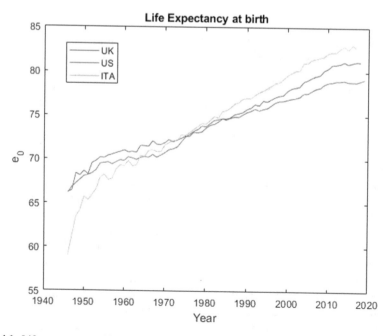

Fig. 4.1 Life expectancy at birth in UK, US and Italy from 1946 to 2019. *Source* Human Mortality Database

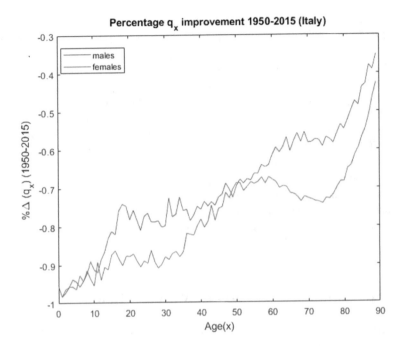

Fig. 4.2 Improvements in annual death probabilities at ages 0–90 in Italy between 2015 and 1950. *Source* Human Mortality Database

portraying the time-evolution of life expectancy at birth for Italy, the United Kingdom and the United States from 1946 to 2019. Nonetheless, the figure shows that the extent of such improvements varied temporally and by country. Mortality decline has been heterogeneous across ages and sexes as well, as testified by Fig. 4.2, which displays the change in conditional yearly death probabilities by age and sex for Italy between 1950 and 2015.

The year 2020, however, brought a worldwide-shared setback to mortality rates' decrease. In the absence of yearly official 2020 data, to describe the impact of the COVID-19 pandemic on mortality rates of several countries, we use weekly data from the STMF (Short-Term Mortality Fluctuations) dataset provided by the Human Mortality Database. First we consider the changes in the total death rates. Figure 4.3 shows their 2020 weekly series for 6 OECD countries (Spain, Italy, France, Germany, the United Kingdom and the United States) and compares their value with the average of the previous 5 years. The figure displays remarkable deviations from the average for all the countries, starting from the spread of the COVID-19 diseases in March. It highlights also that the pandemic shock affected the mortality rates in different countries differently, both in terms of timing and severity. As for timing, with the exception of Germany, we see total death rates spike in coincidence with the first wave of contagions in March. While during the summer death rates were in line with the average of the previous 5 years in all countries, with the exception of the United

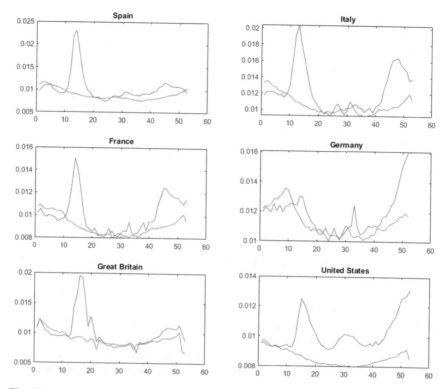

Fig. 4.3 Weekly measured total death rate for 6 countries in 2020. *Source* SMTF Dataset

States, we see the effects of the second wave of contagions in the fall after week 40. As for severity, total death rates, which have been steadily decreasing overtime in the last decades,[1] increased in 2020 by 18.6% in US, 18% in Spain, 15.9% in Italy, 12.2% in Great Britain, 9.89% in France, and 5.3% in Germany compared to their 2015–2019 averages. In Table 4.1 we report the percentage increase in the death rate for the six countries considered for the three age groups—65–74, 75–84 and 85+—which displayed the highest lethality level to the disease. We also distinguish between males and females. The table highlights that, with few exceptions, males death rates for all age groups deteriorated more than females'. The evidence collected in this section is material to our discussion on stochastic mortality models. Indeed, it allows us to stress that the pandemic shock due to the spread of the COVID-19 disease had heterogeneous severity on different countries and age groups. Andrasfay and Goldman (2021) document the disproportionate impact of the shock on the hispanic America and African America sub-populations in the US, for whom the estimated drop in life expectancy in 2020 is far more severe than for the general population. A very important aspect which cannot be ascertained from the data as of yet is to

---

[1] The average annual change in death rates ranges between −1% and −2% for all age classes and countries in the sample.

Table 4.1 Percentage increase in the 2020 death rate versus 2015–2019 average. *Source* Authors' elaboration from the SFTM Dataset

Country	Males			Females		
	65–74 (%)	75–84 (%)	85+ (%)	65–74 (%)	75–84 (%)	85+ (%)
Spain	12.29	9.72	13.00	15.38	8.3	11.73
Italy	14.29	14.20	9.99	8.33	10.48	11.71
France	8.12	4.94	8.12	7.40	3.29	7.75
Germany	0.53	4.32	2.13	−0.83	3.42	0.5
Great Britain	10.9	9.24	9.49	7.16	6.05	6.91
United States	16.49	12.61	11.06	12.59	10.09	11.45

what extent the shock to mortality rates we observed in 2020 will cause long-lasting effects, possibly producing a change in the observed mortality trend. Some countries, including Germany and the United Kingdom, experienced the peak of contagions in the first months of 2021, and hence the 2021 mortality figures are expected to be affected by the pandemic shock as well as the 2020 one. How much of the shock will affect mortality rates in future years is at present hard to predict. All in all, the stylised facts and considerations presented in this section make the case for a multi-population analysis and modelling of mortality when pandemic shocks are specifically considered.

4.3 Stochastic Mortality Models and Pandemics: Single-Population Models

The previous section describes the characteristics of the sudden worldwide shock to mortality rates that occurred in 2020. This section reviews the stochastic mortality models proposed in the literature to account for such features. We consider single-population models, distinguishing between contributions in which mortality rates are modelled in discrete-time and continuous-time.

4.3.1 Discrete-Time Single Population Models

The Lee-Carter (1992) model was among the first stochastic models proposed for mortality, and today it is probably the most well-known. Mathematically can be represented as follows:

$$ln(m_{x,t}) = \alpha_x + \beta_x k_t + e_{x,t}. \tag{4.1}$$

In the equation above, the central death rate for age x in year t, $m_{x,t}$, is a function of a time-varying factor k_t, called the mortality index, which is common across ages but impacts them differently through the coefficients $\beta_x \in \mathbb{R}$. α_x is a constant age-dependent term which represents the baseline level of mortality for age x. Model estimation, which is traditionally achieved by applying a two-stage procedure, captures sudden jumps in mortality through the changes in the k_t values. However, at least two considerations are in order. First, mortality changes captured through k_t affect the different ages through β_x independently of the magnitude of the change. This prevents the model from capturing the age-specificities of pandemic or war-related shocks, as pointed out by Liu and Li (2015). Second, when forecasting, it is necessary to select a model for the k_t series. The modelling choice will obviously impact how past observations shape the distributional properties of future mortality rates. The standard solution is to employ a random walk with drift to model k_t:

$$k_t = \mu + k_{t-1} + \sigma Z_t, \tag{4.2}$$

where μ is the drift term and Z_t is a standard normal. This choice rules out the presence of jumps and their occurrence in projections. Usually, in line with this reasoning, outliers, such as the 1918 Spanish flu-related spike, are either excluded by the estimation process or they are dealt with using ad-hoc interventions (see Li and Chan 2005). Some works have generalised the Lee-Carter model and its most notable extensions (such as the one proposed by Renshaw and Haberman (2006) accounting for cohort effects) to specifically account for jumps in annual mortality rates. Milidonis et al. (2011) introduced a regime switching model, which can capture both jumps and changes in the mortality volatility pattern via the different regime states. Wang et al. (2013) extended the Renshaw and Haberman (2006) model, considering jump processes for the error term. Chen and Cox (2009), building on the continuous-time model proposed by Cox et al. (2006), modified the process k_t in Eq. (4.2) to include a jump component. They proposed two specifications, both based on the inclusion of a jump occurring with probability p and producing a shock $Y_t \sim N(m, s^2)$. One assumes that jumps generate permanent shifts to the mortality index, which is thus modelled as:

$$k_t = k_{t-1} + \mu - pm + \sigma Z_t + Y_t N_t, \tag{4.3}$$

where N_t is a Bernoulli distribution which takes value 1 with probability p and 0 otherwise. An alternative specification assumes that jumps have transitory effects, lasting one period only:

$$k_t = k_{t-1} + \mu + \sigma Z_t + Y_t N_t - Y_{t-1} N_{t-1}, \tag{4.4}$$

The authors claimed that this second choice is more appropriate, because the most severe shocks observed over the last century, such as the pandemic flu of 1918 and the tsunami of 2004 had negligible permanent effects. Whether this will remain true for the COVID-19 pandemic shock is an open question. While epidemiological similarities with the Spanish flu may suggest that development, the severe downturn experienced by the economies worldwide may produce long-term consequences jeopardising the steady decrease in the mortality index that we have observed so far over previous decades. From the modelling point of view, this may imply changes in the mortality trends, as discussed by Sweeting (2011). The models proposed by Chen and Cox (2009) can be estimated via Conditional Maximum Likelihood. If applied to the US data from 1900 to 2003, the models estimate the probability of observing a jump in k_t to be around 4% under both specifications. When a jump occurs, in the transitory-effect model, the sudden increase in k_t it produces is 4 times greater (in absolute value) than its annual negative drift. The importance of accounting for jumps in the mortality index is confirmed by Özen and Şahin (2020), who improved the model fit by allowing for a non-constant mean time between jump arrivals using renewal processes. In the works we have reviewed so far, jumps affected the mortality index. Thus, jump-driven shocks do not have age-specific effects different from the Brownian shock captured by the no-jump equivalent mortality index. To overcome this issue, Liu and Li (2015) specify the model as follows:

$$\ln m_{x,t} = \alpha_x + \beta_x k_t + N_t J_{x,t} + e_{x,t}, \tag{4.5}$$

where $J_{x,t}$ is a time-dependent response to the presence of a jump. $J_{x,t}$ can then be specified as being an age-dependent response to a common shock or as being fully age-specific. The authors provide evidence that accounting for age-specific jump responses is important to improve the model fit. Given the evidence of Sect. 4.2, such a feature appears crucial when including the COVID-19 shock in the data.

4.3.2 Continuous-Time Single-Population Models

Before the Lee-Carter model was extended to account for jumps, jumps had already been introduced in continuous-time mortality models. Indeed, although introduced by Milevsky and Promislow (2001) almost 10 years later than Lee and Carter (1992), continuous-time stochastic mortality models with jumps were proposed just few years later by Biffis (2005) and by Luciano and Vigna (2008). In the continuous-time framework, the time to death of individuals is modelled as the first jump time of a time-inhomogeneous Poisson process with stochastic intensity. The main advantage of such a framework is that, if the mortality intensity process is chosen within the class of affine processes—as in Biffis (2005) and Luciano and Vigna (2008)—survival probabilities are available in closed form. Moreover, the mortality model can be coupled with standard financial risk models to obtain prices of fairly evaluated life insurance policies and their hedges (see Luciano et al. 2012). More formally, a mortality intensity μ_t^x for an individual aged x at time 0 is affine if it is assumed to be an affine function of a jump-diffusive affine process X_t such that:

$$dX_t = \delta(t, X_t)dt + \sigma(t, X_t)dW_t^x + dJ_t^x, \tag{4.6}$$

where W_t^x is a standard Brownian motion, J_t^x is a pure-jump process, and $\delta(\cdot)$, $\sigma(\cdot)\sigma^T(\cdot)$, and the jump-arrival intensity have an affine dependence on X_t. The symbol ^{T} denotes matrix transposition. Within this set-up, the survival probability up to time T, $S_x(t, T)$, under some technical conditions, can be obtained as:

$$S_x(t, T) = \mathbb{E}\left[e^{-\int_t^T \mu_s^x ds} \right] = e^{\alpha(t,T)+\beta(t,T)\cdot X_t}, \tag{4.7}$$

where $\alpha(\cdot)$ and $\beta(\cdot)$ solve a system of Ordinary Differential Equations (ODEs) which depend on the model specification. These processes are very flexible and easy to calibrate. They have been extensively applied in the pricing (Wills and Sherris 2010), hedging (Luciano et al. 2012) and portfolio choice (Menoncin and Regis 2020) contexts. Luciano and Vigna (2008) showed that for several generations, in the Italian population, accounting for jumps improves the model fit, in particular when the intensity follows a Gaussian process. However apart from the analysis in Luciano and Vigna (2008), most of the literature restricts the application of the affine framework to purely diffusive processes (Schrager 2006; Jevtić et al. 2013, for instance). An exception is Luciano et al. (2008), where the dependent lives of couples are considered. Hainaut and Devolder (2008) took instead a different approach, based on the use of pure-jump Lévy processes.

4.4 Stochastic Mortality Models and Pandemics: Multi-population

Pandemic shocks have the crucial characteristic of manifesting themselves globally in a short time frame, affecting the mortality patterns of many populations (countries) and sub-groups within populations. Capturing the heterogeneity and the dependence of such impacts across different groups is a non-trivial task, which requires the use of multi-population models.

Multi-population models have started emerging in the mid-2000s to capture the joint evolution of mortality dynamics across countries and/or socio-economic groups within the general population. In this section, we review the most prominent, highlighting that only two contributions, up to our knowledge, included jumps in a multi-population setting.

4.4.1 Discrete-Time Models

Li and Lee (2005) applied the original Lee-Carter model to describe the mortality rates of multiple countries jointly. The model they proposed assumes a common mortality index driving the mortality rates of different populations. Several papers extended this set-up to account for issues such as (semi)-coherence (i.e. long-run convergence of the mortality rates of several populations, see Li et al. 2017), Bayesian estimation (see Antonio et al. 2015), cointegration (see Yang and Wang 2013; Jarner and Jallbjørn 2020), factor-based approaches (see Chen et al. 2015). To our knowledge, only Zhou et al. (2013) and Özen and Şahin (2021) considered two-population models with jumps in the discrete-time framework. Zhou et al. (2013) modelled jointly the dynamics of two populations, assuming that:

$$\ln m_{x,t}^j = \alpha_x^j + \beta_x^j k_t^j, \, j = 1, 2, \tag{4.8}$$

where $m_{x,t}^j$ denotes the central death rate for the individual aged x at time t and belonging to population j. They define

$$k_t^j = \hat{k}_t^j + N_t^j Y_t^j, \tag{4.9}$$

where $\hat{k}_t^j$ is the stochastic effect free of jumps, N_t^j is the counting process for jumps at time t in the mortality of population j, whose range is $\{0, 1\}$ and whose distribution may be dependent across populations, and Y_t^j is the severity of a jump at time t for the j-th population. Jumps have transitory (one-period only) effects. While imposing a stationary $AR(1)$ process on $\hat{k}_t^{(1)} - \hat{k}_t^{(2)}$, the framework allows for different—but possibly dependent—jump times, frequencies and severities. The model, applied to the populations of Sweden and Finland, is shown to fit the data well, better than the corresponding no-jump process, because it is able to better capture the presence of outliers, such as the 1918 spike due to the Spanish flu. Recently, Özen and Şahin (2021) introduced jumps in a two-population model, modelling jumps in one of the two populations through a renewal process and linking the other population using a Common Age effect model.

4.4.2 Continuous-Time Models

In continuous-time, only a few papers have focused on two-population or multi-population models. Recently, building on the seminal contribution by Dahl et al. (2008), three papers have proposed multi-population continuous-time mortality models. De Rosa et al. (2021) proposed a two-population model that describes the dependence structure among populations and cohorts within them. The stochastic mortality intensities there follow a square-root process, which guarantees non-negativity. Similarly, Sherris et al. (2020) introduced a model in which common and idiosyncratic Gaussian factors drive the mortality surface of two populations. Jevtić and Regis (2019) described a general set-up for multi-population continuous-time stochastic mortality models using affine processes. All these works considered purely diffusive processes. To our knowledge, no multi-population model in continuous-time proposed so far includes jumps. In the next section, to fill this gap, we present an extension of the general framework in Jevtić and Regis (2019) to jump-diffusive affine processes.

4.5 A Continuous-Time Multi-population Model with Jumps

We propose here a continuous-time multi-population model with jumps. Consider the following setting. The mortality intensity of an individual aged $x + t$ at time t belonging to population j, with $j = 1, ..., M$, is defined as:

$$\mu_{x+t}^{j} := g_0^{j}(x + t, t, x) + g_1^{j}(x + t, t, x)X_t^{j}, \tag{4.10}$$

where X_t^{j} is a stochastic process in $\mathbb{R}^N$, $g_0^{j}(x + t, t, x)$ is a base-line level of mortality for the individual and $g_1^{j}(x + t, t, x)$ is the vector of responses of X_t^{j}. Here, X_t^{j} is described by the following stochastic differential equation:

$$dX_t^{j} = A^{j}(X_t^{j}, t)dt + \Sigma(X_t^{j}, t)dW_t + dZ_t, \tag{4.11}$$

where W_t is a vector of standard Brownian Motions and Z_t is a pure-jump process. Notice that W_t and Z_t are in principle the same across populations. In Jevtić and Regis (2019) the different populations j and ages within them are modelled jointly by assuming that they respond differently to a set of common Brownian noises. To that setting, we add a jump component. Notice that the age-specific and population-specific responses g_1^{j} allow for a very rich description of the heterogeneous effects of Brownian and jump shocks. If functions $A^{j}(\cdot)$, $\Sigma\Sigma^T(\cdot)$ and the jump intensity $\lambda(X_t^{j})$ of process Z_t have an affine dependence on X_t^{j}, the survival probabilities $S_x^{j}(t, T)$ can be computed analytically. Suppressing time-dependence for notational convenience, we assume that:

- $A^{j} = K_0 + K_1 \cdot X^{j}$;
- $(\Sigma\Sigma^T)_{ij} = (H_0)_{ij} + (H_1)_{ij} \cdot X^{j}$;
- $\lambda = l_0 + l_1 \cdot X^{j}$;
- $\theta(\cdot)$ defines the "jump transform" that determines the jump-size distribution.

Then, following Duffie et al. (2000), the survival probability of an individual aged x at time t can be computed analytically as:

$$S_x^{j}(t, T) = \mathbb{E}\left[e^{-\int_t^T \mu_{x+t+s}ds}\right] = e^{\alpha(t,T)+\beta(t,T)\cdot X_t^{j}}, \tag{4.12}$$

where $\alpha(t, T)$ and $\beta(t, T)$ solve the following system of ODEs:

$$\dot{\beta}(t) = g_1^j(x + t, t, x) - K_1^T \beta(t) - \frac{1}{2}\beta(t)^T H_1 \beta(t) - l_0(\theta(\beta(t)) - 1) \quad (4.13)$$

$$\dot{\alpha}(t) = g_0^j(x + t, t, x) - K_0 \cdot \beta(t) - \frac{1}{2}\beta(t)^T H_0 \beta(t) - l_1(\theta(\beta(t)) - 1). \quad (4.14)$$

The solution to this system of equations depends on the specification taken by Eqs. (4.10) and (4.11). Under particular conditions, the solution is available in closed form. For instance, this applies to multi-cohort extensions of the jump processes proposed in Luciano and Vigna (2008) for the single-cohort case. Indeed, when the continuous part of (4.11) follows a multi-dimensional Ornstein-Uhlenbeck-process or a Feller-process and jumps are compound Poisson with constant intensity l and exponentially distributed size, $g_0^j = 0$ and $g_1^j = \mathbf{1}$, the system of equations above has an analytical solution. More generally, such a solution may be approximated analytically. A pure-jump factor through (4.11) may be considered, in particular, to allow for population-specific and age-specific responses to sudden shocks, which may not otherwise be well captured by purely-diffusive model specifications.

4.6   Conclusions

This chapter reviewed the literature on stochastic mortality models featuring multiple populations and jumps, which can better capture the effects of pandemic shocks and their heterogeneous intensity across countries, cohorts, socio-economic groups. On top of that, the chapter proposes a modelling framework based on continuous-time jump-diffusive processes. The framework is flexible and rich enough to describe the observed differential mortality across groups and capture the cohort-specific and population-specific effects of jumps. It has the advantage of providing closed-form expressions for the survival probabilities of ages within sub-populations. Additionally, it can be coupled with standard financial risk models to provide valuation and management tools for insurance and reinsurance products affected by mortality risk. However, the affine framework may be restrictive when it comes to the jump component modeling relative to pure-jump specifications, such as the one proposed by Hainaut and Devolder (2008). Also, model estimation, depending on the model specification, may require particular care. We postpone it to further research and future data availability.

Acknowledgements Luca Regis acknowledges support from the "Dipartimenti di Eccellenza 2018–2022" grant from the Italian Ministry of University and Research (MIUR).

References

T. Andrasfay, N. Goldman, Reductions in 2020 US life expectancy due to Covid-19 and the disproportionate impact on the Black and Latino populations. Proc. Nat. Acad. Sci. **118**(5) (2021)

K. Antonio, A. Bardoutsos, W. Ouburg, Bayesian Poisson log-bilinear models for mortality projections with multiple populations. Eur. Actuarial J. **5**(2), 245–281 (2015)

E. Biffis, Affine processes for dynamic mortality and actuarial valuations. Insur. Math. Econ. **37**(3),
443–468 (2005)

H. Chen, S.H. Cox, Modeling mortality with jumps: applications to mortality securitization. J. Risk
Insur. **76**(3), 727–751 (2009)

H. Chen, R. MacMinn, T. Sun. Multi-population mortality models: a factor copula approach. Insur.
Math. Econ. **63**, 135–146 (2015)

S.H. Cox, Y. Lin, S. Wang, Multivariate exponential tilting and pricing implications for mortality
securitization. J. Risk Insur. **73**(4), 719–736 (2006)

M. Dahl, M. Melchior, T. Møller, On systematic mortality risk and risk-minimization with survivor
swaps. Scand. Actuarial J. **2008**(2–3), 114–146 (2008)

C. De Rosa, E. Luciano, L. Regis. Geographical diversification and longevity risk mitigation in
annuity portfolios. ASTIN Bull., page forthcoming (2021)

D. Duffie, J. Pan, K. Singleton, Transform analysis and asset pricing for affine jump-diffusions.
Econometrica **68**(6), 1343–1376 (2000)

D. Hainaut, P. Devolder, Mortality modelling with Lévy processes. Insur. Math. Econ. **42**(1), 409–
418 (2008)

S.F. Jarner, S. Jallbjørn, Pitfalls and merits of cointegration-based mortality models. Insur. Math.
Econ. **90**, 80–93 (2020)

P. Jevtić, L. Regis, A continuous-time stochastic model for the mortality surface of multiple populations. Insur. Math. Econ. **88**, 181–195 (2019)

P. Jevtić, E. Luciano, E. Vigna, Mortality surface by means of continuous time cohort models. Insur.
Math. Econ. **53**(1), 122–133 (2013)

R.D. Lee, L.R. Carter, Modeling and forecasting U.S. mortality. J. Am. Stat. Assoc. **87** (419),
659–671 (1992). ISSN 01621459

J.S.-H. Li, W.-S. Chan, R. Zhou, Semicoherent multipopulation mortality modeling: the impact on
longevity risk securitization. J. Risk Insur. **84**(3), 1025–1065 (2017)

N. Li, R. Lee, Coherent mortality forecasts for a group of populations: an extension of the Lee-Carter
method. Demography **42**(3), 575–594 (2005)

S.-H. Li, W.-S. Chan, Outlier analysis and mortality forecasting: the United Kingdom and Scandi-
navian countries. Scand. Actuarial J. **2005**(3), 187–211 (2005)

Y. Liu, J.S.-H. Li, The age pattern of transitory mortality jumps and its impact on the pricing of
catastrophic mortality bonds. Insur. Math. Econ. **64**, 135–150 (2015)

E. Luciano, E. Vigna, Mortality risk via affine stochastic intensities: calibration and empirical
relevance. Belg. Actuarial Bull. **8**(1), 5–16 (2008)

E. Luciano, J. Spreeuw, E. Vigna, Modelling stochastic mortality for dependent lives. Insur. Math.
Econ. **43**(2), 234–244 (2008)

E. Luciano, L. Regis, E. Vigna, Delta–gamma hedging of mortality and interest rate risk. Insur.
Math. Econ. **50**(3), 402–412 (2012)

F. Menoncin, L. Regis, Optimal life-cycle labour supply, consumption, and investment: the role of
longevity-linked assets. J. Bank. Finan. **120**, 105935 (2020)

M. Milevsky, D. Promislow, Mortality derivatives and the option to annuitise. Insur. Math. Econ.
29(3), 299–318 (2001)

A. Milidonis, Y. Lin, S.H. Cox, Mortality regimes and pricing. North Am. Actuarial J. **15**(2), 266–
289 (2011)

S. Özen, Ş Şahin, Transitory mortality jump modeling with renewal process and its impact on
pricing of catastrophic bonds. J. Comput. Appl. Math. **376**, 112829 (2020)

S. Özen, Ş Şahin, A two-population mortality model to assess longevity basis risk. Risks **9**(2), 44 (2021)

A.E. Renshaw, S. Haberman, A cohort-based extension to the Lee–Carter model for mortality reduction factors. Insur. Math. Econ. **38**(3), 556–570 (2006)

D. Schrager, Affine stochastic mortality. Insur. Math. Econ. **38**(1), 81–97 (2006)

M. Sherris, Y. Xu, J. Ziveyi, Cohort and value-based multi-country longevity risk management. Scand. Actuarial J., 1–27 (2020)

P.J. Sweeting, A trend-change extension of the Cairns-Blake-Dowd model. Ann. Actuarial Sci. **5**(2), 143–162 (2011)

C.-W. Wang, H.-C. Huang, I.-C. Liu, Mortality modeling with non-Gaussian innovations and applications to the valuation of longevity swaps. J. Risk Insur. **80**(3), 775–798 (2013)

S. Wills, M. Sherris, Securitization, structuring and pricing of longevity risk. Insur. Math. Econ. **46**(1), 173–185 (2010)

S.S. Yang, C.-W. Wang, Pricing and securitization of multi-country longevity risk with mortality dependence. Insur. Math. Econ. **52**(2), 157–169 (2013)

R. Zhou, J.S.-H. Li, K.S. Tan, Pricing standardized mortality securitizations: a two-population model with transitory jump effects. J. Risk Insur. **80**(3), 733–774 (2013)

Open Access This chapter is licensed under the terms of the Creative Commons Attribution 4.0 International License (http://creativecommons.org/licenses/by/4.0/), which permits use, sharing, adaptation, distribution and reproduction in any medium or format, as long as you give appropriate credit to the original author(s) and the source, provide a link to the Creative Commons license and indicate if changes were made.

The images or other third party material in this chapter are included in the chapter's Creative Commons license, unless indicated otherwise in a credit line to the material. If material is not included in the chapter's Creative Commons license and your intended use is not permitted by statutory regulation or exceeds the permitted use, you will need to obtain permission directly from the copyright holder.

Chapter 5
A Mortality Model for Pandemics and Other Contagion Events

Gary Venter

Abstract The crisis caused by COVID-19 has had various impacts on the mortality of different sexes, age groups, ethnic and socio-economic backgrounds and requires improved mortality models. Here a very simple model extension is proposed: add a proportional jump to mortality rates that is a constant percent increase across the ages and cohorts but which varies by year. Thus all groups are affected, but the higher-mortality groups get the biggest increases in number dying. Every year gets a jump factor, but these can be vanishingly small for the normal years. Statistical analysis reveals that even before considering pandemic effects, mortality models are often missing systemic risk elements which could capture unusual or even extreme population events. Adding a provision for annual jumps, stochastically dispersed enough to include both tiny and pandemic risks, improves the results and incorporates the systemic risk in projection distributions. Here the mortality curves across the age, cohort, and time parameters are fitted using regularised smoothing splines, and cross-validation criteria are used for fit quality. In this way, we get more parsimonious models with better predictive properties. Performance of the proposed model is compared to standard mortality models existing in the literature.

5.1 Introduction

Probabilistic mortality models usually assume that deaths are independent, identically distributed Yes/No events, which makes the number of Yes events in a period binomially distributed. For low-probability events such as deaths in a year, binomial distributions are very close to Poisson, which is more convenient for modelling. It is common, however, for heavier-tailed distributions, such as negative binomial, to give better fits, which suggests that there are unmodelled correlated effects. Population

G. Venter (✉)

Actuarial Science, School of Professional Studies, Columbia University, 203 Lewisohn Hall, 2970 Broadway, MC 4119, New York, NY 10027, USA

e-mail: gv2112@columbia.edu

© The Author(s) 2022

M. C. Boado-Penas et al. (eds.), *Pandemics: Insurance and Social Protection*, Springer Actuarial, https://doi.org/10.1007/978-3-030-78334-1_5

events, such as disease outbreaks, extreme weather and natural disasters produce year-to-year jumps and contribute to such effects on mortality rates. This is likely to be a reason the mortality models have heavier-tailed residuals. Including model provisions for such population events can improve the fit of the theoretically appropriate Poisson models, while also providing enough flexibility to account for pandemic experience.

In this chapter, a simple model with mortality jumps every year, which are negligible in ordinary years, is proposed: a single factor, drawn annually from a fixed distribution, increases all the modelled mortality means proportionally. In the example, the jump distribution fit to historical data improves on typical models, and also gives reasonable probabilities for events like the COVID-19 pandemic. The other model parameters are fit to smoothing splines, which have several advantages enumerated below.

A few related papers on pandemic mortality rates have appeared recently. For example, Özen and Şahin (2020) postulate occasional population jump events to better predict the market behaviour of mortality catastrophe bonds. Chen and Cox (2009) use a similar idea for mortality securities in general. Zhou et al. (2013) generalise this to two-population modelling, which they point out is often used for mortality securities. Alijean and Narsoo (2018) test a number of mortality models over an extended period for quality of fit, and find that a jump model like these works better than the well-known continuous models, particularly in that it is able to account for the effects of the Spanish flu epidemic around 1918. Barigou et al. (2021) fit smoothed surfaces to mortality rates, with and without pandemic effects, using methodology very similar to the smoothing splines used here. Cairns et al. (2020) model COVID-19 mortality in greater demographic detail. O'hare and Li (2017) test several standard mortality models of log rates for normality of residuals, and find they almost universally fail their tests. This is actually to be expected from Poisson-distributed mortality rates, and the models here use log links to fit the actual mortality counts. They also find evidence of correlation of residuals, which does indicate the type of systematic effects that the jump model here is aiming to avoid.

Section 5.2 provides a summary of the methodology and findings. Section 5.3 lays out methodology to do parametric regression by fitting smoothing splines across all of the age, period, and cohort variables in MCMC, using Bayesian shrinkage for the smoothing, and discusses the predictive advantages of such smoothing. This is done for both cubic and linear splines, optimising the degree of smoothing by conditional expectations instead of the typical cross-validation methods. Details of the mortality models used and the fitting process are given in Sect. 5.4. Section 5.5 contains the results, as well as a discussion of possible generalisations of the models. Section 5.6 concludes, and some computer code is in the Appendix.

5.2 Highlights of Methodology and Findings

5.2.1 Summary of Methodology

The approach is to apply smoothing splines to fit standard mortality models to the French data, using both Poisson and negative binomial residuals. For the best models tried, the negative binomial fit better for both male and female populations. The models were extended to add an annual mortality multiplier factor drawn from a stochastic process, while using only Poisson residuals.

Mortality data generally comes in annual blocks by calendar age at death, and year of death minus age approximates the year of birth cohort. One class of models has factors for age, period (i.e., year), and cohort. If these factors simply multiply, this is called an APC model. These are usually estimated in log form, so the model is additive, and it can be fit by regression. More complicated models multiply the period log factors by an age modifier, to reflect the fact that medical advances, etc., affect some ages more than others. The Lee and Carter (1992) model has age effects and age-modified period effects. Renshaw and Haberman (2006) add cohorts, and this is called the RH model below. The ages that trend faster can change, however, especially for models over wide age and period ranges. Developing economies, for instance, can initially have sharp improvements in pre-teen mortality, then later shift to the more usual pattern from improved treatments for diseases affecting older people. Hunt and Blake (2014) allow multiple trends over various periods, each with its own age modifiers. Venter and Şahin (2018) used this for US males, and found an additional trend for HIV-related mortality in the 1980s and 1990s for young adults, and another complex trend for ages in the 40s, perhaps related to substance abuse.

Parametric curves are often used across the age, period, and cohort log factors. For instance, Perks (1932) introduces a four-parameter mortality curve that fits pretty well across ages. However the parameters do change from time to time. Xu et al. (2019) use curves from multifactor bond-yield models to fit mortality by age for individual cohorts or cohort groups. These come with automatic projections to older ages, which are like long bond yields in these models.

Cubic splines are often fit across the factors instead of using parametric curves. They are smoother and use fewer parameters than models with parameters at every age, period, and cohort. A recent refinement is to use smoothing splines, which typically are cubic splines with a penalty included to constrain the average second derivative across the curve. Linear splines have generally been considered too jagged for such modelling, but Barnett and Zehnwirth (2000) introduce a form of linear smoothing splines that constrain second differences, which is analogous to how smoothing is done for cubic splines, and these can be fairly smooth as well. Using smoothing splines across regression variables is called semiparametric regression. A detailed source is Harezlak et al. (2018), and Venter and Şahin (2021) apply it to an actuarial model.

5.2.2 Summary of Findings

For the female population, adding the annual jump factors made this Poisson model better than the best previous (negative binomial) model. For the males, doing this was almost as good, but the contagion was not captured completely. More complex possible population event models are discussed in Sect. 5.5.

Figure 5.1 shows the fitted mortality multipliers for both populations. These are estimated in logs as additive terms. The male and female factors are generally similar. Their peaks are at 2003, a year with about 14,000 extra deaths from an extreme heat wave. 2018 also has high factors, and it also had a bad heat wave, but preparations for it were better, so it is not quite as high. The 2020 COVID-19 pandemic had about 5 times as many deaths as in 2003. Under the estimated distribution of factors, for males 2003 was about a 1 in 9 year event and 2020 about 1 in 50 years. As such population events can occur over adjacent years, these frequencies are not unreasonable.

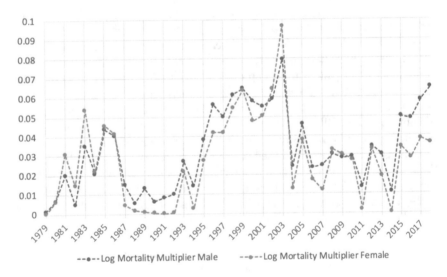

Fig. 5.1 Fitted mortality multipliers

5.3   Semiparametric Regression in MCMC

MCMC is an effective tool for fitting shrinkage splines. In this section, first the theoretical background for this is presented, with Bayesian and frequentist interpretations. Then the design matrices for estimating splines by regression are specified, followed by the hows and whys of parameter shrinkage, including some history. Cross validation is used in semiparametric regression to compare the goodness of fit of models from a predictive viewpoint, and this is relatively simple to do in MCMC. Cross validation is also widely used to estimate the best degree of shrinkage for a model, but this is problematic. A better alternative is available when using MCMC estimation.

5.3.1   MCMC Parameter Shrinkage

Markov Chain Monte Carlo (MCMC) is a method for estimating probability distributions through stochastic sampling. "Markov chain" means that when generating a sequence of samples, each one uses the previous one, without reference to earlier history. It was developed by physicists in the late 1940s as an efficient numerical integration method to model distributions of particle interactions. After enough samples, it has distributional stability—new samples will all be from the same distribution.

In Bayesian statistics, MCMC is used to sample parameter sets from the joint distribution of the parameters and the data. This can be computed as the product of the prior distribution of the parameters and the conditional distribution of the data given the parameters. The latter is the likelihood, so the product is called the joint likelihood. By the definition of conditional distribution, it is also the product of the conditional distribution of the parameters given the data with the probability of the data. The latter is not known, but it is a constant. Thus the joint likelihood is proportional to the conditional distribution of the parameters given the data, and sampling from it with MCMC will provide an estimate of that conditional distribution.

The same thing can be done with random effects. In classical statistics, parameters are unknown constants which do not have distributions. Random effects are like parameters, but they have distributions. These are not subjective distributions but are part of the model specification, just like residual distributions are. They can be evaluated based on the fit and revised as necessary. Random-effects estimation typically just computes the mode of the joint likelihood, but it could estimate the conditional distribution of the effects given the data by MCMC like Bayesians do. Many contemporary Bayesians are abandoning subjective probabilities and consider the priors to be specified distributions that can be revised as needed, just like is done in random effects.

The terminology used here is somewhat intermediate. In the models used, there are no parameters in the frequentist sense, just random effects. But that is not a widely understood term, so it is simpler to just call them parameters. As their postulated distributions are not subjective, the terms "prior" and "posterior" can be misleading. Modern Bayesians insist that these terms do not imply subjective probabilities, but their subjective use is so ingrained in the vocabulary that trying to define them out can lead to confusion.

Smoothing splines involve constraints that limit the second derivatives or second differences of the curves. This can be done in MCMC for each parameter individually by postulating shrinkage distributions. Here, a shrinkage distribution is any distribution with mode at or asymptotic to zero. They give more weight to parameters closer to zero, so favour such parameters. Examples include the standard normal distribution and double-exponential distributions, which mirror the exponential on the negative reals. Gamma distributions with shape parameter less than 1 have mode asymptotic to zero, and are used here for the distribution of the annual event frequency jumps.

The purpose of parameter shrinkage is not just to make smoother curves. It actually can improve the predictive accuracy of models. This will be easier to discuss once the models are formally specified.

5.3.2 Spline Regressions

The APC models are a good place to start, as in log form they are purely linear models. The log model value at any data point is just the sum of the age, period, and cohort parameters for that point.

Mortality data for the illustration is taken from Human Mortality Database (2019). Deaths are in a column listed by year then age within year. In this case this is French female and male data for years 1979–2018 and ages 50–84. These are from 74 year-of-birth cohorts, 1929 to 1968, each with 1–35 ages. The column of death counts is the dependent variable for each model. Also the population exposure for each observation is provided as another column.

A design matrix is then constructed giving dummy variable values in columns for each age, period, and cohort for each observation. For specificity, the first age, period, and cohort variables are not included. There is also a constant term in the model that the code computes outside of the design matrix. The ages, periods, and cohorts are not independent of each other and further constraints are needed, as discussed later. The design matrix thus has 34 columns for age variables, 39 for period variables, and 73 for cohort variables. To fill out the design matrix, reference columns are created for each data observation to specify the age, period, and cohort it comes from.

For a straight regression APC model, the variables are 0,1 dummies. Age i has value 1 in just those rows for which the observation is from age i, and similarly for the period and cohort variables. For linear splines, an alternate regression could be done where all the variables are for the slopes of the line segments connecting the log factors, which are the first differences of the log factors. The age i variable for this would still be 0 for observations from ages less than i and 1 for observations at age i. Later observations would add the later slopes to the existing log factors before those points, so the age i variable would also be included in them. Thus the age i variable would be 1 for observations from ages i and greater, and the same for periods and cohorts.

For the linear splines here, the variables are the second differences of the final age, period, and cohort parameters. Then the variable for age i is still 1 for observations with age i, and 0 for earlier observations. Its value increases the final log factor at ages $k \geq i$ and so is in all those points, just like the slope parameters were. But it also increases the slope (first difference) at i, and so is in all the later first differences as well. Thus it gets added in an additional time for each subsequent point. This means that for a cell from age k, the dummy for age i is $(1 + k - i)_+$. The same is true for periods and cohorts.

Cubic splines have similar design matrices. To define these, consider one direction of variables, like age, with spline-segment variables f_1 to f_n. Then the design matrix value $f_i(z)$ for an observation at age z is a function defined for any real $0 \leq z \leq n$ so the curves can be interpolated. They are: $f_1(z) = 1$, $f_2(z) = z$, and for $i > 2$, $f_i(z) = (z - (i - 2))_+^3 - (n - (i - 2))(z - (n - 1))_+^3$. The second term is nonzero only for $z > n - 1$. The design matrix just has values for integer k. For a cell from age k, the value for the variable for age i is $f_i(k)$. These basis functions, from Hastie et al. (2017),[1] assume that the splines are linear outside of $1 < z < n$. Here they are modified by constant factors by column that change the fitted coefficients but simplify the formulas and put the fitted coefficients more on the same scale.

The average second derivative of the spline is a complicated but closed-form function of the parameters. A smoothing spline minimises the NLL plus a selected constant λ times the average second derivative. This is a penalised regression similar to ridge regression and lasso. It can be readily estimated with a nonlinear optimiser for any given λ. Some experimentation with this found that increasing λ usually shrinks all the spline parameters to a similar degree. This suggests an alternative way to define cubic smoothing splines in MCMC: use shrinkage distributions on the spline parameters to penalise large segment parameters. This also reduces the average second derivative very similarly to the usual smoothing splines, but does not map exactly to a specific shrinkage λ. It does make the splines estimable in MCMC, which has advantages discussed below.

5.3.3 Why Shrinkage?

With these design matrices, the spline models are just regression. Without parameter shrinkage they would give the same fits as the regular 0,1 dummy variables. The reason for shrinkage is based in reduced error variances. This traces back to the Stein (1956) paper, which showed that when estimating three or more means, the error variance is reduced by shrinking the estimates all towards the overall mean to some degree. Actuaries have been doing this heuristically with credibility weighting since

---

[1] A derivation is at https://stats.stackexchange.com/questions/172217/why-are-the-basis-functions-for-natural-cubic-splines-expressed-as-they-are-es.

Mowbray (1914). The same thing was extended to regression in Hoerl and Kennard (1970) with ridge regression. That minimises the negative loglikelihood NLL plus a proportion of the sum of the squared parameters. For parameters β_j this estimation minimises:

$$NLL + \lambda \sum \beta_j^2$$

They showed that there is always some positive λ that reduces the error variance from that of maximum likelihood estimation. They also called their method "regularisation" as it was derived from a slightly more general method from Tikhonov (1943) which is often translated from the Russian using that non-descriptive term.

Ridge regression is actually a form of shrinking the fitted means towards the grand mean. In practice, all the variables are standardised to have mean zero, variance one before going into the design matrix. Then each fitted value is the constant plus terms of mean zero, so shrinking the parameters shrinks the fitted values towards the overall mean. Just like in credibility, this biases the estimated means towards the overall mean while improving estimation accuracy. The problem is in selecting λ. In practice this is usually done by cross validation. The data is divided into a number of smaller subsets, and the model is estimated by fitting the data many times for different values of λ with each of the subsets excluded separately. Then the NLL can be computed for each subset using the models that excluded it, and the results compared among the different λs.

Ridge regression eventually led to lasso—least absolute shrinkage and selection operator, following Santosa and Symes (1986) and Tibshirani (1996). This replaces the square of the parameters with their absolute values, and so minimises

$$NLL + \lambda \sum |\beta_j|$$

Lasso shrinks some parameters to exactly zero—more as λ increases—which is why it is both a shrinkage and a variable selection operator. There is good software available, e.g., in R, for this, and it quickly does extensive cross validation as well. That is why lasso has become more popular than ridge regression.

Ridge regression and lasso constraints are summed over the parameters but can be approximated in MCMC by postulating shrinkage distributions for the parameters individually. For instance, if β is normal in 0 and σ, its log density, ignoring constants, is:

$$-\log(\sigma) - 0.5(\beta/\sigma)^2$$

Then minimising the joint negative loglikelihood would mean minimising $NLL + \log(\sigma) + 0.5\beta^2/\sigma^2$. For any fixed σ, the $\log(\sigma)$ term is a constant, so this reduces to ridge regression with $\lambda = 0.5/\sigma^2$.

If β is Laplace (double exponential) distributed in 0 and s, its log density is:

$$-\log(2s) - |\beta|/s$$

For a fixed s, then, minimising the negative joint loglikelihood comes down to minimising $NLL + |\beta|/s$, so is lasso with $\lambda = 1/s$.

There is a problem with parameter shrinkage for APC models, however. Each parameter represents the effect of a given age, period, or cohort. These can rarely be taken out of the model, which would effectively happen if they were shrunk too much towards zero. That is where the splines become useful. Shrinking a slope change or cubic segment parameter to zero just extends the existing curve one step further at that point. Thus with shrinkage the smoothing splines become more compressed and do not skip any age, period, or cohort effect. The points where the curves change are called knots, and shrinkage simultaneously optimises which knots to use as well as the fitted accuracy. These are separate steps in non-MCMC spline fitting and are often not coordinated enough to optimise both simultaneously.

5.3.4 Cross Validation in MCMC

The usual way of measuring goodness-of-fit of models is penalised loglikelihood. The likelihood is penalised because of sample bias—bias in the likelihood arising from measuring it on the sample that the parameters were fit to. A bias penalty is calculated, usually as a function of number of parameters and sample size. Subtracting the penalty from the loglikelihood gives the measure. These are attempting to estimate what the loglikelihood would be on a new, independent sample. The penalty from the small sample AIC, AICc, for instance, has been shown to give an unbiased estimate of the sample bias. Of course there is still some error in such estimates.

Models with shrinkage, like ridge regression, lasso, and smoothing splines, do not have parameter counts that are appropriate for these measures. Due to shrinkage, the parameters do not use up the same degrees of freedom as they do in other models. Cross-validation is a way to estimate the sample bias when there are not good parameter counts available. The sample is divided up into subsamples and the likelihood measured on each subsample with the parameters fit without it, and the results are compared for various values of the shrinkage parameter. Thus cross validation produces a penalised likelihood. How many subsamples and how they are selected influence this process, and it also has estimation error for the sample bias.

MCMC can efficiently produce an estimate of the unbiased loglikelihood, using an approach called leave-one-out cross validation, or loo. Each sample point is taken as an omitted subsample in computing the cross-validation likelihood. Because there are many parameter sets generated in MCMC estimation, the likelihood of the point can be computed under each one of those. Gelfand (1996) showed that a weighted average of those likelihoods, with more weight given to the worse values, gives an unbiased but noisy estimate of the likelihood of a left-out point. The sum of those estimates is the loo likelihood, and as a sum it has a lower average estimation error. This method is called importance sampling, and in fact the weights are inversely proportional to the likelihood. Thus the loo estimate for a point is the harmonic mean of its likelihoods across the parameter sets.

This produces an unbiased estimate of the sample bias, but unfortunately it has turned out to have a high estimation error. Recently, Vehtari et al. (2019) improved upon it, with something akin to extreme value theory. They fit a Pareto to the tail of the distribution of likelihoods for a point, and for the 20% most extreme likelihoods use the Pareto fit instead of the actual likelihood. The sum of these over the data points gives a more accurate estimate of the likelihood for the model excluding sample bias. After the MCMC estimate is finished, this can be done quickly in an R application itself called loo.

The shrinkage parameter can then be estimated by running MCMC several times with various trial shrinkage values, and picking the one with the highest loo. This is how cross validation is used to select the shrinkage parameter in other forms of estimation as well. As with any estimation based on optimising penalised likelihood, this runs the risk of actually choosing the parameter that produces the greatest understatement of the sample bias instead of the most predictive λ.

There is an alternative approach under MCMC: also postulate a distribution for the shrinkage parameter, so it too becomes a random effect. This is called the fully Bayesian method, as now there are no parameters in the model in the frequentist sense of being an unknown constant. Usually you would want to use a distribution for this that has minimal impact on the conditional mean of this parameter. The estimates below were done with the Stan MCMC package. That assumes an initial distribution for a parameter, unless otherwise specified, as uniform on $\pm1.7977\mathrm{e}{+}308$, which is the range of real numbers in the double-precision system most computers use. That could be postulated for the log of the shrinkage parameter, as assuming a wide range for a positive parameter usually biases it upwards. I usually find that ±8 is wide enough for that log. MCMC would of course produce a range of values for the parameter, and the whole sample would be used in risk analysis. The conditional mean is usually presented as the estimate. This is easier than cross validation in that you do not need runs for several different trial values, and it is better in that it does not have the problem of favouring shrinkage parameters with understated sample bias. MCMC sample distributions of parameters represent possible parameter sets that could have generated the data along with the distribution of those sample sets. The mean of this gives the most accurate estimate for each parameter.

5.4 Model Details

The models above were fit using semiparametric regression for Poisson and negative binomial residuals and with and without the gamma-distributed jumps. The Poisson distribution is very tight for large means, which forces those fits to emphasise the larger cells. But this is the distribution that independent deaths should follow. Comparing fits to the negative binomial thus provides an indicator of missing systematic effects in the model.

5.4.1   *Formulas*

The form of the negative binomial used is the one in General Linear Models. The mean is μ and the variance is $\mu + \mu^2/\phi$. An indicator of how much spread a distribution can have is stdv./mean or its square, variance/mean^2. For the Poisson, that is $1/\mu$, and that can get small for large cells, which forces the model to fit very well at those cells. For the negative binomial, it is $1/\mu + 1/\phi$, which puts a minimum on that ratio, and increases model flexibility. Still, if there are no contagion effects, the Poisson should work well.

A gamma distribution is used for the annual mortality multipliers. The gamma in a, b has mean ab and variance ab^2. The density at zero is asymptotic to infinity for $a < 1$. The models here assume $a = 0.1$. For females, b is estimated to be close to 0.4. That gamma then has mean 0.04, and its standard deviation is about 0.125. A large part of this distribution is close to zero, with the median at 0.00024. The largest log factor for this sample, at 2003, is about 0.1, which is at the 90th percentile of the distribution. Probably for the 2020 pandemic year the log factor will be more like 0.5, which is close to the 98.5th percentile, which seems plausible.

All the parameters except the constant and the annual log factors are assumed to be Laplace (double exponential) distributed, with mean zero and the same scale parameter s. The Laplace variance is $2s^2$, so the smaller s is, the tighter is the range around zero, so the more shrinkage it produces. Also *log s* is assumed to be uniform on ± 6. For males, sampled values of it range from -3 to 1.5, and for females they are in a tight range around -5. The constant term, which is eventually exponentiated, is assumed uniform on ± 8. it ends up near 0.5 for females and 2.5 for males.

For the APC model, let y be the column of death counts, x be the design matrix, v be column of parameters, c be the constant term, which here is not in the design matrix, and q the column of exposures. Using "$\times$" for pointwise vector multiplication, the Poisson or negative binomial mean μ is:

$$\mu = q \times \exp(c + xv)$$

This is more complicated for models with age modifiers on the period trend, as these are not purely linear. For this, first let z be x with just the age and cohort variables, and let t be the design matrix with the period variables. Let A be a spline design matrix for all the age variables, including the first age. Now v is the vector of just the age and cohort parameters. Let u be the corresponding column of period parameters, so tu is the vector of period factors by observation, and let w be another column of age parameters for the age weights. Then set $r = Aw$, the raw age weights on trend for each observation. These need another constraint for identifiability, so let $m = \max(r)$ and set $p = \exp(r - m)$. These are the final age weights on trend for each observation, and are positive with a maximum of 1. With all of this, the mean is now:

$$\mu = q \times \exp(p \times tu + c + xv).$$

For the contagion model, with mortality multipliers by year, another design matrix B is needed with 0,1 variables for each year, including the first year, to pick out the year each observation is from. This is multiplied by the column h of log age weights selected from the gamma distribution, and then added to the log mortality. Then:

$$\mu = q \times \exp(Bh + p \times tu + c + xv)$$

5.4.2 Fitting Process

The overlaps among the three directions in APC models require additional constraints. In parameter shrinkage, some parameters are shrunk to zero, which takes those variables out of the model, and this is enough of a constraint to keep the design matrix from being singular. The parameters then are not readily interpretable as to what the period and cohort trends mean individually, but since they come from finding the most predictive model, it sometimes could be reasonable to take them at face value. That would be a place for actuarial judgement. Some argue that since the trends are not separately identifiable without constraints, there are no real individual effects. But there are drivers of the trends in other data. For instance, health-related behaviours, like diet, smoking, and exercise, tend to be more cohort effects than period effects. Change in demographic mix can also affect mortality, and this can have both period and cohort effects driven by immigration and demographic differences in family size.

The models in he example use linear splines, but most of the fitting procedures below work for cubic splines as well. Limited experience suggests that more parameters go to zero with cubic splines, making those models somewhat more parsimonious. That could lead to better or worse loo fit measures.

MCMC sample sets rarely have any parameters that are zero in every sample, but it can simplify the model and improve loo a bit by eliminating some of the small ones. Setting parameters to zero is like setting variables with low t-statistics to zero in regression models. That is done in regression to eliminate parameters that have a good chance of actually being zero. But in spline models, nearby parameters can be highly negatively correlated, and can have opposite signs and move simultaneously. So a parameter having zero in its sample range could still be contributing to the model. The rule followed in the fitting here was to eliminate parameters with absolute values of 0.001 or less, as long as they had ranges approximately symmetric around zero. But if that reduced the loo fit measurement, they were put back in. Sometimes offsetting adjacent parameters were both eliminated in this way.

MCMC can have difficulties in finding good parameter sets when there is a large number of highly negatively-correlated variables. This can be shortcut by first using lasso with minimal shrinkage to eliminate the least-needed variables, with the rest used in MCMC. Generally the fitting procedure above will eliminate more of these. The glmnet R package for lasso is very efficient, and includes a cross-validation routine that can indicate the minimal reasonable degree of shrinkage. Glmnet has some difficulties with the cubic spline design matrix, and may look too parsimonious. If so, lasso can easily be fit for still lower selected values of λ by a nonlinear optimiser.

Glmnet has a Poisson option that uses a log regression and exposure multipliers. The approach taken was to first use this just on age and period variables. After eliminating some of those, the cohort variables were added in lasso to give an APC model. It is not possible to input all the APC variables initially, as the design matrix is singular until some variables are eliminated. From the initial set of 146 age, period, and cohort variables, 50 made it into the female model, with 45 for the male model. Then the APC models were fit in MCMC.

The RH model was fit with the surviving variables from the APC fit, along with all the age variables for the age weights. Poisson and negative binomial versions of the RH model were fit to both populations. The negative binomial fit better for both, and had 65 variables in the female model and 48 in the male. But due to shrinkage, the female model had fewer effective parameters, according to the loo penalty. Those parameters were then used in the contagion model, with 40 additional annual jump variables, some ending up negligible.

The individual-year log jump factors in the contagion model combined can pick up some of the overall trend. Steps were taken here to inhibit that, so that they could be interpreted as jumps with no trend in them. For instance, using a low gamma a parameter pushes the jumps more towards zero, which makes it less likely that they would pick up larger trends. Also the period variable for age 1 was put into the model. It had been left out initially for identifiability reasons, but with other parameters no longer in the model, that was not necessary. Even with higher values of the gamma a, if a trend appears in the jump parameters, the jumps could still be displayed as residuals to a line fit through them. The interaction of this with the other trends could produce an overall better model in some cases, but that would have to be tested by trying different values of the gamma a, Also these parameters together could interact with the constant term. The gamma shrinkage here was enough to prevent that, but one of them could be left out as well to prevent such overlap. That could not be any of them, though, as they all are positive, with some close to zero and some large. The year with the lowest residuals to the RH model would be a logical choice to leave out in this situation.

5.5 Results

Table 5.1 summarises the fit statistics, and Fig. 5.2 graphs the resulting parameter curves for the event jump model. Loo is a penalised loglikelihood, so higher is better. It comes with a degree-of-freedom type penalty. In a good model, the penalty will be not too different from the parameter count, and ideally will be less, due to parameter shrinkage. Being much higher is an indication that the model is not working well. Probably some sample parameter sets work well for some of the observations and other sets for other observations. This would make the loo measure worse.

Table 5.1 Fit summary statistics

	Female			Male		
	Loo	Penalty	Parameters	Loo	Penalty	Parameters
APC Poisson	−9420	155	42	−9998	133	37
RH Poisson	−9053	181	78	−10029	142	73
APC NB	−8328	24	33	−8830	25	33
RH NB	−8191	34	32	−8750	29	50
Event jump Poisson	−7912	103	106	−8818	125	95

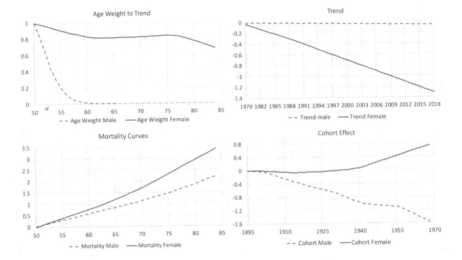

Fig. 5.2 Fitted parameter curves

The Poisson APC and RH models do not fit well for either population. The negative binomial versions have much better loo measures. The standard deviation/mean gives an indication of how much variability from the data is allowed for the model means. With this data, for the Poisson that can be as small as 1% for the largest cells. This widens to about 2.5% for the negative binomial fitted model. It is more like 3% for both distributions for the smaller cells. This forces the Poisson models to fit very tightly at the largest cells.

The models all fit better in the female population. For the males, the APC model fits better than the RH model for Poisson residuals. The event jump model is clearly the best fit for the female data, even though it uses quite a few more parameters. For the males, it is by far the best Poisson model, but it is not quite as good as the negative binomial RH model. Thus there are still some systematic elements that this model is not picking up. The trend is small for the male population, and it does not affect most ages, as seen in the top two graphs in Fig. 5.2. Leaving out the trend and trend weight variables might seem to save parameters without hurting the fit much. This turns out not to be the case. The trend and weight variables actually do not use many parameters. The entire trend for males shows up in the cohort parameters and so does not apply to the older ages. For females, the time trend is similar to the male cohort trend, but is somewhat offset by an increase for cohorts after 1940. The journal NIUSSP has done various related studies.[2] They find higher smoking rates in France than in most countries, and this increased substantially for females in recent generations, particularly for those with lower education levels. It has come down a bit for males.

5.5.1 Extensions: Generalisation, Projections and R Coding

The event-jump model here is very simple. All the ages get the same proportional increase in mortality. A generalisation would be to have the excess deaths follow a parametric age curve, like Gompertz or others, as in Perks (1932). This would have 2–4 parameters. These parameters could even be put on semiparametric curves and allowed to evolve over time. As they are positive, their logs could be postulated to follow such curves. They might be slowly evolving, not using many change parameters, with occasional larger shifts. The population events themselves tend to extend for a few years, perhaps due to weather conditions or diseases. It might be more efficient to put the logs of the jump parameters on semiparametric curves to fit this kind of jump correlation and to save parameters. The gamma distribution for jumps has heavier-tailed options still with mode zero, such as even lower a, or heavier-tailed distributions like the log-logistic or its generalisation, the Burr.

Projected future mortality risk scenarios would include the event jump parameters. Each simulation would draw a jump for each year from the gamma distribution. MCMC program output is good for the parameter risk part of the simulation, as all the sampled parameter scenarios are available. These come with built-in dependence among the different parameters. Future projections of the trend and cohort parameters can be generated from the double-exponential distributions for the slope changes. This should use the correlation matrix of the parameters, which can be computed from the scenario output. The normal copula would be a convenient way to simulate these parameters. Also, putting the jumps on semiparametric curves would give the simulations groups of years that are fairly high simultaneously, which would be more realistic if simulated losses in nearby years is a risk of concern. Compared to other jump models, an advantage of this one is that the low-median high-mean jump distribution produced reasonable probabilities for extreme outcomes even though it did not have extreme events in the historical data used.

R uses the command "fit_ss = extract(fit, permuted = FALSE)" to get the samples from a Stan run with an output called "fit," and put them into an object called "fit_ss." MCMC produces samples that are serially correlated, so some users prefer to permute them before using them. This does not help with most analyses, however. The non-permuted option in the extract command gives a convenient three-dimensional

---

[2] E.g., see https://www.niussp.org/category/article/?postyear=2017.

output: sample parameters by variable for each chain. These can be output to disk by "write.csv(fit_ss, file="samples_fit.csv")." It is often more convenient to do this one chain at a time to preserve the variable headings. Some sample code is included in the Appendix.

5.6 Conclusions

This chapter introduced a mortality model with a jump term for contagion to fit pandemic mortality experience as well as smaller population-mortality events. It was estimated by fitting Poisson and negative binomial distributions to mortality counts with known exposures, using a log link function, and was found to be an improvement to standard continuous mortality models. The models were all fit by semiparametric regression in MCMC, with shrinkage distributions for each parameter.

The number of deaths for a homogeneous group will be binomially distributed if they are independent. This can be closely approximated by a Poisson. If model residuals are heavier-tailed, like negative binomially, some systematic drivers are missing in the model. This was found to be the case for French female and male populations. One reason could be population mortality events in some periods. These were added to the models as gamma-distributed annual mortality jumps that affect mortality for all ages proportionally. This made the female Poisson model better than the previous negative binomial model, and improved the male Poisson models but there were apparently some mortality drivers still missing for males. Possible extensions were discussed. The gamma distribution used for the jumps is wide enough to capture pandemic effects such as COVID-19 in 2020.

Semiparametric regression with linear smoothing splines is a way to improve models by reducing overfitting. This worked well here, and gave simple parameter curves across all the variable dimensions. An advantage of including the event jumps is that simulated risk scenarios can include their probabilities. It is possible that using mortality trend without shrinkage could pick up all the historical jumps, but that would lose the advantages of shrinkage and could be more difficult to project in a way that would keep this risk element.

Appendix

First is R code for running Stan for the RH model plus the event jumps, followed by the Stan code. Simpler models would use only some of these lines.

The R code starts by loading in the needed packages, and then the data. This includes columns for the deaths and exposures, and then the various design matrices. The data is combined in a list file to pass to Stan, then Stan is run. Loo is run on the output, and parameters printed and sent out to disk.

```
setwd("~/OneDrive/R/Mortality/Contagion")
library("loo")
library(rstan)
rstan_options(auto_write = TRUE)
options(mc.cores = parallel::detectCores())
library(readxl)

  y = as.integer(scan('Le_mort.txt'))
  expo = scan('Le_expo.txt')
    #scan turns a column file into a vector
  x = as.matrix(read_excel("Le_AC_nb.xlsx"))
    #age and cohort only
  x = x[,-c(4,5,10,14,22)]
    #eliminates some variables in later runs
  trendmat = as.matrix(read_excel("Le_Y_nb.xlsx"))
    #year for trend
  wghtm = as.matrix(read_excel("age_all.xlsx"))
    #ages for weights
  yearm = as.matrix(read_excel("Y01.xlsx")) #0,1 trend matrix
  wghtm = wghtm[,-c(5,10,16,17,22,23,31,34)]
  trendmat = as.matrix(trendmat[,-c(2,3)])
  yearm = yearm[,-1]
  n <- ncol(trendmat)
  u <- ncol(wghtm)
  t <- ncol(yearm)
  N = length(y)
  U = ncol(x)
    c(N,U,n,u,t)
dx = list(N=N,U=U,expo=expo,y=y,x=x,n=n,u=u,t=t,
            trendmat=trendmat,wghtm=wghtm,yearm=yearm)
# this is to send input data to MCMC model
# now run model
  set.seed(77)
  Sys.time()
  fit = stan(file = 'rh_mort_pois_double.stan',data=dx,
            verbose = FALSE, chains = 3, iter = 20000, warmup =
            15000, control = list( max_treedepth = 12))
  Sys.time()
#now compute loo
log_nb1 <- extract_log_lik(fit)
loo_nb1 <- loo(log_nb1)
loo_nb1
#show parameters and write means to disk
print(fit, pars=c("cn","b", "v","w", "d","s","m2"),probs=c(.025,
            0.25, 0.5, 0.75, 0.975), digits_summary = 3)
out <- get_posterior_mean(fit)
write.csv(out, file="out.csv")
```

Now the Stan code. This is translated into C++ then compiled for more efficient execution. Most of it is just defining the variable types and dimensions for C++. It comes in required code blocks. The means are computed in the transformed-parameters block, and the model block gives the distributions assumed for the data

and the parameters. The last block computes the log likelihoods for use in the loo
code later.

```
data {
    int N;      // number of obs
    int U;      // number of age+cohort variables
    vector[N] expo;
    int y[N];
    matrix[N,U] x;   //age and cohort matrix
    int n;              //number of trend params
    int u;              //number of wghts params
    int t;
    matrix[N,n] trendmat;  //trendmat design matrix
    matrix[N,u] wghtm;      //age wght dsgn mat
    matrix[N,t] yearm;      //year 0,1 dummy dsgn mat
}
parameters {  // all except v will get default uniform prior
  real<lower=-8, upper=8> cn;     //constant, in assumed range
  real<lower=-6, upper = 6> logs; //log of s, related to lambda
  real logb;  //log of gamma b parameter
  vector[U] v;  //age and cohort parameters
  vector[u] w;  //trend-wghts parameters
  vector[n] d;  //trend parameters
  vector<lower=0, upper = 1>[t] m2; //factor by age
}
transformed parameters {
  real s;  // shrinkage parameter
  vector[N] mu;
  real m;  // max weight
  real b;  // gamma b for factor
  vector[N] colwghts;
  vector[N] trend;
  vector[N] mean2nd;
  s = exp(logs);
  b = exp(logb);
  mean2nd = yearm*m2; // 2nd mean by observation
  colwghts = wghtm*w;
  m = max(colwghts);
  colwghts = exp(colwghts-m); //makes max = 1, all >0
  trend = trendmat * d;
  for (j in 1:N) trend[j] = trend[j]*colwghts[j];
  mu = exp(trend+x*v+cn+mean2nd);
  for (j in 1:N) mu[j]=mu[j]*expo[j];
}
model { // gives priors for those not assumed uniform.
Choose this one for lasso.
    for (i in 1:U)  v[i] ~ double_exponential(0, s);  //
    more weight to close to 0
    for (i in 1:u)  w[i] ~ double_exponential(0, s);
    for (i in 1:u)  w[i] ~ double_exponential(0, s);
    for (i in 1:t)  m2[i] ~ gamma(0.1, b);
    y ~ poisson(mu);
}
```

```
generated quantities { //outputs log likelihood for
  testing purposes   vector[N] log_lik;
  for (j in 1:N) log_lik[j] = poisson_lpmf(y[j] | mu[j]);
}
```

References

M.A.C. Alijean, J. Narsoo, Evaluation of the Kou-modified Lee-Carter model in mortality forecasting: Evidence from French male mortality data. Risks **6**(123) (2018)

K. Barigou, S. Loisel, Y. Salhi, Parsimonious predictive mortality modelling by regularization and cross-validation with and without Covid-type effect. Risks **9**(5) (2021)

G. Barnett, B. Zehnwirth, Best estimates for reserves. Proc. Casualty Actuarial Soc. **87**, 245–303 (2000)

A.J.G. Cairns, D. Blake, A.R. Kessler, M. Kessler, The impact of Covid-19 on future higher-age mortality. Pensions Institute (2020). http://www.pensions-institute.org/wp-content/uploads/wp2007.pdf

H. Chen, S.H. Cox, Modeling mortality with jumps: applications to mortality securitization. J. Risk Insur. **76**(3), 727–751 (2009)

A.E. Gelfand, Model determination using sampling-based methods, in *Markov Chain Monte Carlo in Practice*, ed. by W.R. Gilks, S. Richardson, D.J. Spiegelhalter (Chapman and Hall, London, 1996), pp. 145–162

R.J. Harezlak, D. Rupert, M.P. Wand, *Semiparametric Regression with R* (Springer, 2018)

T. Hastie, R. Tibshirani, J. Friedman, *The Elements of Statistical Learning* (Springer, Corrected 12th Printing, 2017)

A.E. Hoerl, R. Kennard, Ridge regression: biased estimation for nonorthogonal problems. Technometrics **12**, 55–67 (1970)

Human Mortality Database, University of California, Berkeley (USA), Max Planck Institute for Demographic Research (Germany) and United Nations (2019). https://www.mortality.org/

A. Hunt, D. Blake, A general procedure for constructing mortality models. North Am. Actuarial J. **18**(1), 116–138 (2014)

R. Lee, L. Carter, Modelling and forecasting U.S. mortality. J. Am. Stat. Assoc. **87**, 659–675 (1992)

A.H. Mowbray, How extensive a payroll exposure is necessary to give a dependable pure premium? Proc. Casual. Actuarial Soc. **1**, 24–30 (1914)

C. O'hare, Y. Li, Models of mortality rates—analysing the residuals. Appl. Econ. **49**(52), 5309–5323 (2017)

S. Özen, Ş. Şahin, Transitory mortality jump modelling with renewal process and its impact on pricing of catastrophic bonds. J. Comput. Appl. Math. **376**(1) (2020)

W. Perks, On some experiments in the graduation of mortality statistics. J. Inst. Actuar. **63**(1), 12–57 (1932). https://www.cambridge.org/core/journals/journal-of-the-institute-of-actuaries/article/on-some-experiments-in-the-graduation-of-mortality-statistics/F3347914B8A72BAECF3AFD3D95E5FF50

A.E. Renshaw, S. Haberman, A cohort-based extension to the Lee-Carter model for mortality reduction factors. Insur. Math. Econ. **38**, 556–570 (2006)

F. Santosa, W.W. Symes, Linear inversion of band-limited reflection seismograms. SIAM J. Sci. Stat. Comput. **7**(4), 1307–1330 (1986)

C. Stein, Inadmissibility of the usual estimator of the mean of a multivariate normal distribution. Proc. Third Berkeley Symp. **1**, 197–206 (1956)

R. Tibshirani, Regression shrinkage and selection via the lasso. J. Royal Stat. Soc. Ser. B (Methodol.) **58**(1), 267–288 (1996)

A.N. Tikhonov, On the stability of inverse problems. Doklady Akademii Nauk SSSR **39**(5), 195–198 (1943)

A. Vehtari, D. Simpson, A. Gelman, Y. Yao, J. Gabry, Pareto smoothed importance sampling (2019). https://arxiv.org/abs/1507.02646

G. Venter, Ş Şahin, Parsimonious parametrization of age-period-cohort models by Bayesian shrinkage. Astin Bull. **48**(1), 89–110 (2018)

G. Venter, Ş. Şahin, Semiparametric regression for dual population mortality. North Am. Actuar. J. (2021). https://doi.org/10.1080/10920277.2021.1914665

Y. Xu, M. Sherris, J. Ziveyi, Continuous-time multi-cohort mortality modelling with affine processes. Scand. Actuar. J. **2020**(6), 526–552 (2020)

R. Zhou, J.S.-H. Li, K.S. Tan, Pricing standardized mortality securitizations: a two-population model with transitory jump effects. J. Risk Insur. **80**(3), 733–774 (2013)

Open Access This chapter is licensed under the terms of the Creative Commons Attribution 4.0 International License (http://creativecommons.org/licenses/by/4.0/), which permits use, sharing, adaptation, distribution and reproduction in any medium or format, as long as you give appropriate credit to the original author(s) and the source, provide a link to the Creative Commons license and indicate if changes were made.

The images or other third party material in this chapter are included in the chapter's Creative Commons license, unless indicated otherwise in a credit line to the material. If material is not included in the chapter's Creative Commons license and your intended use is not permitted by statutory regulation or exceeds the permitted use, you will need to obtain permission directly from the copyright holder.

Chapter 6
Risk-Sharing and Contingent Premia in the Presence of Systematic Risk: The Case Study of the UK COVID-19 Economic Losses

Hirbod Assa and Tim J. Boonen

Abstract Motivated by macroeconomic risks, such as the COVID-19 pandemic, we consider different risk management setups and study efficient insurance schemes in the presence of low probability shock events that trigger losses for all participants. More precisely, we consider three platforms: the risk-sharing, insurance and market platform. First, we show that under a non-discriminatory insurance assumption, it is optimal for everybody to equally share all risk in the market. This gives rise to a new concept of a contingent premium which collects the premia ex-post after the losses are realised. Insurance is then a mechanism to redistribute wealth, and we call this a risk-sharing solution. Second, we show that in an insurance platform, where the insurance is regulated, the tail events are not shared, but borne by the government. Third, in a competitive market we see how a classical solution can raise the risk of insolvency. Moreover, in a decentralised market, the equilibrium cannot be reached if there is adequate sensitivity to the common shock events. In addition, we have applied our theory to a case where the losses are calibrated based on the UK Coronavirus Job Retention Scheme.

6.1 Introduction

The recent COVID-19 crisis increased the collective need for risk management tools for financial institutions, and such tools require the management of systematic losses, illiquidity, default, and the need for financial aid by the government. Most of the credit by borrowing became uncertain, and there is an imminent demand by the clients and the regulator to manage and mitigate the credit risk.

H. Assa (✉)
Kent Business School, University of Kent, CT2 7FS Kent, UK
e-mail: h.assa@kent.ac.uk

T. J. Boonen
Amsterdam School of Economics, University of Amsterdam, Roetersstraat 11, 1018 WB Amsterdam, The Netherlands
e-mail: t.j.boonen@uva.nl

© The Author(s) 2022
M. C. Boado-Penas et al. (eds.), *Pandemics: Insurance and Social Protection*,
Springer Actuarial, https://doi.org/10.1007/978-3-030-78334-1_6

Motivated by the large economic loss due to the recent COVID-19 pandemic, the management of the macroeconomic risk has become the subject of new research. The economic impact of COVID-19 not only emphasises the need for risk management tools to deal with the economic losses for each single country, but it has also shown the need for the global measures to overcome the economic impact. In this paper we look at this problem from an insurance perspective. We consider three risk management platforms, namely, risk-sharing, insurance and the market platform.

In the risk-sharing platform we consider a machinery that redistributes the wealth of the policyholders. We will see that among the three platforms, the risk-sharing platform is the only one that can give the *perfect pooling* solution, which is the most optimal solution. This means that everybody's wealth is the average of the society's (or insurance cohort's) wealth. Furthermore, this solution gives rise to the concept of a *contingent premium*, which means that the premium cannot be collected ex-ante but ex-post. As a wealth distribution mechanism, this platform is to some extent resembling the state fiscal and monetary policies, which will be discussed in the next section.

In the second platform we consider a regulated insurance company that offers insurance policies. The optimal solution in this platform is a *partial pooling* that shares the wealth in the non-extreme events. In this platform, the policyholders do not pay for the extreme events which necessitates the existence of a protection to the policyholder by providing a bail-out plan.

In the third platform, we consider two different markets: a competitive and a decentralised market. In the competitive market, the optimal premium is the mean of the losses that is identical to classical solutions. However, as we will discuss, this platform significantly increases the insolvency risk. In the decentralised market, the main objective is to reach the market equilibrium. We will see that if the systematic event is not a tail event,[1] which means that large losses occur with sufficiently large probability, then the market equilibrium does not exist despite avoiding the risk of insolvency. On the contrary, if the systematic event is a tail event, despite of rising the insolvency risk, there exists an equilibrium in the market.

To study the implications of our paper, we have constructed an example based on the economic impact of COVID-19 in the UK and particularly we use a calibration arising from the Coronavirus Job Retention Scheme (UK furlough scheme during the COVID-19 pandemic). In this way we can measure several things, for instance the magnitude of the insolvency risk in a competitive market or the magnitude of the pricing gaps in a decentralised market due to a difference in the market demand and supply prices. We have made some observations; given the magnitude of the economic risk generated during the COVID-19 pandemic in the UK, traditional insurance products with ex-ante premia are not sufficient to cope with such huge financial losses. Therefore, a new paradigm needs to be adopted where a contingent premium can readily be incorporated to deal with substantial systematic events.

---

[1] In this paper, the tail event and the systematic event are not necessarily identical, however, they usually can overlap. See also footnote 2.

This paper is organised as follows. In Sect. 6.2 we discuss five risk levels in insurance and introduce the concept of systematic risk. In Sect. 6.3 we introduce the mathematical setup, the preliminaries, and a classical insurance platform. In Sect. 6.4 we define three platforms: the risk-sharing, insurance, and market platform. In this section we obtain the optimal solution in each platform and discuss some policy implications. In Sect. 6.5 we study the three platforms in the presence of three common shock models and study the results of particular examples. Especially, we construct our examples for a case that is calibrated based on the UK Coronavirus Job Retention Scheme. In Sect. 6.6 we conclude.

6.2 Risk Levels and Systematic Risk in Insurance

The general idea in risk management is to dilute the risk by splitting it into smaller risks, sharing, and spanning it over time. This can be called 'smoothing' and risk management 'tools' help to do this. We can identify two approaches to risk management: diversification and risk-sharing. Diversification means that the risk of a portfolio of assets is less than the sum of the risks of the single assets. In mathematical terms this can be formulated as follows: any convex combination of two assets carries lower risk than single ones. Despite all the controversies about the definition of diversification, and that it always can reduce the risk, the majority has widely accepted this approach. On the aggregate risk, financial institutions need to hold capital to be allowed to bear any remaining risk.

On the other hand, the underlying idea of risk-sharing is to exchange risk with counterparties such as an insurer, reinsurer, or via market securitisation. Risk-sharing with expected utility by maximising the agents preferences has been studied extensively in actuarial science and economics, see for instance Borch (1962) and Wilson (1968). The most popular concept of efficiency is Pareto optimality. A risk-sharing contract is called Pareto optimal if there does not exist another feasible contract that is better for all agents and strictly better for at least one agent. For instance, if agents are endowed with exponential utilities, then it is Pareto optimal to share the aggregate risk in the market in a proportional way. On the other hand, if coherent distortion risk measures, such as the well-known Expected Shortfall, are used, it is Pareto optimal to share the aggregate risk in the market via tranches (Boonen 2015). Risk-sharing can be related to the concept of hedging or reinsurance. This is closely related to market principles, like Arrow-Debreu equilibria, no-arbitrage, and no-good-deal principles (Assa and Karai 2012). This also has been discussed by Albrecht (1991), and practical considerations as pricing principle are discussed by Wang et al. (1997), where an axiomatic characterisation of the insurance prices is shown.

To see how the two approaches can be used in the real application we need to know about the risk levels. We have identified different levels of risks by the risk trading markets as we assume the existence of a market is the main indication of the willingness to introduce a risk management product. As such we can identify five different levels.

- **The retail-level**: At this level individuals usually share their risk with another entity; for example, they can buy insurance policies for sharing the risk (a *risk transfer*). The insurance retail market is where the risk is managed.
- **The corporate level**: At this level companies use a combination of diversification and risk-sharing approaches to manage the risk e.g., an insurance company pools the risk of its client while at the same time buys reinsurance. The (re)insurance market is where the risk is managed.
- **The catastrophic level**: The main characteristic of catastrophic risk is that their impact is in most cases independent of the pool size, and therefore cannot be managed easily in insurance markets. Natural disasters or cyber risks are among the most known examples of catastrophic risks. The usual ways of managing such risks include introducing a risk transfer platform to share the risk with the financial markets, through for example CAT bonds; or in general to introduce Alternative Risk Transfers (ART) (Banks 2004; Olivieri and Pitacco 2011). These arrangements can transfer the risk of systematic events and can help the managing of the catastrophe risk by lowering the cost of reinsurance, and or the need for capital allocation. Cyber risk includes the costs involved in the event of a data breach or ransomware (Eling 2020).
- **The systemic level**: The risk of system failure is endogenous, and a result of system failure due to the connectedness of insurers' or banks' lending relations. At the theoretical level there have been some discussions how to manage the systemic risk by Merton (1990) and Shiller (2007).
- **The macro-level**: The examples can include world wars and pandemics where no market for trading risk can be considered even at the theoretical level. The underlying probability distribution and the corresponding losses are exogenous and hard to accurately predict. In a macro-level risk, the challenge is that since the insurance market is not feasible even at a theoretical level, it seems there is no efficient way for risk management. From the insurance perspective, we need to have a deeper understanding of insurance principles as well as the time direction of an insurance's risk management.

At all five levels, *systematic risk* may exist, which can be interpreted as a market risk factor that affects the risk variables in the same direction. Systematic risk is usually attributed to a common shock. Examples of systematic risk include financial market indices, but also the aging population, and epidemics. Given the magnitude and the systematic nature, there are lots of discussions on how and in which market to manage the risk of systematic events.

- Bilateral GDP income swaps by Merton (1990) or GDP linked securities by Shiller (2007) are among the major "theoretical" solutions. The idea here is to introduce some securities that can share the risk among the countries in the world. These solutions without a doubt need international collaboration to run markets that trade such instruments.
- Another way of managing the systematic risks is risk-sharing, which is claimed to be happening by globalisation (Flood et al. 2012) and peer-to-peer insurance (Denuit 2020; Feng et al. 2020).

- Another option is to run a deficit and essentially transfer the systematic risk to the same society that needs to bear the risk. This is an example of market failure as society would not be willing to take a rather expensive and uncertain product. But the state, including the government and the central banks, have the authority and tools to sell the risk by running state policies such as monetary and fiscal policies.

It seems that with the current risk management tools a systematic risk has been considered manageable, at the practical or theoretical level, as long as it does not belong to the category of macro risks. That also includes any insurance solution. So, it is necessary to have a closer look into the insurance risk management mechanisms and approaches to see the problem.

In insurance, a stronger approach than diversification is risk pooling, which under the assumption of independence of losses implies the "insurance principle". The insurance principle is explained as follows by Albrecht (1991):

> *for a growing collective the relative (i.e. divided by the number of risks) safety loading and the relative security capital required to maintain a certain security level of the insurance company is decreasing.*

Pooling means sharing losses by aggregating the accident costs and then split this equally. However, if there is a risk of common shocks, it is no longer clear that pooling can imply the principle of insurance. Common shock is modelled to include systematic risk, and heavy losses on the common shock lead to heavy losses of the insurers. Any ex-ante risk management approach needs to render a present value that represents the risk. This is necessary to make a correct risk assessment. This brings us to the concept of risk valuation. Pooling can help to make such an assessment under the independence assumption of losses. However, as we relax the independence assumption, which means there is a risk of common shocks, pooling no longer will be able to easily render a deterministic value. The (wide) dependency relation can properly be related to the concept of systematic risk. The meaning of this is that under a systematic risk circumstance, pooling would not mitigate all randomness and the remaining randomness needs to be measured for valuation.

In a historical context there have been two major risk management institutions i.e., the insurance and the banking industry, with more than 300 years of modern history. While insurance leverages against insurable risk by collecting premia ex-ante, banking leverages against the credit risk by collecting interest ex-post. In the past 300 years, the world has witnessed a handful number of macro risks, including pandemics, world wars, etc. However, insurance never was part of the solution whereas (central) banking always has played a crucial role. By far, the most well-known ways to encounter macro risk is to introduce fiscal and monetary policies, that can be regarded as ex-post policies.

It seems at a macro level risk, insurance may not be sustainable solution. For insurance, we are often using a future extreme loss as a benchmark (say "once in 100 years"), which makes us rely on the modelling aspect of a quantitative risk assessment. The modelling of rare systematic events usually result in a lack of robustness and huge risk assessment errors. However, even with reliable models, the aggregate nature of common shocks, that is one of the major characteristics of systematic events, is a major challenge for introducing sound insurance solutions. The alternative solutions include the monetary and fiscal policies that are implemented by the state institutions like central banks and the governments. Given that in a macro-level event the whole system will be impacted, the governments will need to borrow the risk capital from the future generations and the contributions from the past (the buffer) are generally insufficient. So, an important point here is to distinguish the differences in the time direction of a sound risk management solution for macro risks. This will be more discussed in the following which constitutes part of the main contribution of this paper by introducing insurances with contingent premia as an ex-post rather than ex-ante policies.

In this paper we model systematic events that have an impact on a large part of the pool of policyholders. The closest concept in the literature to this is probably the concept of common shocks. The idea is to introduce an independent random variable (independent from losses), that represents the losses with a common cause. This variable can either be summed up or multiplied to the other random variables or it can change the severity and frequency of losses. Lindskog and McNeil (2003) use the common Poisson shock processes to model dependent event frequencies and examine these models in the context of insurance loss modelling. Meyers (2007) discusses an approach where the common shock variable is a multiplier of a pool of independent losses. In this paper, the author describes some more general models involving common shocks to both the claim count and claim severity distributions. Avanzi et al. (2018) consider a model where the common shock is additive with the loss variables. For the estimation of diversification benefits, they develop a methodology for the construction of large correlation matrices to any dimension.

The concepts of macroeconomic risk and systematic risk have been thoroughly studied in the literature. However, to the best of the authors knowledge they are not covering the type of risk we discuss in this paper. In the literature, macroeconomic risk is usually referred to as the risk generated by macroeconomic factors and political decisions which can include GDP, inflation, unemployment and central bank interest rate. The main objective is to manage the risk of the macroeconomic factors on financial stock returns or global investment. For instance, Majumder and Majumder (2002) consider the volatility of GDP as the most common problem worldwide whose risk can be shared through the trading of GDP growth rate-related bond, to obtain a mutually preferable allocation of aggregate income.

On the other hand, systematic risk in the literature is usually known as a cause of insurance failure, that is associated with many losses that are positively correlated. Beyond the common shock model that we study in this paper, other popular ways to model systematic risk is via vine copulas (Aas et al. 2009) or positive dependence constraints (Bignozzi et al. 2015).

The focus in this paper is on systematic risk rather than on systemic risk. Systemic risk involves the modelling of the potential collapse of a system and the corresponding default events and is typically modelled via interbank lending networks (see, e.g., Eisenberg and Noe 2001). Our focus is the modelling of common shocks in (insurance) loss variables.

6.3 Mathematical Setup

6.3.1 Probability Space

Let us consider a non-atomic probability space $(\Omega, \mathcal{F}, P)$, where Ω is the states of the world, $\mathcal{F}$ is a sigma algebra and P is a probability measure. All random variables are measurable with respect to $\mathcal{F}$. We assume our probability space is rich enough to introduce any sequence of i.i.d. random variables with a given distribution. Consider a set of bounded losses $X_1, X_2, \ldots$ and assume that the losses are identically distributed with a shared distribution $X_i \sim X$. Let us assume that X_i is non-negative and satisfies $\mathrm{esssup}(X_i) = M$. We also denote the cumulative distribution function of random variable X by F_X and the expectation is denoted by E. We consider a framework with two time steps 0, 1.

Consider a set of policyholders $\{1, 2, \ldots, n\}$. We assume that the policyholders are homogeneous in that they have the same initial wealth w_0 at time 0 and have loss variable $X_1, X_2, \ldots, X_n$ that are identically distributed at time 1. The final wealth of policyholder i at time 1 is given by W_i and is a measurable random variable. In absence of purchasing insurance, the final wealth of each single policyholder i is $W_i = w_0 - X_i$. The policyholders are endowed with Von Neumann-Morgenstern expected utility (see Varian 2019) functions denoted by u_i, $1 = 1, 2, \ldots, n$, which are assumed to be increasing and concave.

In the following, we consider a risk tolerance parameter $\eta \in (0, 1)$, that is usually very close to 0. This parameter will be used to measure the sensitivity against the tail (usually unfavourable) events. This parameter specifies the tail events as events $A \in \mathcal{F}$ such that $P(A) \leq \eta$. This definition is motivated by Liu and Wang (2021).[2]

6.3.2 Insurance Preliminaries

We assume two major types of economic agents in our setup: policyholders (also called insureds) and an insurer. The policyholders and the insurer have different attitudes towards risk and insurance.

Policyholders. Policyholders are endowed with risk-averse preferences. To model the behaviour of such agents we could consider suitable utility functions that are applied to the agents' final wealth, which is random. Alternatively, one can also consider the policyholders as risk-neutral (expected profit maximising) agents; this is often assumed in reinsurance where the reinsurance buyer is a firm itself or when the insurance is traded as part of business risk management, for instance in a supply

---

[2] This definition of the tail event in this paper must not be mistaken by the tail event that is introduced in the probability theory which consists of events that can be determined if an arbitrarily finite segment of the sequence is removed (like in Kolmogorov's 0–1 theorem).

chain (Assa et al. 2021). Risk aversion for expected utility maximising agents implies that the utility function is concave. All policyholders maximise their own utility. A choice is objective if all agents agree on the same choice regardless of the utility functions they have. It is well known that a policy is objectively chosen i.e., it is the best choice for all utility functions if and only if it is second-order stochastic dominant. Considering two random variables X and Y, we say that X dominates Y in second stochastic dominance if $E(u(Y)) \leq E(u(X))$ for all increasing and convex utility functions u, and this is denoted by

$$Y \preceq_{SSD} X.$$

In some cases, one also assumes that all policyholders have the same utility function. In this case, one also can appeal to the well-known concept of a representative agent.

The demand for insurance with alternative (not expected utility) preferences has been extensively studied in the literature, see for instance Schlesinger (2000) for an overview. An interesting type of utility that is very useful for investigating the insurance demand is the one that is promoted by prospect theory of Kahneman and Tversky (1979) or rank-dependent utility theory of Quiggin (1993). In our work we consider insuring against low probability events, like events that are expected to happen only once in 100 years, but that have a very large impact in the insurable losses in the market. To better understand the aspects of the problem, note that from the demand side we are dealing with policyholders. For a policyholder, 1% is an ultra-low probability. It is known in the literature that prospect theory and rank-dependent utility theory provide better assessment of the demand for this situation compared with other measures, as the policyholder is expected to overweight the probability of 1% in their mental assessment of the risk. A good study about this subject is Schmidt (2016), who explains empirical evidence that show that people are unwilling to insure rare losses at subsidised premia. There is also a discussion around the importance of risk aversion and loss aversion for assessing small-probability losses (Eeckhoudt et al. 2018).

Insurer. The insurer is an entity that offers the risk management tool to the policyholders. The insurer can be a firm (e.g., an insurance company), government or a (guarantee) fund. The insurer's concern is either to maximise the expected profit or to reach a particular business objective; for instance, to reach a targeted loss ratio. If the insurer is a government, the objective of the insurer is given by a social welfare function, and the insurer can bail-out the pool of policyholders with taxpayers' money in case of high aggregate losses. The insurer is here modelled as a mutual insurer or a stock insurer, and we assume that the insurers are concerned with the welfare of the policyholders or with a non-negative expected profit condition per policy. Moreover, we assume that the insurer does not "over-insure" a risky position. In fact, a special case of over-insuring is double insurance, and this is generally not legally allowed. There are different objectives for the insurer in our platforms that can be seen later in (6.1), (6.2), (6.5) and (6.11).

Risk management platforms. In this paper we use three risk platforms: (a) risk-sharing, (b) insurance and (c) market platforms. In the risk-sharing platform the policyholders share the risk by introducing a fund. There is no other entity like government or the insurance company. In insurance platform we consider there is an insurance company that issues the insurances for the policyholders. The policyholders are making their decision by maximising their preferences and the insurance company is concerned with the welfare of all policyholders. Finally, we consider two market platforms, one that sets for a competitive market solution and another one where in a decentralised market the equilibrium needs to be reached.

In all our platforms, we assume that the following assumption holds, which we denote by NDI (non-discriminatory insurance).

Assumption NDI The insurance will treat all the policyholders equally so that their final wealth has the same distribution.

NDI explicitly tells us that there are no claim hierarchies or priority claims, and all policyholders are treated the same. Any re-labelling (also called permutation) of the policyholder set $\{1, 2, \ldots, n\}$ leads to the same insurance contract. Under NDI, if all policyholders use the same utility function, then the utility after purchasing insurance is the same for every policyholder. This utility can then be interpreted as the utility of a representative policyholder. However, as we will see we do not need to assume that the policyholders have the same utility. In the literature, optimal insurance contracts often satisfy NDI (see, e.g., Albrecht and Huggenberger 2017; Boonen 2019), but we are the first to impose NDI ex-ante as a property for "desirable" insurance contracts. In addition, from a technical point of view and unlike most of the literature, NDI is not concerned about the joint distribution of the policyholders' final wealth i.e., the final wealth joint distributions of the same number of policyholders do not need to be identical. In addition, we do not need to assume that the policyholders have the same utility functions.

As an immediate implication of NDI, we have the following useful lemma.

Lemma 6.1 *Consider a set of individual identically distributed wealth, $W_1, \ldots, W_n$ then for all $j = 1, \ldots, n$, $W_j \preceq_{SSD} \frac{\sum_{1 \le i \le n} W_i}{n}$.*

Proof Let u be an increasing, concave function that is twice differentiable and let $W = \sum_{1 \le i \le n} W_i$. We prove this for W_j, for a given $j \in \{1, \ldots, n\}$. Using Taylor's theorem, we get for some $\zeta_i \in [\min\{W_i, W/n\}, \max\{W_i, W/n\}]$:

$$E(u(W_j)) = n \times \frac{1}{n} \times E(u(W_j)) = \frac{1}{n} \sum_i E\left(u(W_i)\right)$$

$$= \frac{1}{n} \sum_i \left(E\left(u\left(\frac{W}{n}\right) + u'\left(\frac{W}{n}\right)\left(W_i - \frac{W}{n}\right)\right) + \frac{1}{2}u''(\zeta_i)\left(W_i - \frac{W}{n}\right)^2 \right)$$

$$= E\left(u\left(\frac{W}{n}\right)\right) + \frac{1}{n}E\left(u'\left(\frac{W}{n}\right)\sum_i\left(W_i - \frac{W}{n}\right)\right)$$

$$+ \frac{1}{2n}E\left(\sum_i u''(\zeta_i)\left(W_i - \frac{W}{n}\right)^2\right)$$

$$\leq E\left(u\left(\frac{W}{n}\right)\right) + \frac{1}{n}E\left(u'\left(\frac{W}{n}\right)\left(\sum_i\left(W_i - \frac{W}{n}\right)\right)\right) = E\left(u\left(\frac{W}{n}\right)\right).$$

By a simple approximation, the same is true if we consider that the function u is concave and not necessarily two times differentiable. Since the function u has been chosen arbitrarily it follows that $W_j \preceq_{SSD} \frac{W}{n}$.

Definition 6.1 An insurance scheme s is a set of non-negative random variables,

$$s = (\lambda_1, \lambda_2, \ldots, \lambda_n, f_1, f_2, \ldots, f_n),$$

where λ_i is policyholder i's liability to the insurer known as the premium and f_i is the insurer's liability to policyholder i known as the insurance indemnity.

According to this definition, under the scheme s, the policyholder i final wealth is given by

$$W_i = w_0 - \lambda_i - X_i + f_i.$$

If the premia are deterministic, then it is an ex-ante policy. However, in this paper we also consider the case where both the insurance indemnity and the premium to be random variables. For that reason, we may also use the term contingent premium instead of premium.

An interesting example is the perfect pooling insurance scheme: we call the insurance scheme $s = (\lambda_1, \lambda_2, \ldots, \lambda_n, f_1, f_2, \ldots, f_n)$ a perfect pooling scheme if $\lambda_1 = \lambda_2 = \cdots = \lambda_n = \frac{\sum_i X_i}{n}$ and $f_i = X_i$ for all $i = 1, 2, \ldots, n$. In that case, all risk variables in the market are aggregated, and all policyholders bear an equal share of the aggregated risk. Here, the premium is stochastic. We will show that if there is infinitely many policyholders, then deterministic premia appear when the policyholders are endowed with i.i.d. loss variables.

6.4   Risk Management Platforms

In this section, we study three different risk platforms. First, in Sect. 6.4.1, we propose a pure risk-sharing platform. In Sect. 6.4.2, we propose an insurance platform, in which there is a mutual insurer that can opt to default and is protected by a government. Finally, in Sect. 6.4.3, we propose a market platform and consider a competitive and a decentralised model.

6.4.1   Risk-Sharing Platform

In this part, we consider a risk-sharing platform where there is no role for the insurer, but individuals share their risk directly with each other (for instance, via a peer-to-peer network). Lemma 6.1 can be used to show that under NDI the best allocation is the perfect pooling. As we have mentioned before, based on the individual's final wealth we can consider that the total wealth is given by $W = nw_0 - \sum_i X_i$. Suppose the risk-sharing platform is designed to solve[3]

$$
\begin{cases}
\max & \sum_i E\left(u_i(w_0 - \lambda_i - X_i + f_i)\right), \\
\text{s.t.} & \forall j, 0 \leq f_j \leq X_j, \sum_i \lambda_i = \sum_i f_i, \\
& \text{NDI holds,}
\end{cases}
\tag{6.1}
$$

where $\sum_i \lambda_i = \sum_i f_i$ is a budget constraint that guarantees that the aggregate premia are equal to the aggregate insurance indemnities. From Lemma 6.1 it follows that the optimal allocation of total wealth is given by $\frac{W}{n} = w_0 - \frac{\sum_i X_i}{n}$. Then, the final wealth of individual j after risk-sharing is given by $w_0 - \lambda_j - X_j + f_j = w_0 - \frac{\sum_i X_i}{n}, j = 1, \ldots, n$, where λ_j is the premium and f_j is the coverage for policyholder j. This risk exposure after risk-sharing is obtained by choosing full coverage of the losses, $f_j = X_j$, and $\lambda_j = \lambda = \frac{\sum_i X_i}{n}$. This is a general rule for the premium of insurance with full risk coverage. So we have the following theorem.

Theorem 6.1 *Consider a set of identically distributed risk variables, $X_1, X_2, \ldots, X_n$, then the perfect pooling insurance scheme is a solution to the problem (6.1).*

Note that with infinite number of policyholders if the losses are i.i.d then the premium will converge to the mean, but in the case that we do not have the i.i.d assumption this may no longer hold.

---

[3] Under NDI, the objective of this problem and the other ones in the sequel can be replaced by only a representative agent's utility function.

6.4.2 Insurance Platform

Another way of risk management is to introduce an insurance platform. We first take the perspective of a mutual insurer. In mutual insurance, all policyholders share the insurable risk with each other, and the insurer itself has no profit objective. For a typical insurer, the major consideration is the risk of insolvency. This to some extent hints for a different answer than the risk-sharing which is characterised in the following theorem. But before that let us introduce the partial pooling insurance scheme as an insurance scheme $s = (\lambda_1, \lambda_2, \ldots, \lambda_n, f_1, f_2, \ldots, f_n)$ where

$$\lambda_1 = \cdots = \lambda_n = \frac{\sum_i X_i}{n} 1_{\left\{\frac{\sum_i X_i}{n} \leq \mathrm{VaR}_{1-\eta}\left(\frac{\sum_i X_i}{n}\right)\right\}},$$

for some $\eta \in (0, 1)$, and $f_j = X_j$, $j = 1, 2, \ldots, n$, where 1_A is defined as the indicator function of the event $A \in \mathcal{F}$.

Theorem 6.2 *Consider a set of identically distributed risk variables, $X_1, X_2, \ldots,$ X_n, then the partial pooling insurance scheme is a solution to the following problem[4]:*

$$\begin{cases} \max & \sum_i E\left(u_i(w_0 - \lambda_i - X_i + f_i)\right), \\ \mathrm{s.t.} & \forall j, 0 \leq f_j \leq X_j, \, P\left(\sum_i f_i \leq \sum_i \lambda_i\right) \geq 1 - \eta, \\ & NDI \text{ holds.} \end{cases} \quad (6.2)$$

Proof Consider an insurance scheme $s = (\lambda_1, \lambda_2, \ldots, \lambda_n, f_1, f_2, \ldots, f_n)$, where $\forall j, 0 \leq f_j \leq X_j, P\left(\sum_i f_i \leq \sum_i \lambda_i\right) \geq 1 - \eta$ and NDI holds i.e., $(w_0 - \lambda_j - X_j + f_j)_{(j=1,2,\ldots,n)}$ are identically distributed. Let $\lambda'_j = \lambda_j + (X_j - f_j)$. Observe that

$$\begin{aligned} w_0 - \lambda_j - X_j + f_j &= w_0 - \left(\lambda_j + (X_j - f_j)\right) - X_j + \left(f_j + (X_j - f_j)\right) \\ &= w_0 - \lambda'_j - X_j + X_j = w_0 - \lambda'_j. \end{aligned}$$

First, by NDI this shows that λ'_j, $j = 1, 2, \ldots, n$ have the same distribution. Second, we have

$$\sum_i \lambda_i \geq \sum_i f_i \Leftrightarrow \sum_i (\lambda_i + (X_i - f_i)) \geq \sum_i X_i \Leftrightarrow \sum_i \lambda'_i \geq \sum_i X_i.$$

Given the two points above, the insurance scheme $s' = (\lambda'_1, \ldots, \lambda'_n, X_1, \ldots, X_n)$ respects the NDI assumption that has the same objective value as (6.2). So, we can replace f_j by X_j and λ_j by λ'_j, and rewrite the problem as follows:

---

[4] The participation condition of the insurer is here resembling the reduction of the ruin probability that is often used in the literature on risk theory in a dynamic insurance framework.

$$\begin{cases} \max & \sum_i E\left(u_i\left(w_0 - \lambda_i'\right)\right), \\ \text{s.t.} & P\left(\sum_i X_i \leq \sum_i \lambda_i'\right) \geq 1 - \eta, \\ & \text{NDI holds.} \end{cases} \quad (6.3)$$

Now let $\lambda = \frac{\sum_i \lambda_i'}{n}$. Consider the insurance scheme $s'' = (\lambda, \lambda, \ldots, \lambda, X_1, X_2, \ldots, X_n)$. It is clear that s'' satisfies NDI. On the other hand, based on Lemma 6.1, it holds that $\sum_i E(u_i(w_0 - \lambda_i')) \leq \sum_i E(u_i(w_0 - \lambda))$. Given the last two points and given that $\sum_i X_i \leq \sum_i \lambda_i' = \sum_i \lambda$ we get that the solution must satisfy:

$$\begin{cases} \max E\left(u(w_0 - \lambda)\right), \\ P\left(X \leq \lambda\right) \geq 1 - \eta, \end{cases} \quad (6.4)$$

where $u = u_1 + \cdots + u_n$ and $X = \frac{\sum_i X_i}{n}$.

Let us introduce $A = \left\{ X = \frac{\sum_i X_i}{n} \leq \lambda^* \right\}$ where λ^* is a solution to (6.4). As all utility functions are increasing, the maximum on A happens at $\lambda^* = \frac{\sum_i X_i}{n}$ and on A^C at 0. This means $\lambda^* = \frac{\sum_i X_i}{n} 1_A$. Now, let us look at A^C. Since a utility function is increasing then the values of $\frac{\sum_i X_i}{n}$ on A must be smaller than the values of $\frac{\sum_i X_i}{n}$ on A^C. This implies that $A = \left\{ \frac{\sum_i X_i}{n} \leq \text{VaR}_{1-\eta}\left(\frac{\sum_i X_i}{n}\right) \right\}$. So, the solution is given by, $\lambda^* = \frac{\sum_i X_i}{n} 1_{\left\{\frac{\sum_i X_i}{n} \leq \text{VaR}_{1-\eta}\left(\frac{\sum_i X_i}{n}\right)\right\}}$. This proves the theorem.

There are a few points that need to be discussed. First, by NDI, the risk exposure after purchasing the optimal insurance scheme does not depend on the individual policyholders. Second, by setting $\eta = 0$, meaning a perfect solvency condition $\sum_i f_i \leq \sum_i \lambda_i$, we get the perfect pooling solution, which is also optimal for the risk-sharing platform. Third, if we moreover assume that the random losses are independent, then for both the risk-sharing and the insurance platform, we get the same solution in the limit: $\lambda^* \to E(X)$, as $n \to \infty$. However, the most important point is the difference between the optimal value of the risk-sharing and the insurance platform. In the insurance platform, part of the risk can be forgiven. More precisely, the average risk above the value $\text{VaR}_{1-\eta}\left(\frac{\sum_i X_i}{n}\right)$ is not covered by the policyholders, and the government provides protection to the policyholders. Note that such a so-called bail-out happens with probability η, that is usually small. This is to some extent a huge difference and is not at all desirable for the system. This is an important point to observe that the existence of the government will shelter the agents against the part of the risk that is the most harmful. However, at a large scale it seems the government needs to borrow enough funds to manage the risk which indicates the necessity of sponsorship. The sponsorship can either be in the form of a guarantee fund or can be a social reinsurance.

Remark 6.1 With a similar proof, we get that if we assume that λ is deterministic (an ex-ante policy), then a solution is given by $\lambda_1 = \cdots = \lambda_n = \text{VaR}_{1-\eta}\left(\frac{\sum_i X_i}{n}\right)$.

Remark 6.2 In light of Theorem 6.2 one may consider the insurance scheme given by $s = (\lambda_1, \lambda_2, \ldots, \lambda_n, f_1, f_2, \ldots, f_n)$ where

$$\lambda_1 = \cdots = \lambda_n = \min \left\{ \frac{\sum_i X_i}{n}, \text{VaR}_{1-\eta} \left(\frac{\sum_i X_i}{n} \right) \right\}$$

and $f_i = X_i, i = 1, 2, \ldots, n$. This insurance scheme satisfies the following condition:

If $\sum_i f_i > \sum_i \lambda_i$, then $\sum_i \lambda_i = \text{VaR}_{1-\eta} \left(\sum_i f_i \right)$. This condition states that the government only covers losses beyond the deductible threshold, which is the threshold for a bail-out.

Remark 6.3 In this paper, the insurance policies rely on the credit worthiness of the policyholders. Doherty and Schlesinger (1990) and Cummins and Mahul (2004) study insurance policies under conditions of default risk. In addition, Boonen (2019) considers a limited liability framework where the multivariate risk of the policyholders is exchangeable and focuses on the optimal allocation of losses in default. An interesting finding of this work is that a protection fund can be welfare-improving. A protection fund charges levies to policyholders with low realised losses, and this is used to compensate policyholders with high losses in case of a default of the insurer. While limited liability in the existing literature is considered when the insurers can default on their obligations, in this paper our focus is on a mutual insurer that can opt to default only with a sufficiently small probability. In case of default, the insurance claims will be covered by a government.

6.4.3 Market Platform

So far, we have discussed the risk-sharing and the insurance platforms. However, let us look at the problem from a market perspective. We have chosen two different market platforms; one where we consider a competitive market and another one where we consider a decentralised market.

6.4.3.1 Competitive Market

In this part, we present a model in which a "classical" solution appears, and the premium is deterministic. There is a competitive insurance market, in which the policyholders seek optimal insurance contracts with a (stock) insurer that is faced with the participation constraint to make a non-negative expected profit on each insurance policy.

Theorem 6.3 *Consider a set of identically distributed risk variables, $X_1, X_2, \ldots, X_n$ then the insurance scheme $s = (\lambda_1, \lambda_2, \ldots, \lambda_n, f_1, f_2, \ldots, f_n)$ where $\lambda_1 = \cdots = \lambda_n = E(X_i)$, and $f_j = X_j, j = 1, 2, \ldots, n$ solves*

$$\begin{cases} \max & \sum_i E\left(u_i(w_0 - \lambda_i - X_i + f_i)\right), \\ \text{s.t.} & \forall j, 0 \leq f_j \leq X_j, E(\lambda_j) \geq E(f_j), \\ & NDI \text{ holds.} \end{cases} \tag{6.5}$$

Proof From NDI, it follows that $(w_0 - \lambda_j - X_j + f_j)_{j=1,2,...,n}$ are identically distributed, and thus we have $w_0 - \lambda_j - X_j + f_j = w_0 - \lambda_j'$ for $\lambda_j' = \lambda_j + (X_j - f_j)$. By NDI, it follows that λ_j', $j = 1, 2, \ldots, n$ have the same distribution. Moreover, we have

$$E(\lambda_j) \geq E(f_j) \Leftrightarrow E(\lambda_j + (X_j - f_j)) \geq E(X_j) \Leftrightarrow E(\lambda_j') \geq E(X_j).$$

Given the two points above, the insurance scheme $s' = (\lambda_1', \lambda_2', \ldots, \lambda_n', X_1, X_2, \ldots, X_n)$ respects the NDI assumption and thus we have the same objective value as (6.5). So, we can replace f_j by X_j and λ_j by λ_j', and rewrite the problem as follows:

$$\begin{cases} \max & \sum_i E\left(u_i\left(w_0 - \lambda_i'\right)\right), \\ \text{s.t.} & E(\lambda_j') \geq E(f_j), \\ & NDI \text{ holds.} \end{cases} \tag{6.6}$$

Now let $\lambda_j'' = \tilde{\lambda} = \frac{\sum_i \lambda_i'}{n}$, $j = 1, 2, \ldots, n$. Consider the insurance scheme $s'' = (\lambda_1'', \lambda_2'', \ldots, \lambda_n'', X_1, X_2, \ldots, X_n)$. It is clear that s'' satisfies NDI. On the other hand, from Lemma 6.1 we get $\sum_i E\left(u_i\left(w_0 - \lambda_i'\right)\right) \leq \sum_i E\left(u_i\left(w_0 - \lambda_i''\right)\right)$. Given the last two points and given that $\sum_i X_i \leq \sum_i \lambda_i' = \sum_i \lambda_i''$ we get that the solution must solve:

$$\begin{cases} \max & \sum_i E\left(u_i(w_0 - \lambda_i)\right), \\ \text{s.t.} & \forall j, E(\lambda_j) \geq E(f_j). \end{cases} \tag{6.7}$$

Since the utility function is concave, it holds by Jensen's inequality that $E(u_i(w_0 - E(\tilde{\lambda}))) \geq E(u_i(w_0 - \tilde{\lambda}))$. From this and the fact that the utility function is increasing, (6.7) is solved by $\lambda_j = \tilde{\lambda} = E(X_i)$. This proves the theorem.

Theorem 6.3 describes a classic situation where deterministic premia are optimal. The key assumption here is that the insurer is a separate firm that is able to pool and manage risk in absence of a regulator. In absence of such an unregulated firm, the policyholders can still decide to share risk via a variety of platforms, and then stochastic premia may become optimal.

Since there is no regulator it is important to understand the risk associated with such policies. We look at the probability of insolvency: $P(\sum_i X_i > n\tilde{\lambda})$. So, in the limit for an infinite number of policyholders, we look at the following quantity:

$$P\left(\lim_n \frac{\sum_i X_i}{n} > \tilde{\lambda}\right). \tag{6.8}$$

Moreover, we can also measure the average relative magnitude of the losses in default, i.e.,

$$\frac{E\left(\lim_n \frac{\sum X_i}{n} - \tilde{\lambda} \,\middle|\, \lim_n \frac{\sum X_i}{n} > \tilde{\lambda}\right)}{\tilde{\lambda}}. \tag{6.9}$$

This also can be interpreted as the average violation from solvency:

$$E\left(\lim_n \frac{\sum X_i}{n\tilde{\lambda}} - 1 \,\middle|\, \lim_n \frac{\sum X_i}{n\tilde{\lambda}} > 1\right). \tag{6.10}$$

Now, let us compare the solutions of (6.1) and (6.5) in light of the criteria in (6.8)–(6.10). Note that if we consider the case when the losses are independent, then the contingent premium and the deterministic premium are equal if there is an infinite number of policyholders, and the criteria in (6.8)–(6.10) will suggest the use of the deterministic premium. The important issue is that in the presence of a common shock or systematic risk the expected contingent premium $\tilde{\lambda} = E(\lambda)$ cannot truly be representative of the real macro-level impact. We will show examples of this with common shock models in Sect. 6.5.

6.4.3.2 Decentralised Market

In the decentralised market, our focus is on deterministic (ex-ante) premia, and we solve the supply and the demand problem separately. Our main concern is to observe if an equilibrium exists.

Let us first consider the supply side of the market. We use a very popular method in the industry, by keeping the loss ratio below a targeted loss ratio. So, let consider the insurance company, which is the supply side of the market, wants to keep the loss ratio at a given level $\beta \in (0, 1]$. Based on Assa and Wang (2020) this value is around 65 per cent for the industry. In the case that the only aim of the insurer is the insurance solvency, we can consider $\beta = 1$. For instance, in the case that the insurance is run as a state back scheme or when it is run based on a fund, we can assume that $\beta = 1$. In this case the insurance is not supposed to make profit.

In what follows, we need to do the valuation of insurance products. We have adopted a very simple approach by finding the premium λ as follows:

$$\lambda^S = \inf\left\{\tilde{\lambda} : P\left(\frac{\sum_i X_i}{n\tilde{\lambda}} \geq \beta\right) \leq \eta\right\}, \tag{6.11}$$

where the fraction $\frac{\sum_i X_i}{n\tilde{\lambda}}$ is called the loss ratio.

Note that here we have considered the full coverage $f_j = X_j$ to follow the results of the previous sections. It is not very difficult to see that this can be done through the following assessment:

$$\lambda^S = \frac{1}{\beta} \text{VaR}_{1-\eta} \left(\frac{\sum_i X_i}{n} \right).$$

On the other hand, we need to explore the market demand price. So, let us consider a representative agent with a utility u. Using a utility indifference approach for losses with identically distributes as X, we know that the premium for an insurance contract with full coverage is given as follows:

$$u(w_0 - \lambda^D) = E(u(w_0 - X)),$$

or,

$$\lambda^D = w_0 - u^{-1}(E(u(w_0 - X))),$$

where the inverse utility function u^{-1} exists because u is increasing.

We say that an equilibrium exists if $\lambda^D \geq \lambda^S$, and an equilibrium price is given by $\lambda \in [\lambda^S, \lambda^D]$. In the case that we have linear utility (i.e. risk-neutral agents) we have that $\lambda^D_{Lin} = E(X)$. On the other hand, by Jensen's inequality we have:

$$\lambda^D = w_0 - u^{-1}(E(u(w_0 - X))) \geq E(X) = \lambda^D_{Lin}.$$

So as a sufficient condition for the existence of an equilibrium we can check the following condition:

$$E(X) = \lambda^D_{Lin} \geq \lambda^S.$$

We also can find the following demand price for the random variable $\frac{\sum_i X_i}{n}$:

$$\lambda^D_{Ave} = w_0 - u^{-1} \left(E \left(u \left(w_0 - \frac{\sum_i X_i}{n} \right) \right) \right).$$

We claim that $\lambda^D \geq \lambda^D_{Ave}$. To see this note that by using the Lemma 6.1 we have that

$$u(w_0 - \lambda^D) = E(u(w_0 - X)) \leq E \left(u \left(w_0 - \frac{\sum_i X_i}{n} \right) \right) = u(w_0 - \lambda^D_{Ave}),$$

and so $\lambda^D \geq \lambda^D_{Ave}$. So, this can also be used to introduce the following sufficient condition for the existence of an equilibrium:

$$\lambda^D_{Ave} \geq \lambda^S.$$

For instance, in the case that we are only concerned with the insurance solvency i.e., $\beta = 1$, then in the limit we can see that for i.i.d losses we get:

$$\lambda_{Ave}^{D} \to E(X), \text{ and}$$
$$\lambda^{S} \to E(X), \text{ as } n \to \infty.$$

This readily justifies the existence of the insurance market for the assumptions above.

Remark 6.4 As one can see in all three platforms we studied above the independence assumption with an infinite number of policyholders will result in the classical solution. This explains why in many standard models when we are not facing the risk of common shock a (non-contingent) actuarial premium can be the optimal answer. This can be regarded from the so-called "principle of insurance" perspective, as it was discussed in the introduction. It is not very difficult to see that by using the central limit theorem we have that

$$\zeta = P(S_n > n(E(X_i) + z_n)) = P\left(\frac{\frac{S_n}{n} - E(X_i)}{\frac{\sigma}{\sqrt{n}}} > \frac{z_n}{\frac{\sigma}{\sqrt{n}}}\right) \approx 1 - \Phi\left(\frac{z_n}{\frac{\sigma}{\sqrt{n}}}\right)$$

$$\Rightarrow z_n \approx \frac{\sigma}{\sqrt{n}}\Phi^{-1}(1 - \zeta) \to 0,$$

for any $\zeta \in (0, 1)$, where z_n is the relative risk loading $S_n = \sum_{i=1}^{n} X_i$, σ the standard deviation of X_i, and Φ is the CDF of a standard normal distribution.

6.5 Systematic Risk Model and Common Shocks

In this section we focus on the common shock models and will study different risk management framework in the presence of a common shock. For each framework, we also consider specific examples to better understand the impact of the common shocks on the risk management frameworks.

For simplicity, we have chosen the Bernoulli distribution for losses and the constant absolute risk aversion (CARA, or exponential) utility function, i.e.,

$$u(x) = \frac{1 - e^{-ax}}{a}, \text{ for } a > 0 \text{ and } u(x) = x \text{ for } a = 0.$$

Here, a is the risk aversion parameter. Advantages of using CARA utility are the possibility of using negative wealth, the price invariance to the initial wealth and also additivity w.r.t independent losses. A disadvantage of using the CARA utility is that the risk aversion parameter a depends on the currency (a scaling problem), which makes the calibration of the parameter a challenging. In the literature the risk aversion parameter a is a number very close to 0. However, in very particular cases it can even reach values close to $a = 1$ (see, e.g., Babcock et al. 1993). In our numerical assessment we consider $a \in [0, 1]$.

With the CARA utility we readily derive that

$$\lambda^D = \frac{\log(E(e^{aX}))}{a}.$$

In the market platform, we must essentially make sure that $\lambda^S \leq \lambda^D$. With CARA utility when the losses are i.i.d, based on what have been discussed with infinite number of policyholders, we can verify this relation by the Jensen inequality for $a \neq 0$ as follows:

$$\lambda^S = E(X) = \frac{E(aX)}{a} = \frac{\log(e^{E(aX)})}{a} \leq \frac{\log(E(e^{aX}))}{a} = \lambda^D.$$

The case $a = 0$ is obvious.

However, if the above condition does hold, we also want to see the degree of violation of this condition. In a market setup, as mainly the demand is the driving force of the market, we scale everything with the market demand prices and assess the relative pricing gap, i.e.,

$$RPG = \frac{\lambda^S - \lambda^D}{\lambda^D} = \frac{\lambda^S}{\lambda^D} - 1.$$

This value shows how large the gap is between the supply and demand sides of the market. Another benefit of using this quantity is that it is dimension free, so we do not have problem with scales.

In terms of the tolerance probability of the insurance company, i.e., the parameter η, we generally consider two cases, one where the probability of the common shock event is greater and one where the common shock is less than the VaR-parameter. In our case study, since we assume a probability of common shock γ to be equal to 0.01, we consider $\eta = 0.005$ and $\eta = 0.015$. In this way we can see the impact of the risk management by checking if the insurance risk tolerance parameter is sensitive to common shock or not. We also consider the effect of the probability of the non-systematic (alternatively called idiosyncratic) event which we denote by p.

In the following, in three sections we consider three different common shock examples, including the additive, multiplicative and risk rate common shock models. Then in each section we consider the risk-sharing, insurance, and market platforms. We use the examples of the Bernoulli loss variables along with the CARA utility and $n \to \infty$. One of the benefits of using a Bernoulli distribution is that we can easily associate the common shock to the systematic event. In the following, we will specify the systematic event in each case. It is very important to realise if the systematic event is regarded as a tail event. More precisely, if the systematic event probability is less than the parameter η, then it is also a tail event.

For the common shocks, we focus on an example of a wide-spread pandemic event like the Spanish flu and the recent COVID-19, which we assume to be a macro event of roughly 1 per cent of probability ("once every 100 years"). We also follow the example of the UK Coronavirus Job Retention Scheme presented by Assa (2020) as a case-study with systematic risk. The calibration in this paper helps us to set suitable values for the parameters. For instance, for the idiosyncratic event we can consider $p = 0.06$; so for completeness in this paper we can consider a wider range of $0.05 \leq p \leq 0.1$. We also consider $\gamma = 0.01$.

6.5.1 Additive Common Shock Model

Let us consider an i.i.d. sequence of risk variables Y_j, $j = 1, 2, \ldots, n$ and another non-negative random variable Z, independent from all Y_j, $j = 1, 2, \ldots, n$. Let us introduce the loss variables as $X_j = Y_j + Z$. We assume all random variables Y_i have the same distribution as Y. This additive common shock model is proposed by Avanzi et al. (2018) and Boonen (2019) for the modelling of insurable losses. In this setup the common shock is represented by the random loss Z. The systematic event needs to be introduced for each case. However, in this framework a natural suggestion is an event $S \in \sigma(Z)$, where $\sigma(Z)$ is the sigma-field generated by Z. In our examples we consider Bernoulli distributions for Y and Z. Let us consider an idiosyncratic loss variable $Y = 1_{\mathcal{L}_Y}$ and a systematic loss variable $Z = 1_S$, where $P(\mathcal{L}_Y) = p$ and $P(S) = \gamma$. So, naturally the systematic event that we consider is S.

1. **Risk-sharing platform.** As we have seen in Sect. 6.4.1, for risk-sharing we need to be knowledgeable about the average. This is given by:

$$\frac{\sum_i (Y_i + Z)}{n} = \frac{\sum_i Y_i}{n} + Z \to E(Y) + Z.$$

This is the best allocation for a risk-sharing platform because of Lemma 6.1. As one can see the average still includes the common shocks.

If we consider our example we can see the contingent premium to be given as $p + 1_S$. It is very interesting to compare this solution to the classical problem we considered in (6.5), where the solution is just the average of the loss variable which here is given by $p + \gamma$. So, there is a trade-off in the contingent premium, while in the non-systematic event the contingent premium is less i.e., $p < p + \gamma$, but once the systemic event happens the contingent premium is much higher i.e., $p + \gamma < p + 1$. One can see that there is a chance of $\gamma = 0.01$ that a big loss would hit all. This is generally not sufficient for any insurance scheme to be considered sustainable.

2. **Insurance platform.** As we have seen in Sect. 6.4.1, for mutual insurance we need to understand the event $\left\{ \frac{\sum_i X_i}{n} \leq \mathrm{VaR}_{1-\eta} \left(\frac{\sum_i X_i}{n} \right) \right\}$ for $n \to \infty$. This limit is given by:

$$\left\{\frac{\sum_i X_i}{n} \leq \text{VaR}_{1-\eta}\left(\frac{\sum_i X_i}{n}\right)\right\} \rightarrow \{E(Y) + Z \leq E(Y) + \text{VaR}_{1-\eta}(Z)\}$$
$$= \{Z \leq \text{VaR}_{1-\eta}(Z)\},$$

and as a result, we get in limit that

$$\lambda_i = \lambda^* \rightarrow (E(Y) + Z)1_{\{Z \leq \text{VaR}_{1-\eta}(Z)\}},$$

for the solution $\{\lambda_1, \lambda_2, \ldots, \lambda_n\}$ of Problem (6.2).

Now let us consider our special case of Bernoulli distributions. Here we have two cases. First let us consider $\eta = 0.005 < 0.01 = \gamma$. In this case we have $\text{VaR}_{1-\eta}(Z) = 1$, and as a result we get $\{Z \leq \text{VaR}_{1-\eta}(Z)\} = \Omega$, and $\lambda^* = E(Y) + Z$. This essentially means that in the case that the risk tolerance parameter is set at a value that is smaller than the probability of the systematic event, which means that the systematic event is not perceived as a tail event, then, the solution is identical to the one from the risk sharing platform. Second, let us consider $\eta = 0.015 > 0.01 = \gamma$. In this case we have $\text{VaR}_{1-\eta}(Z) = 0$, and then we get $\{Z \leq \text{VaR}_{1-\eta}(Z)\} = S^C$ and $\lambda^* = E(Y)1_{S^C}$. Thus, if the risk tolerance parameter is larger than the probability of the systematic event, then we may end up with a solution that will put a large burden on the government due to tail risk.

3. **Market platform.** In the market platform we consider the competitive and decentralised markets.

Competitive market. It is very easy to see that $\tilde{\lambda} = E(Y + Z)$. For the Bernoulli example this is easily given by $p + \gamma$. In order to see the risk impact of the policy we need to find out about the probability of default generated by (6.5), as in (6.8) which is $P(Z > E(Z))$. For the Bernoulli model this value is equal to $P(S) = \gamma$. The average relative magnitude of the losses in default is given by:

$$E\left(\frac{1_S + p}{\gamma + p} - 1 \Big| S\right) = \frac{1 - \gamma}{\gamma + p}.$$

If we consider the parameter values of our calibration based on the UK Coronavirus Job Retention Scheme, the average relative magnitude of the losses in default is at least $\frac{1-0.01}{0.01+0.1} = 9 = 900\%$, which is quite high.

Decentralised market. We need to find $\text{VaR}_{1-\eta}\left(\frac{\sum_i X_i}{n}\right)$ in the limit to find the supply price as follows:

$$\lambda^S = \frac{1}{\beta}\text{VaR}_{1-\eta}\left(\frac{\sum_i X_i}{n}\right) \rightarrow \frac{1}{\beta}\text{VaR}_{1-\eta}(E(Y) + Z) = \frac{1}{\beta}(E(Y) + \text{VaR}_{1-\eta}(Z)),$$

which holds true by continuity of the VaR risk measure (see Proposition 4.11 in Marinacci and Montrucchio 2004). We also need the demand price

$$\lambda^D = w_0 - u^{-1}(E(u(w_0 - Y - Z))).$$

Our aim is to see when we can verify the market condition $\lambda^S \leq \lambda^D$.
By considering the CARA utility we have that,

$$\lambda^D = \frac{\log(E(e^{aX}))}{a} = \frac{\log(E(e^{aY}))}{a} + \frac{\log(E(e^{aZ}))}{a}, \quad a > 0,$$

and,

$$\lambda^D = E(Y) + E(Z), \quad a = 0.$$

Now, let us look at the pricing gaps:

$$\lambda^S - \lambda^D = \begin{cases} E(Y) + \text{VaR}_{1-\eta}(Z) - \left(\frac{\log(E(e^{aY}))}{a} + \frac{\log(E(e^{aZ}))}{a} \right), & a > 0, \\ \text{VaR}_{1-\eta}(Z) - E(Z), & a = 0. \end{cases}$$

Computing this gap is not an easy job in general. When we use the Bernoulli loss distribution, we derive with our assumptions the following:

$$\lambda^S - \lambda^D = \begin{cases} p + 1_{\gamma > \eta} - \left(\frac{\log(1-p+pe^a)}{a} + \frac{\log(1-\gamma+\gamma e^a)}{a} \right), & a > 0, \\ \text{VaR}_{1-\eta}(Z) - E(Z), & a = 0. \end{cases}$$

As explained in the beginning of this section motivated by the COVID-19 case study, we use $\gamma = 0.01$. Next, we study the pricing gaps. In Fig. 6.1 we show the relative pricing gaps for two cases, one when the risk assessment is sensitive to the systematic risk ($\eta < \gamma$), and one otherwise. As one can see that if the systematic risk probability parameter γ is "captured" by the risk confidence parameter η ($\eta < \gamma$), then the relative price gaps in Fig. 6.1(left) are always positive and greater than 5.5 (meaning 550%) and can become as large as 16 times (1600%) the demand price. On the other hand, one can see that if the risk confidence parameter η is larger than the systematic risk parameter γ, then the prices gap is negative which means that an equilibrium price exists. The VaR may however not be an adequate measure to determine the riskiness of the loss variable as the common shock is not measured. The main problem is the common shock that has a great impact on the valuation. Some side-observation includes the reduction of the gap with the increase of the risk aversion parameter which makes economic sense. However, the behaviour of the pricing gap is different with respect to the changes in the non-systematic risk parameter p.

6.5.2 Multiplicative Common Shock Model

We consider a model, where the common shock is multiplicative, and the individual risks are given by: $X_j = ZY_j$, where $Z \geq 0$, $E(Z) > 0$, and the risk variables $Y_j, j = 1, 2, \ldots, n$, are i.i.d. and independent of Z. Similar to Sect. 6.5.1 we consider

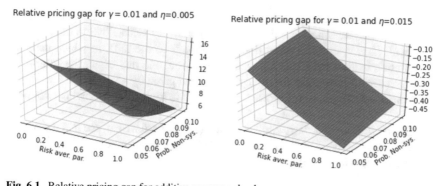

Fig. 6.1 Relative pricing gap for additive common shocks

Bernoulli losses. Let us consider loss $Y = 1_{\mathcal{L}_Y}$ and $Z = 1 + z1_S$, where $P(\mathcal{L}_Y) = p$ and $P(S) = \gamma$ and $z > 0$ is the magnitude of the systematic losses. For simplicity we assume $z = 2$ and 4 for two different values of the common shock effect. We again consider S as the systematic event.

Similar to Sect. 6.5.1, one can get that for the risk-sharing platform it holds that:

1. **Risk-sharing platform.** Based on discussions in Sect. 6.4.1, we need to find the average in limit:

$$\lambda = \frac{\sum_i (ZY_i)}{n} = Z\frac{\sum_i Y_i}{n} \to ZE(Y).$$

   This is the best allocation for a risk-sharing platform because of Lemma 6.1. As one can see the average still includes the common shock Z.

   Now let us again consider Bernoulli distributions for the losses. We then get that $\lambda = ZE(Y) = p + pz1_S$. Similar to the comparison we made in Sect. 6.5.1 with the solution $\tilde{\lambda} = E(Z)E(Y) = p + p\gamma z$ in (6.5), one can realise the trade-off between the two solutions. It is again important to note that the solution in (6.5) cannot be truly representative of the real macro-level impact due to the systematic risk.

2. **Insurance platform.** As we have seen in Sect. 6.4.2, for mutual insurance we need to understand the event $\left\{ \frac{\sum_i X_i}{n} \le \mathrm{VaR}_{1-\eta}\left(\frac{\sum_i X_i}{n} \right) \right\}$ for $n \to \infty$. This limit is given by:

$$\left\{ \frac{\sum_i X_i}{n} \le \mathrm{VaR}_{1-\eta}\left(\frac{\sum_i X_i}{n} \right) \right\} \to \{ZE(Y) \le \mathrm{VaR}_{1-\eta}(Z)E(Y)\}$$

$$= \{Z \le \mathrm{VaR}_{1-\eta}(Z)\},$$

   and as a result, we get in the limit that

$$\lambda_i = \lambda^* \to E(Y)Z1_{\{Z \le \mathrm{VaR}_{1-\eta}(Z)\}},$$

   for the solution $\{\lambda_1, \lambda_2, \ldots, \lambda_n\}$ of Problem (6.2).

Now let us consider the Bernoulli losses. Similarly to Sect. 6.5.1 we distinguish two cases. One where the risk tolerance parameter is sensitive to the common shock (i.e., systematic event does not belongs to the tail), in which case the solution is identical to the risk-sharing solution; and one where the solution is much less impactful for the policyholders and given by $\lambda^* = E(Y)z1_{S^C}$.

3. **Market platform.** Like before, here we need to look at the following two models.

Competitive market. Very similar to the previous case we can see that the probability of insolvency in (6.5) is equal to $P(S)$, and the average relative magnitude of the losses in default is given as

$$E\left(\frac{p(1+z1_S)}{p(1+z\gamma)} - 1 \Big| S\right) = \frac{z(1-\gamma)}{1+z\gamma} = \frac{1-\gamma}{\frac{1}{z}+\gamma}.$$

For our case study on the UK Coronavirus Job Retention Scheme we see that for $\gamma = 0.01$, we get $\frac{99}{\frac{100}{z}+1} \geq 1.94 = \%194$. This is very high value.

Decentralised market. Now we want to see if we can verify the market condition $\lambda^S \leq \lambda^D$. In this case we have that

$$\lambda^D = \frac{\log(E(e^{aX}))}{a} = \frac{\log(E(e^{aZY}))}{a}, \qquad a > 0, \text{ and}$$
$$\lambda^D = E(Y)E(Z), \qquad a = 0.$$

Let us consider the example with Bernoulli-distributed factors. With our assumptions we can see that for $\beta = 1$

$$\lambda^S = E(Y)\text{VaR}_{1-\eta}(Z) = p1_{\{\gamma > \eta\}}.$$

In addition, it holds that

$$e^{aYZ} = e^{a1_{\mathcal{L}_Y}+az1_{\mathcal{L}_Y \cap S}} = \begin{cases} e^{a(1+z)} & \text{on } S \cap \mathcal{L}_Y, \\ e^a & \text{on } \mathcal{L}_Y \backslash S, \\ 1 & \text{on } \mathcal{L}_Y^C, \end{cases}$$

which is equal to

$$\begin{cases} e^{a(1+z)} & \text{with prob. } \gamma p, \\ e^a & \text{with prob. } (1-\gamma)p, \\ 1 & \text{with prob. } 1-p. \end{cases}$$

This implies that

$$\lambda^D = \frac{\log(E(e^{aYZ}))}{a} = \frac{\log(p\gamma e^{a(1+z)} + p(1-\gamma)e^a + (1-p))}{a},$$

and

$$\lambda^S - \lambda^D = \begin{cases} p(1 + z1_{\gamma > \eta}) - \dfrac{\log\left(p\gamma e^{a(1+z)} + p(1-\gamma)e^a + (1-p)\right)}{a}, & a > 0, \\ p((1 + z1_{\gamma > \eta}) - \alpha), & a = 0. \end{cases}$$

Like before, motivated by the UK COVID-19 case study we use the calibration $\gamma = 0.01$. We next examine the pricing gaps.

The results in Fig. 6.2 are very similar to what we have observed in Fig. 6.1, which means there is a trade-off between correctly covering the risk via a VaR and the existence of the market equilibrium. This observation holds true almost regardless of the values of the risk aversion parameter, the non-systematic risk and even the value of z.

6.5.3 Risk Rate Common Shock Model

In this example we discuss different levels of the systematic risk in the market. As it has been discussed in the beginning of the paper, risk management solutions emerge to take care of different level of risk. Markets that deal with less risky aggregate losses are more likely to form and the risk would more perfectly be managed. On

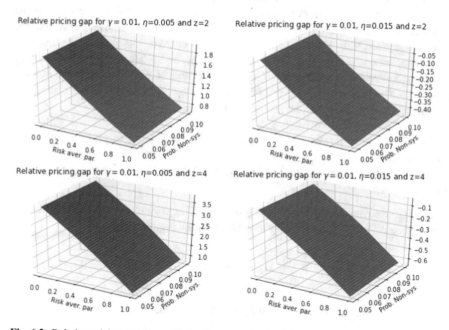

Fig. 6.2 Relative pricing gap for multiplicative common shocks

the opposite side we have the markets that are dealing with larger aggregate losses, which are much harder to create. Here, we use a very simple setup to demonstrate the different levels of systematic risk while still focussing on identically distributed individual loss variables. The major difference is the aggregate risk rather than the individual risk. This differentiates this example from the other two examples where the aggregate risk is not the essential differentiator.

Let us consider a partition $\Omega_1, \ldots, \Omega_m$ of Ω where $p_k := P(\Omega_k)$ for all $k = 1, \ldots, m$. The elements of the partition represent different layers of the risk management market. We assume that $p_1 > p_2 > \cdots > p_m > 0$, associating smaller probabilities with events with larger impact. Let us consider a sequence of loss variables $\{X_n\}_{n \in \mathbb{N}}$ so that $\{X_n | \Omega_k\}_{n \in \mathbb{N}}$ is i.i.d. in the probability space (Ω_k, P_{Ω_k}), where P_{Ω_k} is the conditional probability on Ω_k for any $k = 1, \ldots, m$. Let us assume that $E(X_i | \Omega_1) \leq \cdots \leq E(X_i | \Omega_m)$. This assumption can reflect the fact that more harmful events cause larger expected losses. By assumption it is clear that $E_k = E(X_1 | \Omega_k) = \cdots = E(X_n | \Omega_k)$, for $k = 1, \ldots, m$. It is also clear that the loss distribution is given as follows:

$$F_X = \sum_k p_k F_{X|\Omega_k}.$$

To better understand the example, we also discuss a special case of this model by assuming only a systematic and a non-systematic event. This means in a probabilistic setup the probability space Ω can be partitioned into S and S^C, for systematic and non-systematic events, respectively. Let $\gamma = P(S)$ be a positive number that is the probability of the macro event (e.g., every 100 years). We assume in the systematic event the probability of the loss distribution will change. The same is assumed for the complement set S^C. Let $Sys = \{\emptyset, S, S^C, \Omega\}$, be the sigma-field generated by systematic event. Let us assume $E(X_j | S) = \delta M$ and $E(X_j | S^C) = \alpha M$. We assume that $\alpha > \delta$ which is reflecting the fact that the magnitude of the losses during the systematic event is larger than non-systematic event.

Using the calibration by Assa (2020) on the UK Coronavirus Job Retention Scheme, we set $\alpha = 0.27$ and $\delta = 0.06$ and $\gamma = 0.01$. However, for completeness we consider a range for $\delta \in [0.05, 0.1]$.

Now let us see what will happen to the optimal strategies given by the propositions we have discussed.

1. **Risk-sharing platform**. By using the law of large numbers on each conditional space we need to specifically look at the following quantities:

$$\lambda = \frac{\sum_i X_i}{n} = \frac{\sum_i X_i}{n} 1_{\Omega_1} + \cdots + \frac{\sum_i X_i}{n} 1_{\Omega_m} \to \sum_k E(X|\Omega_k) 1_{\Omega_k}.$$

Considering the special case of $m = 2$, we have that,

$$\lambda = E(X|S^C) 1_{S^C} + E(X|S) 1_S = \delta M 1_{S^C} + \alpha M 1_S = \delta M + (\alpha - \delta) M 1_S.$$

2. **Insurance platform**. As we have seen we need to find the following quantity

$$\lambda^* = \frac{\sum_i X_i}{n} 1_{\left\{\frac{\sum_i X_i}{n} \leq \text{VaR}_{1-\eta}\left(\frac{\sum_i X_i}{n}\right)\right\}}$$

$$\rightarrow \left(\sum_k E(X|\Omega_k) 1_{\Omega_k}\right) 1_{\left\{\sum_k E(X|\Omega_k) 1_{\Omega_k} \leq \text{VaR}_{1-\eta}\left(\sum_k E(X|\Omega_k) 1_{\Omega_k}\right)\right\}}.$$

Let $m^* = \min\{k : \sum_{k' \leq k} P(\Omega_{k'}) \geq 1 - \eta\}$. Then it is clear that

$$\text{VaR}_{1-\eta}\left(\sum_k E(X|\Omega_k) 1_{\Omega_k}\right) = E(X|\Omega_{m^*}),$$

and that

$$\lambda^* = \frac{\sum_i X_i}{n} 1_{\left\{\frac{\sum_i X_i}{n} \leq \text{VaR}_{1-\eta}\left(\frac{\sum_i X_i}{n}\right)\right\}} \rightarrow \left(\sum_k E(X|\Omega_k) 1_{\Omega_k}\right) 1_{\Omega_1 \cup \cdots \cup \Omega_{m^*}}$$

$$= \sum_{k \leq m^*} E(X|\Omega_k) 1_{\Omega_k}.$$

Now using the example of $m = 2$, we easily derive that if $\eta = 0.005 < 0.01 = \gamma$ then

$$\lambda^* = E(X|S^C) 1_{S^C} + E(X|S) 1_S = \delta M 1_{S^C} + \alpha M 1_S,$$

and if $\eta = 0.015 > 0.01 = \gamma$ we have

$$\lambda^* = E(X|S^C) 1_{S^C} = \delta M 1_{S^C}.$$

Hence, if the systematic risk is not a tail event then the solution is identical to the risk-sharing solution. Otherwise, the systematic risk is fully borne by the government, and the policyholders only pay a premium if the non-systematic (idiosyncratic) risk is realized.

3. **Market platform**. Here we need to look at the following models.

Competitive market. One can realise the following trade-off between the model (6.5) and (6.1) premium and the contingent premium:

$$\delta M \leq \tilde{\lambda} = (1 - \gamma)\delta M + \gamma \alpha M \leq \alpha M.$$

The probability of the insolvency is equal to $p = P(S)$. We also find that the average relative magnitude of the losses in default is given by:

$$E\left(\frac{\delta 1_{S^C} + \alpha 1_S}{\delta(1 - p) + \alpha p} - 1 \Big| S\right) = \frac{(1 - p)(\alpha - \delta)}{\delta(1 - p) + \alpha p} \approx \frac{\alpha}{\delta} - 1 \geq 1.7.$$

Decentralised market. We need to consider the following values.

$$\lambda^S \rightarrow= \frac{1}{\beta}\mathrm{VaR}_{1-\eta}\left(\sum_k E(X|\Omega_k)1_{\Omega_k}\right) = \frac{1}{\beta}E(X|\Omega_{m^*}),$$

$$\lambda^D = w_0 - u^{-1}(E(u(w_0 - X))).$$

Using the Bernoulli distributions for the conditional losses we can also find the loss distribution. Let us consider the following model: for any j let $X_j|\Omega_k = 1_{\mathcal{L}_{jk}}$, where $\mathcal{L}_{jk} \subseteq \Omega_k$, and the sequence of the sets $\{\mathcal{L}_{jk}\}_j$ are independent in (Ω_k, P_{Ω_k}) and $P_{\Omega_k}(\mathcal{L}_{jk}) = E_k$. So, if we consider a loss variable X with the same joint distribution as the losses X_j, it itself has a Bernoulli distribution given by $X = 1_{\mathcal{L}}$, where $P(\mathcal{L}) = \sum_k p_k E_k$. Using the exponential utility, we can find the demand price as follows:

$$\lambda^D = \frac{\log(E(e^{aX}))}{a} = \frac{\log(1 - \sum_k p_k E_k + e^a \sum_k p_k E_k)}{a}, \qquad a > 0,$$

and

$$\lambda^D = E(1_{\mathcal{L}}) = \sum_k p_k E_k, \qquad a = 0.$$

We display the relative pricing gap in Fig. 6.3. Now let us consider our example of the UK Coronavirus Job Retention Scheme for $\beta = 1$. Interestingly, we can see the same results as in Figs. 6.1 and 6.2, which means regardless of all other parameters there is a trade-off between the existence of the market equilibrium and correct risk coverage.

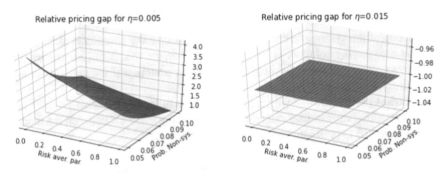

Fig. 6.3 Relative pricing gap for risk rate common shocks and $\gamma = 0.01$

Remark 6.5 In all examples, we can observe an interesting fact that is that the optimal risk-sharing approach in the limit for $n \to \infty$ will result in the following

$$\lambda = E(X|Sys),$$

where Sys is the sigma-field representing the systematic events. More precisely, for the examples with the risk-sharing and insurance platforms, it is $Sys = \sigma(Z)$, and in the example with the market platform, it is $Sys = \sigma(\Omega_1, \ldots, \Omega_n)$. The observation that the average of infinitely many exchangeable random variables reduce to a conditional expectation is not surprising to us, as it is related to De Finetti's theorem with exchangeable risk (see Kingman 1978, for more details on De Finetti's theorem).

6.6 Conclusion

In this paper, we considered an insurance cohort with identically distributed loss variables, and this insurance cohort sought an insurance scheme that keeps everybody's final wealth distributionally the same. This was called the non-discriminatory insurance (NDI) assumption. We have considered a very general setting where there is no need for any dependency structure of the wealth variables; this essentially means that we did not assume that losses, or the wealth distributions, are i.i.d or exchangeable. This general setup is motivated by systematic loss events such as a widespread pandemic (e.g., Spanish flu or COVID-19) with large macroeconomic loss impact. The idea is to provide a platform where we can properly study the risk management of macro-level losses.

We considered three different platforms: a risk-sharing, an insurance and a market platform. There are some general observations from studying these platforms. First, we showed that under NDI the most efficient final wealth is nothing but the average wealth; regardless of the dependency structure. Second, from the first observation we realised that there are benefits of introducing a contingent, ex-post premium. This essentially means that ex-post policies are shown to be more efficient than the current ex-ante insurance policies.

For any specific platform we also have made very interesting observations based on three common shock models, which include the additive, multiplicative and the risk rating common shock model. First, the risk-sharing platform is an efficient platform and any insurance scheme only acts as a wealth re-distributor ex-post. Second, in the insurance platform as the insurance companies are regulated and need to be solvent the optimal answer is a partial risk-sharing scheme. As a result, we observe that in this platform the risk-sharing platform the policyholders do not bear the risk of the tail events which necessitated the existence of a social scheme run by the state or government to bear the tail event risk. Third, we studied the market platform. In a competitive market, we see that we will come up with a deterministic premium, which gives an ex-ante policy. However, we observe that this platform dramatically increases the risk of insolvency. In the decentralised market platform, we realised that there is a trade-off between the tail risk of the insurer (supply side) and the market equilibrium. On one hand, if the common shock is not a tail event then we have no market equilibrium. On the other hand, if the common shock is tail event, even though a market equilibrium exists, there is substantial tail risk. Apparently if there is no common shock then both absence of tail risk and insurance market equilibrium can happen at the same time.

We have calibrated our models to the Coronavirus Job Retention Scheme (the UK furlough scheme during the COVID-19 pandemic), and all the observations are made on that basis. The results of the paper suggest new policy implications; the most important of which is the consideration of ex-post (contingent) premia, where the insurance premium is collected after the observation of the realised losses. Our observation is that in the presence of common shock, the major issue would not be the sophistication of the loss modelling or contracting, but it is the paradigm that may need to be changed. The insurance market needs be sustainable for systematic shock events, which for our case study means that unemployment insurance premia becomes contingent to the occurrence of the systematic risks such as COVID-19. This means that to reach the optimal allocation one adjusts the premium by directing wealth from people who did not lose to the people who have lost. In short, such insurance plans serve as a mechanism to *diversify* idiosyncratic risk and to *share* systematic risk. The systematic risk caused by COVID-19 is hard to insure as discussed by Richter and Wilson (2020).

The observation from the real world to a good extent confirms our conclusions. First, there has been a dispute[5] over the insurance coverage of the pandemic losses in the UK which has emphasised the unwillingness (or inability) of the supply-side (insurers) of the insurance market to settle the claims. Second, the government in the UK has introduced a generous furlough scheme[6] that ran for a few months to cover a large portion of the workforce in the UK. This has been executed in different means but the necessary capital will increase the government deficit and need to be paid back either by direct taxes or inflation. That can be regarded as some kind of contingent premia. Third, the UK government has supported businesses through the Trade Credit Insurance (TCI) guarantee, which again seems to be more like a contingency measure.[7] However, none of these solutions are carefully planned, and they are all based on the short-term assessments. Finding an insurance solution consistent to what we discuss in this paper seems to be a suitable direction to consider for further research.

---

[5] https://www.bbc.co.uk/news/business-55661702.

[6] https://www.ons.gov.uk/employmentandlabourmarket/peopleinwork/employmentandemployeetypes/articles/furloughingofworkersacrossukbusinesses/23march2020to5april2020.

[7] https://www.gov.uk/government/news/government-to-support-businesses-through-trade-credit-insurance-guarantee.

References

K. Aas, C. Czado, A. Frigessi, and H. Bakken, Pair-copula constructions of multiple dependence. Insur. Math. Econ. **44**(2), 182–198 (2009). ISSN 0167-6687.

P. Albrecht, Financial approach to actuarial risks? in *Proceedings of the 2nd AFIR International Colloquium, Brighton*, vol. 4 (1991), pp. 227–247

P. Albrecht, M. Huggenberger, The fundamental theorem of mutual insurance. Insur. Math. Econ. **75**, 180–188, 2017. ISSN 0167-6687.

H. Assa, Macro risk management: An insurance perspective (2020). Available on SSRN: https://ssrn.com/abstract=3670881

H. Assa, K.K. Karai, Hedging, pareto optimality and good deals. J. Optim. Theory Appl. (2012)

H. Assa, M. Wang, Price index insurances in the agriculture markets. North Am. Actuarial J. (2020)

H. Assa, H. Sharifi, A. Lyons, An examination of the role of price insurance products in stimulating investment in agriculture supply chains for sustained productivity. Eur. J. Oper. Res. **288**(3), 918–934 (2021). ISSN 0377-2217.

B. Avanzi, G. Taylor, B. Wong, Common shock models for claim arrays. ASTIN Bull. **48**(3), 1109–1136 (2018)

B. Babcock, E.K. Choi, E. Feinerman, Risk and probability premia for CARA utility functions. J. Agric. Resour. Econ. **18**(1), 17–24 (1993)

E. Banks, *Alternative Risk Transfer: Integrated Risk Management through Insurance, Reinsurance, and the Capital Markets* (Wiley, The Wiley Finance Series, 2004)

V. Bignozzi, G. Puccetti, L. Rüschendorf, Reducing model risk via positive and negative dependence assumptions. Insur. Math. Econ. **61**, 17–26 (2015). ISSN 0167-6687

T.J. Boonen, Competitive equilibria with distortion risk measures. ASTIN Bull. **45**(3), 703–728 (2015)

T.J. Boonen, Equilibrium recoveries in insurance markets with limited liability. J. Math. Econ. **85**, 38–45 (2019). ISSN 0304-4068

K. Borch, Equilibrium in a reinsurance market. Econometrica **30**, 424–444 (1962)

J.D. Cummins, O. Mahul, The demand for insurance with an upper limit on coverage. J. Risk Insur. **71**(2), 253–264 (2004). ISSN 00224367

M. Denuit, Investing in your own and peers' risks: the simple analytics of p2p insurance. Eur. Actuarial J. **10**(2), 335–359 (2020). ISSN 2190-9741

N.A. Doherty, H. Schlesinger, Rational insurance purchasing: Consideration of contract nonperformance. Quart. J. Econ. **105**(1), 243–253 (1990). ISSN 00335533, 15314650

L. Eeckhoudt, A.M. Fiori, E. Rosazza Gianin, Risk aversion, loss aversion, and the demand for insurance. Risks **6**(2), (2018). ISSN 2227-9091

L. Eisenberg, T.H. Noe, Systemic risk in financial systems. Manage. Sci. **47**(2), 236–249 (2001). ISSN 00251909, 15265501

M. Eling, Cyber risk research in business and actuarial science. Eur. Actuarial J. **10**(2), 303–333 (2020). ISSN 2190-9741

R. Feng, C.C. Liu, S. Taylor, Peer-to-peer risk sharing with an application to flood risk pooling. Mimeo (2020). Available on SSRN: https://papers.ssrn.com/sol3/papers.cfm?abstract_id=3754565

R.P. Flood, N.P. Marion, A. Matsumoto, International risk sharing during the globalization era. Canadian J. Econ. Revue Canadienne d'Economique **45**(2), 394–416 (2012). ISSN 00084085, 15405982

D. Kahneman, A. Tversky, Prospect theory: An analysis of decision under risk. Econometrica **47**(2), 263–291 (1979). ISSN 00129682, 14680262

J. Kingman, Uses of exchangeability. Ann. Probab. **6**(2), 183–197 (1978)

F. Lindskog, A.J. McNeil, Common poisson shock models: applications to insurance and credit risk modelling. ASTIN Bull. **33**(2), 209–238 (2003)

F. Liu, R. Wang, A theory for measures of tail risk. Math. Oper. Res. (2021)

N. Majumder, D. Majumder, Measuring income risk to promote macro markets. J. Policy Model. **24**(6), 607 – 619 (2002). ISSN 0161-8938

M. Marinacci, L. Montrucchio, Introduction to the mathematics of ambiguity, in *Uncertainty in Economic Theory: A Collection of Essays in Honor of David Schmeidler's 65th Birthday*, ed. by I. Gilboa (Routledge, Taylor & Francis Group, 2004)

R.C. Merton, The financial system and economic performance. J. Financ. Serv. Res. **4**(4), 263–300 (1990). ISSN 1573-0735

G.G. Meyers, The common shock model for correlated insurance losses. Variance **1**(1), 40–52 (2007)

A. Olivieri, E. Pitacco, *Introduction to Insurance Mathematics: Technical and Financial Features of Risk Transfers* (Springer, Berlin Heidelberg, 2011)

J. Quiggin, *Generalized Expected Utility Theory: The Rank Dependent Model* (Springer, Netherlands, 1993)

A. Richter, T. Wilson, Covid-19: implications for insurer risk management and the insurability of pandemic risk. Geneva Risk Insur. Rev. **45**, 171–199 (2020)

H. Schlesinger, The theory of insurance demand, in *Handbook of Insurance*, vol. 22 of *Huebner International Series on Risk, Insurance, and Economic Security*, ed. by G. Dionne (Springer, 2000)

U. Schmidt, Insurance demand under prospect theory: a graphical analysis. J. Risk Insur. **83**(1), 77–89 (2016)

R.J. Shiller, *Macro Markets: Creating Institutions for Managing Society's Largest Economic Risks*, repr. (Clarendon lectures in economics. Oxford Univ. Press, Oxford, 2007)

H. Varian, *Intermediate Microeconomics: A Modern Approach: Media Update* (W.W Norton & Company, 2019)

S.S. Wang, V.R. Young, H.H. Panjer, Axiomatic characterization of insurance prices. Insur. Math. Econ. **21**(2), 173–183 (1997). ISSN 0167-6687

R. Wilson, The theory of syndicates. Econometrica **36**(1), 119–132 (1968). ISSN 00129682, 14680262

Open Access This chapter is licensed under the terms of the Creative Commons Attribution 4.0 International License (http://creativecommons.org/licenses/by/4.0/), which permits use, sharing, adaptation, distribution and reproduction in any medium or format, as long as you give appropriate credit to the original author(s) and the source, provide a link to the Creative Commons license and indicate if changes were made.

The images or other third party material in this chapter are included in the chapter's Creative Commons license, unless indicated otherwise in a credit line to the material. If material is not included in the chapter's Creative Commons license and your intended use is not permitted by statutory regulation or exceeds the permitted use, you will need to obtain permission directly from the copyright holder.

Chapter 7
All-Hands-On-Deck!—How International Organisations Respond to the COVID-19 Pandemic

María del Carmen Boado-Penas, Gustavo Demarco, Julia Eisenberg, Kristoffer Lundberg, and Şule Şahin

Abstract The COVID-19 pandemic is affecting all countries. Since the World Health Organization declared the COVID-19 outbreak a Public Health Emergency of International Concern on 30 January 2021, governments across the world have mobilised on a tremendous scale and put in place different policies to contain the spread of the virus and its negative effects on society. International organisations have supported these efforts through evidence-based policy recommendations and emergency financing packages. This chapter presents a brief overview of the responses

Gustavo Demarco—Disclaimer: Views in this paper do not represent the official views of the World Bank. "All possible errors, mistakes, and shortcomings are the authors' own" whould be a disclaimer for all authors. Kristoffer Lundberg—-Disclaimer: Views in this paper do not represent the official views of the Swedish Ministry of Health and Social Affairs nor the Swedish Government. "All possible errors, mistakes, and shortcomings are the authors' own" whould be a disclaimer for all authors.

M. C. Boado-Penas (✉) · Ş. Şahin
Department of Mathematical Sciences, University of Liverpool, L69 7ZL Liverpool, UK
e-mail: carmen.boado@liverpool.ac.uk

Ş. Şahin
e-mail: sule.sahin@liverpool.ac.uk

G. Demarco
Pensions Global Solution Group – Social Protection and Jobs, The World Bank, 1818 H St, NW – Washington DC, 20433, USA
e-mail: gdemarco@worldbank.org

J. Eisenberg
Institute of Statistics and Mathematical Methods in Economics, TU Wien, Wiedner Hauptstr. 8-10, 1040 Vienna, Austria
e-mail: jeisenbe@fam.tuwien.ac.at

K. Lundberg
Ministry of Health and Social Affairs, Division for Coordination and Support – Policy Analysis Unit, SE 103 33 Stockholm, Sweden
e-mail: kristoffer.lundberg@regeringskansliet.se

© The Author(s) 2022
M. C. Boado-Penas et al. (eds.), *Pandemics: Insurance and Social Protection*,
Springer Actuarial, https://doi.org/10.1007/978-3-030-78334-1_7

127

made by international organisations and European Union towards COVID-19. Special attention is given to the guidance of these organisations on the changes in social insurance and pension plans to protect the most vulnerable population groups.

7.1 Introduction

On New Year's Eve 2019 the World Heath Organization's (WHO) Country Office in the People's Republic of China picked up a media statement posted by the Wuhan Municipal Health Commission from their website mentioning 27 cases of a new "viral pneumonia" in Wuhan, People's Republic of China, see WHO (2020a), European Centre for Disease Prevention and Control (2021). The WHO officials at the Country Office took immediate action and notified the International Health Regulations (IHR)[1] focal point in the WHO Western Pacific Regional Office about the media statement of the pneumonia cases and provided a translation of it from Chinese to enable further dissemination. WHO's Epidemic Intelligence from Open Sources (EIOS) platform also picked up a media report on ProMED (a programme of the International Society for Infectious Diseases) about the same cluster of cases of "pneumonia of unknown cause", in Wuhan, China (ibid). This promoted a reaction from several health authorities around the world and they, in turn, contacted the WHO seeking additional information, see WHO (2020a). This was the first time the world got to know the virus, Severe Acute Respiratory Syndrome Coronavirus-2 (SARS-CoV-2), that would be associated to cause the disease known as COVID-19 and the pandemic that it would cause.

On 30 January 2020 the WHO Director-General Dr. Tedros Adhanom Ghebreyesus[2] declared the COVID-19 outbreak a Public Health Emergency of International Concern WHO (2021a). This is the WHO's highest level of alarm, and it is a rallying call to all countries to immediately take notice and to act (ibid). As described in Chapter 1, many governments took drastic measures, such as lockdowns, to tackle COVID-19 and launched economic support package to protect the vulnerable groups. Every country response to fight this pandemic has been different, as shown in Fig. 7.1, with some implementing stricter measures earlier than others depending on the health and financial concerns, and public willingness to comply with measures, amongst others. However, in general, the measures taken by individual governments have proven to be inadequate at safeguarding the lives and livelihoods of vulnerable groups.

---

[1] The International Health Regulations (IHR) are a legally binding instrument of international law that aims for international collaboration. The IHR is the international legal treaty which empowers WHO to act as the main global surveillance system.

[2] Dr. Tedros Adhanom Ghebreyesus is an Ethiopian biologist, public health researcher, and official. He has since 2017 served as Director-General of WHO.

COVID-19: Stringency Index

This is a composite measure based on nine response indicators including school closures, workplace closures, and travel bans, rescaled to a value from 0 to 100 (100 = strictest). If policies vary at the subnational level, the index is shown as the response level of the strictest sub-region.

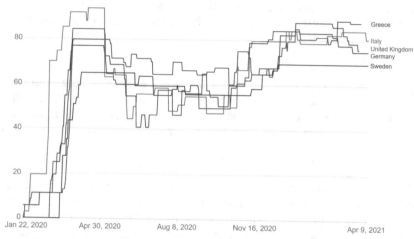

Source: Hale, Angrist, Goldszmidt, Kira, Petherick, Phillips, Webster, Cameron-Blake, Hallas, Majumdar, and Tatlow (2021). "A global panel database of pandemic policies (Oxford COVID-19 Government Response Tracker)." Nature Human Behaviour. – Last updated 11 April, 06:20 (London time)
OurWorldInData.org/coronavirus • CC BY

Fig. 7.1 COVID-19: stringency index for selected countries

International organisations have taken an important initiative to provide a unified guidance and raise awareness of the impact of the pandemic. This chapter aims to give an overview on the roles of the international organisations and how these have responded to COVID-19. With this in mind, the following section provides a description of the EU response together with the roles of the two main European institutions in charge of the pandemic: the European Commission and the European Centre for Disease Prevention and Control (ECDC). Section 7.3 focuses on the actions of the World Bank to COVID-19 pandemic and discusses the Pandemic Emergency Financial Facility created in 2017. Section 7.4 presents the responses from other international organisations: United Nations (UN), WHO, International Labour Organization (ILO) and the Organisation for Economic Co-operation and Development (OECD). Section 7.5 describes the recommendations from international organisations with respect to social insurance and pensions. The chapter concludes with a discussion of the role of the international organisations to fight against the pandemic.

7.2   The EU Response to COVID-19

The European Union is a political and economic union of 27 democracies. There are seven principal decision-making bodies of the EU, and they are listed in Article 13 of the Treaty on European Union, see Treaty on European Union (2010).

On 24 January 2020 the first European case of COVID-19 was reported in France, European Centre for Disease Prevention and Control (2021). Since the onset of the pandemic the EU has been at the centre of the response in Europe. The EU's actions taken towards the COVID-19 pandemic can be summed up in 10 main points, see European Parliament (2020):

- *Slowing the spread of the virus*
  To help limit the virus's spread, the EU has closed its external borders to non-essential travel, while ensuring essential goods keep moving across the EU. The European Centre for Disease Prevention and Control[3] (ECDC), which provides rapid risk assessments and epidemiological updates on the outbreak, has been given additional resources.

- *Providing medical equipment*
  The EU member states have taken action to procure life-saving medical equipment, such as ventilators and protective masks, under the Civil Protection Mechanism. In addition, the EU has set up a huge international tender allowing member states to make joint purchases of equipment and drugs etc. The EU has also organised an online fundraiser for vaccines, medicines, and diagnostics to fight the coronavirus worldwide.

- *Promoting research*
  The EU research programme is funding 18 research projects, and 151 teams across the EU to help find a vaccine against COVID-19 quickly.

- *Boosting European solidarity*
  The European Parliament[4] has supported new rules allowing member states to request financial assistance from the EU Solidarity Fund to also cover health emergencies. This allows for up to EUR 800 million to become available for member states to fight the pandemic.

- *Assuring the EU's recovery*
  To help the EU to recover from the pandemic a new proposal for the EU's long-term budget for 2021–2027 has been agreed upon. This proposal also includes a stimulus package (see the Recovery and Resilience Facility below).

- *Supporting the economy*, see European Council (2021).
  On the 11th of February 2021 the European Council established the Recovery and Resilience Facility, which will make EUR 672.5 billion in grants and loans available for public investment and reforms in the 27 member states to help them address the impact of the COVID-19 pandemic. The aim of the Facility is to

---

[3] The ECDC is an EU agency whose mission is to strengthen Europe's defences against infectious diseases. See https://www.ecdc.europa.eu/en for more details.

[4] The European Parliament is one of three legislative branches of the EU and one of its seven institutions. Together with the Council of the European Union, it adopts European legislation, commonly on the proposal of the European Commission. The Parliament is composed of 705 democratically elected members.

foster the green and digital transitions and to build resilient and inclusive societies. Member states will receive support based on the national recovery and resilience plans, which are under preparation in spring 2021.

- *Protecting jobs*
  The European Commission has unlocked EUR 1 billion from the European Fund for Strategic Investments in guarantees to encourage banks and other lenders to provide up to EUR 8 billion of liquidity in support to some 100,000 European businesses.

- *Repatriating EU citizens*
  With the outbreak tens of thousands of Europeans became stranded around the world but thanks to the EU Civil Protection mechanism they could be returned home.

- *Helping developed countries face the pandemic*
  The European Commission has unlocked EUR 20 billion to help non-EU countries fight the crisis as part of an EU package for a coordinated global relief to tackle the spread of COVID-19.

- *Ensuring accurate information*
  The spread of disinformation about the coronavirus puts people's health at risk. Ensuring that everyone has access to accurate and verified information in their own language has been called upon by members of the European Parliament, and social media companies have been asked to tackle disinformation and hate speech.

In addition to the general EU response, there are two main European institutions in charge of the pandemic, the European Commission and ECDC, whose roles towards the pandemic are described in Sects. 7.2.1 and 7.2.2.

7.2.1   EU Commission Response

The European Commission is the EU's politically independent executive arm and is coordinating a common European response to the coronavirus outbreak, see European Commission (2021a). The Commission is taking resolute action to reinforce the public health sector and mitigate the socio-economic impact in the EU. It is mobilising all means at its disposal to help the Member States coordinate their national responses and are providing objective information about the spread of the virus and effective efforts to contain it. The president of the European Commission Ursula von der Leyen[5] has established a Coronavirus response team at political level to coordinate the Commission response to the pandemic. The European Commission has been

---

[5] Ursula von der Leyen was born 8 October 1958 and is a German politician, prior to becoming president of the Commission she served in the Cabinet of Germany between 2005 and 2019.

negotiating intensely to build a diversified portfolio of vaccines for EU citizens at fair prices and contracts have been concluded with several vaccine developers, securing a portfolio of more than 2.6 billion doses, see European Commission (2021b).

7.2.2 The ECDC Response

The ECDC is an EU agency whose mission is to strengthen Europe's defences against infectious diseases, as such it has a pivotal role in Europe. The ECDC produces the risk assessments, technical reports, advice and information on a regular basis. Since 31 December 2019 when the events in Wuhan was first reported, the ECDC Epidemic Intelligence has reported on this and started close surveillance of the COVID-19 related events. On 7 January 2020 the ECDC issued the Threat Assessment Brief—assessing the risk for travellers, introduction and further spread into the EU, see European Centre for Disease Prevention and Control (2021).

7.3 The World Bank Response to COVID-19

According to the World Bank (WB) the poorest and most vulnerable countries will be hit the hardest by the pandemic. Therefore, a fast response package is needed to save lives and alleviate the consequences of COVID-19. In April 2020, the World Bank approved financial emergency support for developing countries to protect lives and support economic recovery.[6] Through this COVID-19 fast-track facility, the World Bank is making available up to $160 billion over the following 15 months. On April 2, the first group of projects—amounting to 1.9 billion—was released to assist 25 countries. In May 2020, the World Bank Group (WBG) announced its emergency operations in fighting COVID-19 had reached 100 developing countries. The WBG also designs and implements community-driven development (CDD) programmes to respond to urgent needs including access to clean water, rural roads, schools and health clinic constructions or support for micro-enterprises, amongst others.

Additional funding uses the COVID-19 Multiphase Programmatic Approach (MPA) with the aim to support vaccination of 1 billion people globally, see World Bank (2020b). Another financial instrument used by the WBG during the pandemic is the development policy financing (DPF) that provides IBRD loans, IDA credit/grant and guarantee budget support for governments to help achieving sustainable, shared growth and poverty reduction. For three stages of crisis response—Relief, Restructuring and Resilient Recovery, see World Bank (2020a).

---

[6] See WHO (2020b) and Gentilini et al. (2020) for more details about the WHO's response to COVID-19.

The World Bank operates across three stages of intervention, driven by the stages of the crisis, and a four-pillar response to the pandemic crisis tailored to the specific country. The relief stage involves emergency response to the health threat posed by COVID-19 and its immediate social, economic, and financial impact. Once the countries start to have the pandemic controlled and re-open their economies, the restructuring stage focuses on strengthening health system for pandemic readiness, restoring human capital, and restructuring of firms and sectors, amongst others. The resilient recovery stage entails benefiting from new opportunities to build a more sustainable, inclusive and resilient future world transformed by the pandemic.

The four pillars of crisis response comprise:

- Pillar 1: Bank emergency support for health interventions aimed at saving lives threatened by the virus.

- Pillar 2: WBG social response for protecting poor and vulnerable people from the impact of the economic and social crisis triggered by the pandemic.

- Pillar 3: WBG economic response for saving livelihoods, preserving jobs, and ensuring more sustainable business growth and job creation.

- Pillar 4: Support for strengthening policies, institutions, and investments to achieve a resilient, inclusive, and sustainable recovery.

WBG COVID-19 CRISIS RESPONSE			
Eliminate Extreme Poverty and Promote Shared Prosperity in a Sustainable Manner Capital Package Commitments IDA19 Commitments & Special Themes Partnerships UN Agencies IMF & MDBs Private Sector Vaccine Partnerships Civil Society Macroeconomic Stability and Strong Fiscal Framework Flexibility and Adaptive Learning Bridging the Digital Divide			
WBG COVID-19 Crisis Response	**Relief Stage**	**Restructuring Stage**	**Resilient Recovery Stage**
Pillar 1 **Saving Lives**	Public Health Emergency	Restructuring Health Systems	Pandemic-ready Health Systems
	Health MPA & Project restructurings	Health MPA & new IPFs	Health MPA & new IPFs
	DPFs	IFC Health Value Chain Platform	IFC LTF to pvt providers & manufacturers
Pillar 2 **Protecting the Poor & Vulnerable**	Social Emergency	Restoring Human Capital	Building Equity and Inclusion
	Cash/in-kind transfers, CDD Projects, DPFs	Cash/in-kind transfers, CDD projects, DPFs	Cash/in-kind transfers, CDD projects, DPFs
	Project Restructurings	new IPFs	ASA on active labor market policies
	Govt guarantees to MFIs	IFC recapitalization of strategic MFIs	IFC lending to MFIs
Pillar 3 **Ensuring Sustainable Business Growth & Job Creation**	Economic Emergency	Firm Restructuring & Debt Resolution	Green Business Growth & Job Creation
	DPFs, FILs, P4Rs and IPFs	DPFs and IPFs	DPFs and IPFs
	IFC trade & working cap lines, MIGA instruments	IFC restructuring and recapitalization of firms	IFC/MIGA instruments
	PPP Financing Vehicles	PPPs, IFC LTF and MIGA instruments	PPPs
Pillar 4 **Strengthening Policies, Institutions and Investments for Rebuilding Better**	Maintain Line of Sight to Long-term Goals	Policy and Institutional Reforms	Investments to Rebuild Better
	DPFs on fiscal strenthening & service delivery	DPFs on policies and institutional reforms	Full range of WBG instruments
	ASA for understanding COVID-19 related	for restructuring & resilience	with focus on PPP, Upstream project
	transformations, SME & MFI guarantee schemes	ASA for restructuring	development and mobilizing pvt solutions
	ASA for debt sustainability, mgt and transparency	ASA for tracking Twin Goals and SDGs	ASA for tracking Twin Goals and SDGs
WBG FINANCIAL CAPACITY			
IDA Hybrid Model IBRD Financial Stability Framework IFC Financial Model MIGA Financial Model			

Fig. 7.2 WBG COVID-19 crisis response. *Source* World Bank

In Fig. 7.2 we can see the financial instruments that the WBG applies depending on the pillars and the stage of the pandemic.

7.3.1 Pandemic Emergency Financial Facility (PEF)

In 2017, the Pandemic Emergency Financial Facility (PEF) was launched by the World Bank, in consultation with the World Health Organisation, and the private sector to financially help developing nations[7] facing a serious outbreak of infectious disease. In practice, the World Bank collects the premiums and issues bonds and swaps to private investors in return for favourable interest rates (6.5–11.1% above LIBOR). This will compensate investors for the risk that the bonds will need to make pay-outs to fight pandemics under certain conditions. In this way, the PEF, which can be seen as a type of catastrophe bond, is also known as pandemic bonds. It differs from other funds because it draws money from capital markets rather than relying solely on voluntary contributions.[8]

The coverage of PEF that was set up for an initial period of 3 years, from July 2017 to June 2020, has no cost for countries and the funds do not need to be repaid, World Bank (2019). The PEF covers large-scale outbreaks for diseases identified as likely to cause major pandemics.[9] The risk of these large-scale outbreaks is modelled so that the premium cost and coverage are calculated.

The objectives of the PEF are to: (i) make available essential surge financing to respond to an outbreak with pandemic potential and to minimise its health and economic consequences and (ii) help catalyse the creation of a global market for pandemic insurance instruments by drawing on resources from insurance, bonds and/or other private sector financial instruments.

The PEF shares some features of a parametric insurance contract in the sense that there is an objective trigger (threshold) event and a pay-out scheme. Also, the differences between the actual economic losses and the trigger creates basis risk. An epidemic might not trigger a parametric pay-out—for example, if the infection level[10] does not reach a particular level agreed in the specifications of the contract—but still can provoke some damages and losses.

---

[7] All countries that qualify for credits from the Bank's International Development Association are eligible to access PEF funds. In addition, international organisations and non-governmental organisations supporting response efforts in affected countries are also eligible to access PEF funds.

[8] The two main funds, i.e., the UN's Central Emergency Relief Fund (CERF) and the World Health Organization's Contingency Fund for Emergencies (CFE) have struggled with the financing, see Brim and Wenham (2019).

[9] This group of diseases includes pandemic Influenza, Coronaviruses, Filoviruses (i.e., Ebola), Crimean Congo haemorrhagic fever, Rift Valley fever, and Lassa fever.

[10] Other triggers are number of deaths, number of cases in each country, percentage of confirmed cases to total cases and the growth rate of cases, amongst others.

PEF is an innovative solution and was designed to be the perfect complement to traditional insurance to cover pandemic risks. However, the PEF was widely criticised mainly due to the generous returns to investors and difficulty in accessing funding during the early stages of the disease outbreaks when action is crucial, Jonas (2019). Also, in the PEF, there are several activation criteria (i.e. the total infected cases, total deaths, number of countries affected, outbreak growth and spread) that served as a joint trigger and consequently were difficult to reach those simultaneously. In the Kivu Ebola epidemic, the PEF only paid $31 million by the 13th month of the outbreak while the premiums paid to bondholders reached a total of $75.5 million, Brim and Wenham (2019). On April 2020—four months after the start of the outbreak—the first payment of $196 million for the PEF was triggered by the COVID-19 pandemic.

In 2019, the World Bank indicated that they were planning to issue a new set of bonds starting in May 2020. However, after facing significant criticism for delayed payments to developing countries during the COVID-19 pandemic, the World Bank has cancelled the launch of PEF 2.0.

7.4 Other International Responses to COVID-19

In this section, we describe the responses from the key international organisations that have shaped the responses in the EU countries: the UN, the WHO, OECD and ILO.

7.4.1 The UN Response

The UN is an intergovernmental organisation created in 1945 to maintain international peace and security, develop friendly relations among nations, to achieve international cooperation in solving international problems of an economic, social, cultural, or humanitarian character, and to be a centre for harmonizing action in the attainment of these common ends.[11]

The UN regards the COVID-19 pandemic as more than a health crisis, and hence it has triggered a UN Comprehensive Response to COVID-19 launched by the UN Secretary General António Guterres.[12] The response, which is ongoing, aims to, see United Nations (2021):

- Deliver a global response that leaves no-one behind.

- Reduce the vulnerability to future pandemics.

---

[11] Charter of the United Nations and Statue of the International Court of Justice, 1945, San Francisco.

[12] António Manuel de Oliveira Guterres was born 30 April 1949 and is a Portuguese politician serving as the ninth secretary-general of the United Nations since 2017.

- Build resilience to future shocks.
- Overcome the severe and systematic inequalities exposed.

The UN response is divided into three pillars, see United Nations (2020). The first pillar is a large-scale, coordinated, and comprehensive response by the WHO. The second pillar includes a wide-ranging effort to safeguard lives and livelihoods by addressing the devastating near-term socio-economic, humanitarian, and human rights aspects of the crisis. The final pillar aims towards creating a transformative recovery that leads to a better post-COVID-19 world.

7.4.2 The WHO Response

The WHO is a specialised agency of the UN responsible for public health, and as such it is at the heart of the world response towards the COVID-19 pandemic, see WHO (2021a). According to the Constitution of the World Health Organization "Health is a state of complete physical, mental and social well-being and not merely the absence of disease or infirmity".[13] The WHO has a broad mandate that enables the organisation to work with universal health care, risks monitoring, coordinating health emergencies, and promoting health and well-being.

The WHO response is comprehensive, and it is beyond the scope of this chapter to present it here. However, the WHO produces a Weekly Epidemiological Update which provides an overview of the global, regional and country-level COVID-19 cases and deaths, highlighting key data and trends, in addition to other important epidemiological information concerning the COVID-19 pandemic, see WHO (2021b). Furthermore, the COVID-19 Weekly Operational Update reports presents a weekly update on the WHO and its partners' actions in response to the pandemic (ibid).

7.4.3 The OECD Response

The OECD which is an international organisation that works to build better policies for better lives, OECD (2021). The goal of the OECD is to shape policies that foster prosperity, equality, opportunity, and well-being for all (ibid). The OECD works on establishing evidence-based international standards and finding solutions to a range of social, economic, and environmental challenges. The organisation seeks for improvement through knowledge from data and analysis, exchange of experiences, best-practice sharing, and advice on public policies and international standard-setting. The OECD response has aimed at answering questions relating to what impact the coronavirus pandemic would be on individuals and societies. Furthermore, the

---

[13] The Constitution was adopted by the International Health Conference held in New York from 19 June to 22 July 1946 and was signed on 22 July 1946 by the representatives of 61 States.

OECD intends to find solutions to strengthen healthcare systems, secure businesses, maintain jobs and education, and stabilise financial markets and economies.

7.4.4   The ILO Response

ILO has developed a four-pillar policy framework to tackle the economic and social impact of the COVID-19 based on International Labour Standards, see ILO (2020b). The policies emphasise the human dimension of the crisis caused by the pandemic and urge the governments to address the challenges with a human-centred approach. The pillars, listed below, form comprehensive and integrated recommendations on the key areas of policy action.

- Pillar 1: Stimulating the economy and employment.
- Pillar 2: Supporting enterprises, jobs and incomes.
- Pillar 3: Protecting workers in the workplace.
- Pillar 4: Relaying on social dialogue for solutions.

It is not expected that the harmful effect of the pandemic is distributed equally. It will be most damaging in the poorest countries and the poorest neighbourhoods. Pandemic and its aftermath created its unique way of discrimination and disadvantaged mainly the people who are in the informal economy, people with disabilities, migrant workers, indigenous people, women, and people living with HIV. The crisis also gave rise to child labour, forced labour, and human trafficking particularly women and girls. Due to the lack of access to social protection, these vulnerable groups are more affected by income shocks, ILO (2020a). Enhancing and enforcing the laws and policies on equality and non-discrimination are crucial to mitigate the risks.

7.5   Consequences of COVID-19 Responses on Social Security and Pensions

Elderly people and people with disabilities are particularly at risk of COVID-19. Specific actions are needed to protect the lives and health of these groups, so that a sufficient level of income is guaranteed, and they are better prepared for unexpected health-related expenses caused by a pandemic. In theory, current pensioners should not be economically affected by the pandemic as they are recipients of a regular income. However, pension payments—for many the only source of income—can be very low. According to ILO, support to pensioners should focus on protecting the low-income category.

Income support during the COVID-19 pandemic, Apr 11, 2021
Income support captures if the government is covering the salaries or providing direct cash payments, universal basic income, or similar, of people who lose their jobs or cannot work.

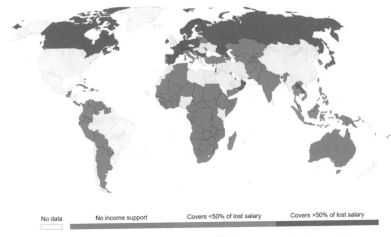

| No data | No income support | Covers <50% of lost salary | Covers >50% of lost salary |

Source: Hale, Angrist, Goldszmidt, Kira, Petherick, Phillips, Webster, Cameron-Blake, Hallas, Majumdar, and Tatlow (2021). "A global panel database of pandemic policies (Oxford COVID-19 Government Response Tracker)." Nature Human Behaviour. – Last updated 11 April, 06:20 (London time)
Note: This income support may not apply to workers in all sectors, and may vary at the sub-national level.
OurWorldInData.org/coronavirus • CC BY

Fig. 7.3 World map of the income support during the COVID-19 pandemic

Financial aid may take the form of additional support to those affected by the infection or a general pension increase. In Egypt, for example, a general pension increase has been implemented due to the pandemic while in Serbia one-off payment was made to all pensioners. Other countries, such as Slovenia, Sri Lanka and Tunisia have made selective top-ups in favour of the lower income pensioners and in Costa Rica, Colombia, Kosovo, or Mexico have made advance pension payments.

However, this kind of financial aid comprises serious dangers. Many countries had financially unsustainable pension systems already before COVID-19. Unexpected and considerable financial expenses, especially if the pension increases are applied to large groups of pensioners for extended periods, worsen the viability of the system even more in the long run. On the other hand, pandemics increase the unemployment and consequently reduce the income from contributions into social security programmes. In addition, some countries have even suspended, or reduced, contributions made by employees and/or employers to provide an incentive to retain workers.

During the first wave of the pandemic social security played a vital role to protect the most vulnerable groups. COVID-19 caused a high incidence of partial unemployment. Some countries extended the eligibility of unemployment benefits to cover not only full unemployment but also partial unemployment. Figure 7.3 gives an overview which countries provide a financial support to those who lose their jobs or cannot work. Contemplating that the significant turmoil on financial markets due to COVID-19 reduces returns on pension investments, the asset and the liability sides in pension insurances gape far apart.

Finally, loosening regulations on pension withdrawals, which some countries with Defined Contribution (DC) schemes have allowed, will compromise the adequacy of DC pensions in the future, if large sums have been withdrawn from individual accounts, unless compensatory measures are set in place.

According to the WB, after a pandemic ends, some of the (highly unpopular) measures to set in place post-crisis compensatory mechanisms and guide the system back onto the road to long-term financial stability are

- Repayment of past unpaid contributions by employers and workers.
- Extending participants working life by raising retirement ages.
- Limiting or eliminating the access to early retirement.

7.6 The Need of a United Action Tactic

The present book chapter gives an overview on how some of the most famous international organisations have responded to COVID-19 as of March 2021. The question arises whether the help and guidance provided by international organisations have been enough and timely.

Since the onset of the pandemic, a lot of critics has been expressed towards international agencies: WHO would have "failed in its basic duty" or the EU would lack consolidation and coordinated effort in responding the challenges of COVID-19. Unfortunately, the question of failing and guilt may easily become a political instrument and can repeatedly arise also in the future, see for instance the discussion in Gasbarri (2020). Debre and Dijkstra (2021) harshly criticise the responses of 75 international organisations, accusing some of them of bureaucracy and of benefiting from cross-border crises. On the other hand, Bill Gates, whose foundation is the second-largest funder of the WHO, stated that

> Halting funding for the World Health Organization during a world health crisis is as dangerous as it sounds. Their work is slowing the spread of COVID-19 and if that work is stopped no other organization can replace them. The world needs WHO now more than ever.

In July 2020, the Pew research center (a nonpartisan American think tank based in Washington, D.C.) conducted a survey across 14 countries about citizen's perception of the COVID-19 responses taken by both the countries' own governments and international organisations such as WHO and EU. The satisfaction scale ranged from 100 (totally satisfied) to 0 (not satisfied at all). Whilst countries such as Denmark (95), Australia (94), Germany (88) or South Korea (86) were quite satisfied with the measures taken by their own countries, the citizens from UK (46) or US (47), where the governments' response was not so harsh at the beginning of the pandemic, did not find the actions taken adequate. The WHO's score ranged between 74 in Denmark and 19 in South Korea, the measures taken by the EU received even a weaker recognition between 68 in Germany and 19 in South Korea, see Fig. 7.4. The

In all other countries surveyed, people rank the U.S. coronavirus response lowest

% who say ___ has done a __good job__ dealing with the coronavirus outbreak

	U.S.	Own country	WHO	EU	China
Spain	20%	54%	67%	65%	49%
Italy	18	74	54	54	51
Canada	16	88	67	65	36
UK	16	46	64	64	37
Sweden	15	71	73	56	33
France	15	59	62	57	44
Japan	15	55	24	34	16
Australia	14	94	54	46	25
Netherlands	14	87	66	68	42
Belgium	11	61	61	51	40
Germany	9	88	66	68	41
Denmark	7	95	74	66	26
South Korea	6	86	19	19	20
13-COUNTRY MEDIAN	15	74	64	57	37

Note: In Australia and Canada, the question was asked about "COVID-19." In Japan, it was asked about "the novel coronavirus," and in South Korea, it was asked about "Corona19."
Source: Summer 2020 Global Attitudes Survey. Q10a-e.
"U.S. Image Plummets Internationally as Most Say Country Has Handled Coronavirus Badly"

PEW RESEARCH CENTER

Fig. 7.4 Survey on the citizen's perception of the COVID-19 responses

survey also concluded that those countries with a favourable opinion of the UN were more likely to value the WHO's response to the pandemic positively.

Inside the EU there has been and, as of March 2021, still exists a big disparity in the level of infections, of testing capacities and endowments in medical supplies, see for instance The Guardian (2020). The lack of coordination by countries being close neighbours unavoidably leads to a reinfection and to multiple epidemic waves as the virus freely crosses the borders. Also, dealing with the aftermath of an economic disaster will strongly differ between the EU member states. When the severity of the COVID-19 pandemic, which is ongoing as of April 2021, ebbs away, it will be important to draw a true and objective balance from what did and did not work in the international and domestic responses.

The question about the next pandemic, the WHO answers with "not if, but when". In a report of the 30 March 2021, see WHO (2021c), the WHO claims that united action is needed to create a resilient international pandemic response in the future. Isolationism and nationalism cannot address the challenges of a global crisis, just fighting together as a united front[14] governments and international organisations can prevent the repeat of a disaster like COVID-19.

---

[14] A united front is a concept from the communistic vocabulary meaning building temporary coalitions for fighting concrete problems.

References

B. Brim, C. Wenham, Pandemic emergency financing facility: struggling to deliver on its innovative promise. BMJ **367** (2019), https://www.bmj.com/content/367/bmj.l5719. ISSN 0959-8138. 10.1136/bmj.l5719

M. Debre, H. Dijkstra, Covid-19 and policy responses by international organizations: crisis of liberal international order or window of opportunity? (2021). https://doi.org/10.13140/RG.2.2.28978. 43203

European Centre for Disease Prevention and Control. Timeline of ECDC's response to COVID-19, 2021, https://www.ecdc.europa.eu/en/covid-19/timeline-ecdc-response

European Commission, Coronavirus response (2021a), https://ec.europa.eu/info/live-work-travel-eu/coronavirus-response_en

European Commission, Safe COVID-19 vaccines for Europeans (2021b), https://ec.europa.eu/info/live-work-travel-eu/coronavirus-response/safe-covid-19-vaccines-europeans_en

European Council, EU recovery package: council adopts recovery and resilience facility (2021), https://www.consilium.europa.eu/en/press/press-releases/2021/02/11/eu-recovery-package-council-adopts-recovery-and-resilience-facility/

European Parliament, 10 things the EU is doing to fight the coronavirus (2020), https://www.europarl.europa.eu/news/en/headlines/society/20200327STO76004/10-things-the-eu-is-doing-to-fight-the-coronavirus

L. Gasbarri, Blog of the European Journal of International Law: the failure to pursue the mandates of international organizations in the midst of the COVID-19 pandemic (2020). https://www.ejiltalk.org/the-failure-to-pursue-the-mandates-of-international-organizations-in-the-midst-of-the-covid-19-pandemic/

U. Gentilini, M. Almenfi, P. Dale, Global database on social protection and jobs 19 (2020), http://documents.worldbank.org/curated/en/590531592231143435/pdf/Social-Protection-and-Jobs-Responses-to-COVID-19-A-Real-Time-Measures-June-12-2020.pdf. Living database, version 14

ILO, ILO policy brief: a policy framework for tackling the economic and social impact of the COVID-19 crisis (2020a), https://www.ilo.org/wcmsp5/groups/public/---dgreports/---dcomm/documents/briefingnote/wcms_745337.pdf

ILO, ILO presents its responses to COVID-19 at Geneva Environment Dialogues (2020b), https://www.oecd.org/about/

O. Jonas, Pandemic bonds: designed to fail in Ebola. Nature **572**(7769) (2019). https://doi.org/10.1038/d41586-019-02415-9

OECD, About the OECD (2021), https://www.oecd.org/about/

The Guardian, Coronavirus has revealed the EU's fatal flaw: the lack of solidarity (2020), https://www.theguardian.com/commentisfree/2020/apr/28/eu-coronavirus-fund-share-crisis-soul-european-parliament-fiscal

Treaty on European Union, Consolidated version of the Treaty on European Union/Title III: Provisions on the Institutions—Wikisource, the free online library (2010), https://en.wikisource.org/wiki/Consolidated_version_of_the_Treaty_on_European_Union/Title_III:_Provisions_on_the_Institutions

United Nations, Comprehensive response to COVID-19 (2020), https://www.un.org/sites/un2.un.org/files/un-comprehensive-response-to-covid-19.pdf

United Nations, COVID-19 response (2021), https://www.un.org/en/coronavirus/UN-response

WHO, A year without precedent: WHO's COVID-19 response (2020a), https://www.who.int/news-room/spotlight/a-year-without-precedent-who-s-covid-19-response/

WHO, World Bank COVID-19 response (2020b), https://www.worldbank.org/en/news/factsheet/2020/10/14/world-bank-covid-19-response?cid=EXT_WBEmailShare_EXT

WHO, Timeline of WHO's response to COVID-19 (2021a), https://www.who.int/emergencies/diseases/novel-coronavirus-2019/interactive-timeline

WHO, Coronavirus disease (COVID-19) weekly epidemiological update and weekly operational update (2021b), https://www.who.int/emergencies/diseases/novel-coronavirus-2019/situation-reports

WHO, COVID-19 shows why united action is needed for more robust international health architecture (2021c), https://www.who.int/news-room/commentaries/detail/op-ed---covid-19-shows-why-united-action-is-needed-for-more-robust-international-health-architecture

World Bank, The pandemic emergency financing facility: operational brief for eligible countries (2019)

World Bank, Saving lives, scaling-up impact and getting back on the track (2020a)

World Bank, COVID-19 strategic preparedness and response program (SPRP) using the multi-phase programmatic approach (MPA) project: additional financing (2020b), http://documents.worldbank.org/curated/en/882781602861047266/World-COVID-19-Strategic-Preparedness-and-Response-Program-SPRP-using-the-Multiphase-Programmatic-Approach-MPA-Project-Additional-Financing

Open Access This chapter is licensed under the terms of the Creative Commons Attribution 4.0 International License (http://creativecommons.org/licenses/by/4.0/), which permits use, sharing, adaptation, distribution and reproduction in any medium or format, as long as you give appropriate credit to the original author(s) and the source, provide a link to the Creative Commons license and indicate if changes were made.

The images or other third party material in this chapter are included in the chapter's Creative Commons license, unless indicated otherwise in a credit line to the material. If material is not included in the chapter's Creative Commons license and your intended use is not permitted by statutory regulation or exceeds the permitted use, you will need to obtain permission directly from the copyright holder.

Chapter 8
Changes in Behaviour Induced by COVID-19: Obedience to the Introduced Measures

Nuria Badenes-Plá

Abstract The pandemic of COVID-19 that has plagued our planet since the beginning of 2020, has disrupted the way of life of society in general. As in other pandemics suffered throughout history, isolation has been a crucial measure to avoid contagion, causing effects beyond health, in many areas of life. How society obtains economic resources, spends them, enjoys leisure, or simply interacts, is now different. The political and economic context has changed, freedom of movements and expectations are also different. All this generates changes in the behaviour of society that does not react uniformly in all countries. This chapter reviews some of the modifications in behaviour caused by the present circumstances, as what will happen in future pandemics is not predictable for sure. The emphasis is placed on obedience observed in different contexts to imposed restrictions. Homes have become workplaces, consumption patterns have changed, and the derived effects are not always beneficial or distributed equally across the social strata.

8.1 Introduction: Pandemics and Isolation

Since human beings created nuclei of coexistence, diseases have been present. The first pandemics began to be documented when the disease spread and affected various regions of the planet. Several key characteristics must be considered to define what is considered a pandemic according to Morens et al. (2009), such as geographic extension, disease movement, high attack rates, minimal population immunity, novelty, infectiousness, contagiousness, and severity. Pandemics have transformed the societies in which they appeared and have decisively influenced the course of history. During the development of pandemics such as the bubonic plague, the death toll exceeded the number of the living. With the fields unworked, the crops rotted, there was a shortage of agricultural products, monopolised by those who could afford them. Prices rose, and so did penalties for the less well off. Historians agree to point

N. Badenes-Plá (✉)
Instituto de Estudios Fiscales, Av del Cardenal Herrera Oria, 378, 28035 Madrid, Spain
e-mail: nuria.badenes@ief.hacienda.gob.es

© The Author(s) 2022
M. C. Boado-Penas et al. (eds.), *Pandemics: Insurance and Social Protection*,
Springer Actuarial, https://doi.org/10.1007/978-3-030-78334-1_8

143

out also positive economic and social effects for survivors that do not mitigate or compensate for the initial economic and social devastation and loss of life. In the case of the bubonic plague, abundant land, higher wages due to the drop in the supply of work, more job opportunities for women in guilds that had previously vetoed them. Neither in the past nor our days the effects of epidemics are not evenly distributed in the economy. Some sectors benefit from the exceptional demand for certain goods and services, while others suffer disproportionately. Inequality is also reflected in disease and mortality: in countries with lack public universal health care systems, the level of income may be decisive.

Apart from the fact that once infected, the worst economically situated citizens (and therefore with less probability of access to health coverage) have a higher risk of dying, COVID-19 does not hit the poorest population with more virulence as has happened with other pandemics in history. The most vulnerable groups in the face of infection are the elderly, and therefore a fall in labour supply or productivity is not expected as a consequence of the costs in terms of human lives. But it is the very efforts to stop the spread of the virus that have contributed to a dramatic slowdown of the global economy, which may affect the worst economically situated to a greater extent, either in terms of the impossibility of confinement, loss of employment, or difficulty to re-join the labour market.

As stated by the ILO[1] the crisis has had a different impact on enterprises, on workers, and on their families, though in each case deepening already existing disparities. Special attention needs to be given to the following groups: women, informal economy workers, young workers, older workers, refugees, and migrant workers, micro-entrepreneurs, and the self-employed. The greater impact of the crisis on workers and micro-enterprises already in a vulnerable situation in the labour market could well exacerbate existing-working poverty and inequalities. OXFAM[2] concludes on the same line, stating that "The coronavirus pandemic has exposed, fed off and exacerbated existing inequalities of wealth, gender and race. This crisis has laid bare the problems with our flawed global economic system and other forms of structural oppression that see a wealthy few thrive, while people in poverty, many women, Black people, Afro-descendants, Indigenous Peoples, and historically marginalized and oppressed communities around the world, struggle to survive". Laborde et al. (2020) estimate that globally, absent interventions, over 140 million people could fall into extreme poverty due to COVID, an increase of 20% from present levels.

Throughout history, humanity has faced numerous pandemics that have generated mortality figures even higher than those caused by COVID-19 that we suffer today. The Antonine plague between 165 and 180 A.D. caused the death of five million people. Four centuries later, the Justinian plague (first bubonic plague) claimed between 30 and 50 million lives in a single year. It is estimated that the Black Dead (second bubonic plague) in the mid-fourteenth century killed 200 million people, being

---

[1] https://www.ilo.org/wcmsp5/groups/public/@dgreports/@dcomm/documents/briefingnote/wcms_745337.pdf.

[2] https://oxfamilibrary.openrepository.com/bitstream/handle/10546/621149/bp-theinequality-virus-250121-en.pdf.

the deadliest pandemic known, and to have killed between 30 and 60% of Europe's population. Smallpox caused 56 million dead in 1520. Cocoliztli epidemic between 1545–1548 was a form of viral haemorrhagic fever that killed 15 million inhabitants of Mexico and Central America. Six cholera outbreaks have killed a million people between the 19th and 20th centuries. The Spanish flu was the deadliest virus of the 20th century, causing between 40 and 50 million deaths, while HIV has produced a death toll of 35 million since 1981.

In recent times, and before the emergence of COVID-19 pandemic, the SARS (Severe Acute Respiratory Syndrome) and MERS (Middle East Respiratory Syndrome) epidemics were experienced, although the impact on the population were of a much smaller nature. The SARS epidemic originated in 2002 in China, spread to 26 countries, but the number of cases reached a total of 8,098 and 774 people lost their lives. From 2012 to 2019, the total number of laboratory-confirmed MERS infection cases that were reported to WHO globally was 2,468, of which 851 were fatal. China was slow to warn of the emergence of the new SARS pathogen. Although the first case was declared in November 2002, WHO was only informed in February 2003. There was also no special transparency with the numbers of infections, deaths and spread of the new virus, which by March 2003 had already been present in Hong Kong, Vietnam, Canada and Singapore. Once the Chinese government changed its policy, it developed an impressive control strategy involving the public which culminated in containment as reported by Ahmad et al. (2009).

Having a significant number of lives lost and a high mortality rate in both cases, past situations could not predict an incidence of the magnitude reached with COVID-19, a pandemic that, until the first month of 2021, has infected more than 95 million people and caused the death of 2 million worldwide. Comparing the three coronaviruses, the one that causes COVID-19 is the least lethal, but the one that is transmitted more easily, and also the one in which patients do not present symptoms before they become contagious.

The quarantine, a measure that has been applied for centuries, also constitutes today a strategy to fight pandemics. The first sanitary cordon in history closed the ports of Genoa and Venice in the fourteenth century in which 10 days of observation were imposed. The origin of the quarantine dates from 1383 in Marseille. In the 18th century, all of Prussia was isolated for six consecutive weeks, and when the plague reached Greece, its neighbouring states were isolated for five weeks. In 1722, a military cordon separated Paris. And the threat of cholera isolated Russia between 1829 and 1832, establishing the death penalty if the border was crossed. In 1576 Milan was declared in quarantine because of the plague, and a single male member of each family could go out to buy food once a day. The so-called Spanish flu spread between 1918 and 1919 and despite its rapid expansion, isolation measures were not taken in all places. In the USA, the example of two opposing decisions led to very different results: St. Louis urged its population to remain confined, while in Philadelphia the activity was maintained, suffering a mortality eight times higher than that of St. Louis. Opting for isolation or quarantine has always been a decision of great importance given the social need to prevent contagion. The isolation is actually a strategy and a way of life adopted by diverse indigenous peoples who, when they were related to external agents in the past, suffered massive deaths due to the contagion of unknown diseases against which his immune system had not developed adequate defences. This epidemiological behaviour is decisive in the prevention of contagion in pandemics, although the most vulnerable groups cannot always comply with it due to the need to continue the work activity that guarantees livelihood.

As Tognotti (2013) points out, quarantine has been an effective way of controlling communicable disease outbreaks for centuries, although it has always been debated, perceived as intrusive, and accompanied at all times and under all political regimes by a current of suspicion, mistrust and riots. The historical perspective helps with understanding the extent to which panic, connected with social stigma and prejudice, frustrated public health efforts to control the spread of disease. Measures involving isolation require vigilant attention to avoid causing prejudice and intolerance. Public trust must be earned through regular, transparent communications that weigh the risks and benefits of public health interventions. If public health measures in the fight against pandemics imply restriction of freedoms and are not applied from a public power in which the citizens' trust, they may be frustrated.

This chapter shows evidence of how the COVID-19 pandemic has changed everyday life behaviours. After this introduction, the second section reviews the factors that explain the acceptance and compliance with the rules imposed to stop contagion. The third section studies the changes related to consumption, bad habits, teleworking and family relationships. The fourth section concludes.

8.2 Obedience to the Introduced Rules After COVID-19 Across Countries

In this section, research and reports from different countries are reviewed analysing what has been the response and the degree of acceptance or compliance of citizens with the government measures to stop the COVID-19 pandemic. These acceptance and compliance depend on the particular characteristics of the citizens, the values of the society in which they are inserted, the toughness and confidence in their effectiveness, as well as the trust in the governments that impose them.

8.2.1 Citizens' Demographic Characteristics

Clark et al. (2020) discovered from a large international sample, the importance of believing that taking health precautions will be effective for avoiding COVID-19 and prioritizing one's health, as predictors of voluntary compliance behaviours. In contrast, age, perceived vulnerability to COVID-19, and perceived disruptiveness of catching COVID-19 were not found significant predictors of health behaviours. Better information might increase voluntary compliance with government rules and recommendations but warning individuals about their vulnerability, providing details about the inconvenience of getting COVID-19, might not.

Roman et al. (2020) rely on the idea that the mass media often reproduce those meanings that obey the dominant interests. Using a questionnaire from different regions in Spain they conclude that gender and age are determining variables in the legitimation and implementation of social control between peers and that there is also a relationship between the way people perceive the role of media and their predisposition to abide by and exercise social norms and control. Women answered more affirmatively than men that the confinement was necessary and that the act of recriminating inappropriate behaviour of the population was beneficial in fighting the pandemic. Higher percentages were found in central age groups (26–55 years) with regard to the perception of social norms. Young people present predictive behaviours of risk because they perceive social norms in a more flexible way. Focused on the obedience during the first 10 days of confinement in Spain, Tabernero et al. (2020) analyse the relationship among personal social values, self-efficacy for self-protection, and the management of social isolation and beliefs in collective efficacy with the development of certain specific behaviours. They show that maintaining beliefs both in an individual capacity and in the ability of the community to carry out actions that protect us from the virus exerts a direct positive influence on the development of protective behaviours. Also, they conclude that Spanish citizens have greater confidence in their own abilities to develop behaviours that help curb the virus than the confidence they place in others. The variable of risk information seeking plays an important role in explaining behaviour, because as citizens become more informed, they develop a greater number of normative physical distancing behaviours.

Drawing on individual panel data from France, Briscese et al. (2020) find that some basic sociodemographic characteristics, as well as personality traits, are relevant predictors of compliance with health measures. In particular, age is positively associated with complying and women are more likely to have changed their behaviour compared to men. Education is not associated with public compliance, and conscientiousness is positively associated with having changed behaviour in line with the recommendations. Extraversion is negatively associated with having changed one's daily behaviours in the light of the pandemic. When considering ideology, the results indicate that ideological extremity is associated with reduced adherence to public health recommendations, and compliance increases when moving from the left to the right end of the ideology scale. They find also strong empirical confirmation over the association between fear and compliance, which is consistent with the study by Harper et al. (2020). Based on a large international community sample to complete measures of self-perceived risk of contracting COVID-19, fear of the virus, moral foundations, political orientation, and behaviour change in response to the pandemic, they found that the only predictor of positive behaviour change (e.g., social distancing, improved hand hygiene) was fear of COVID-19.

Focused on the effect of age, Daoust (2020) uses a survey regarding global insights on citizens' perceptions and responses to the COVID-19 pandemic across 27 countries and concludes that elderly people-even being the most vulnerable population-, are not systematically more responsive in terms of prospective self-isolation and willingness to isolate. Moreover, they are not more disciplined in terms of compliance with preventive measures, especially wearing a face mask when outside their home, which is surprising and quite troubling.

8.2.1.1 Trust in Government, Media and Health System

The conditions under which the COVID-19 pandemic will lead either to adherence to measures put in place by authorities to control the pandemic or to resistance and the emergence of open could be explained by three factors (Reicher and Stott 2020): the historical context of state-public relations, the nature of leadership during the pandemic, and procedural justice in the development and operation of these measures.

The unprecedented confinement situation offers an opportunity to analyse behaviour in such circumstances. Sibley et al. (2020) investigate the immediate effects of a nationwide lockdown by comparing matched samples of New Zealanders assessed before and during the first 18 days of lockdown. The study found that people in the pandemic/lockdown group reported higher trust in science, politicians, and police, higher levels of patriotism, and higher rates of mental distress compared to people in the pre-lockdown pre-pandemic group. Results were confirmed in within-subjects' analyses. The study highlights social connectedness, resilience, and vulnerability in the face of adversity.

The Spanish Government imposed a lockdown of more than three months, that was applied employing six fifteen-day extensions due to the high death rates achieved. As the confinements are prolonged, their enforcement becomes more difficult. The Spanish Sociological Research Centre conducted a survey asking 4,258 respondents about their rating of the response to the situation and their ability to cope with further extensions of the state of emergency. Fernandez-Prados et al. (2020) found that three out of every four Spaniards would cope with extensions of self-isolation at home. The variable of political inclination shows explanatory power in the responses: 40.6% of people with political leanings to the right would not face an extension of the blockade, while 87% of respondents with left leanings could face new extensions to the confinement. The communication strategy, both of leadership and of political measures, seems to be important in fostering social resilience. Other socio-economic characteristics allow tracing the average profile of the citizen most resilient or willing to cope with the prolongation of the state of emergency and lockdown: a woman over 60 years old, living in a town of fewer than 10,000 inhabitants, with a primary level of education, low social class, left-wing political leanings and no religious beliefs.

Van Rooij et al. (2020) attempt to identify the reasons why Americans come to comply with the stay-at-home and social distancing measures using data from an online survey, conducted on April 3, 2020, of 570 participants from 35 states. Their results show that while perceptual deterrence was not associated with compliance, people comply less when they fear the authorities. Instead, compliance operated through two broad processes. First, compliance is shaped by people's capacity to obey the measures, their self-control, and their lack of opportunity to violate. As such, part of compliance is not shaped by people's choice, but rather by their personal abilities and the context in which they lived. And second, compliance is shaped by people's intrinsic motivations, which determined the choices they could make, including substantive moral support, and social norms.

Al-Hasan et al. (2020) compare the responses of 482 citizens from the United States, Kuwait, and South Korea, and they underline the fundamental role of the government when adhering to measures to control the pandemic. Governments need to enhance their efforts on publicizing information on the pandemic, as well as employ strategies for improved communication management to citizens through social media as well as mainstream information sources. Their work uses web-based survey data in May 2020, and the results suggest that overall, perception of government response efforts positively influenced self-adherence and others' adherence to social distancing and sheltering, with some differences across countries, broadly the United States and Kuwait had better effects than South Korea.

Compliance with regulations may be tormented by the influence of the media. Simonov et al. (2020) found for the USA that a tenth increase in Fox News cable viewership ends up in a 1.3 decimal point reduction within the propensity to stay-at-home, while they fail to seek out conclusive effects of CNN viewership on social distancing. Given that Fox News Channel has been widely described as providing biased reporting in favour of conservative political positions, arises the question of whether journalism broadcasts directly influence viewer beliefs or merely function a platform to push the beliefs of political candidates

Misinformation can become a double-edged sword, as shown by Hameleers et al. (2020). Using the responses from the US, UK, Netherlands, and Germany—which experienced relatively high levels of misinformation and disinformation- they conclude that those citizens who experienced misinformation and were willing to seek further information were also more compliant with official guidelines. On the other hand, those individuals perceiving more disinformation and less willing to seek additional information were less compliant.

Briscese et al. (2020) test whether and how intentions to comply with social-isolation restrictions respond to the duration of their possible extension in Italy at three critical points in the COVID-19 pandemic. Italians reported being more likely to reduce, and less likely to increase, their self-isolation effort if negatively surprised by a given hypothetical extension, whereas positive surprises had no impact.

Referring to 38 Eurasian countries, Chan et al. (2020) carry out a study that tries to determine to what extent trust in the health system influences behaviour in response to the crisis. They conclude that societies with low levels of health care confidence initially exhibit a faster response concerning staying home, a reaction that plateaus sooner, and declines with greater magnitude than does the response from societies with high health care confidence. What is more interesting is that they verify that trust in the government prevails over the health system in behavioural decisions, as regions with high trust in the government but low confidence in the health care system dramatically reduces their mobility.

Confinement and restrictions maintained for a long period can lead to non-compliance, not because of the irresponsibility of the population, but because of the inability to work from home. Many Latin American and Caribbean governments implemented stringent lockdown measures, hoping to curb the spread. Despite this, the virus has hit these Latin American Countries. Following the OxCGRT,[3] 14 countries have been under stringent stay-at-home orders for over 150 days. As stated by UNDP,[4] "while initial compliance was high, the amount of time that people spend at home has been on a downward trend in all countries. In some cases, this corresponds with a less stringent stay-at-home orders (i.e. Aruba, Barbados, Trinidad and Tobago, Uruguay). However, in other cases, people have started to spend less time at home despite the fact that "stay-at-home" orders have remained equally strict (i.e. Bolivia, Brazil Colombia, Ecuador, Honduras, Jamaica, Mexico, Peru). Moreover, it's not always clear in which direction the policy change and behaviour change influence one another. For example, in some countries, people started staying at home before strict measures were put in place (i.e. Belize, Costa Rica, Mexico); whereas in others this change in behaviour seems to have taken place after the measures were put in place (i.e. Honduras, Peru, Venezuela). The converse may also be true in terms of easing restrictions—for example, less stringent policies in some countries were instituted in the wake of already declining compliance (i.e. Aruba, Barbados, Belize)".

8.2.2 *Cultural Tradition*

In addition to analysing compliance with the restrictions imposed by governments, cases can be offered in which the responsibility and good work of the population are trusted. Japan, with a population of more than 120 million inhabitants, and a density greater than 330 inhabitants per square kilometre, shows relatively low numbers of infected and deceased (258,000 and 3,800 respectively in the first days of January 2021). One possible explanation would be that Japanese culture is inherently suited for social distancing, and many Japanese wear face masks in the winter to avoid transmission of respiratory infections. Besides, the country has not imposed confinement and the obligation to close a business as in other countries. Japan avoided harsh measures, instead of issuing official requests for self-restraint and voluntary business closures. Japan has taken advantage of the concept of "seken" the power of peer pressure which reflects a particular power dynamic and order that appears whenever Japanese people gather in a group.

A case similar to that of Japan, and also atypical compared to its neighbouring countries is the Swedish. Sweden has taken a lighter-than-most approach to social distancing for COVID-19, relying on people to monitor themselves for symptoms, stay home when ill, practice good handwashing, and avoid crowds. This strategy

---

[3] The Oxford Coronavirus Government Response Tracker (https://www.bsg.ox.ac.uk/research/research-projects/coronavirus-governmentresponse-tracker).

[4] See https://www.latinamerica.undp.org/content/rblac/en/home/presscenter/director-sgraph-for-thought/home-alone---sustaining-compliance-with-prolonged-covid-19-stay-.html.

has been designed by Anders Tegnell, an epidemiologist at Sweden's Public Health Agency, who has not been exempt from criticism.

The anticipated response of the population to the most severe or lax guidelines, conditioned by cultural traditions, has not gone unnoticed as a determining element of the policies to combat the pandemic. Yan et al. (2020) combine two basic dimensions (centralisation of government and national cultural orientation) to analyse the different strategies adopted by governments in the fight against the spread of the virus in Japan, Sweden, China and France. These four countries are chosen as stereotypes representing fundamental differences in institutional arrangements and cultural values. Sweden is a country with a more decentralised regime and looser culture, whereas China has a more centralized regime and tighter culture. On the other opposing pair, France exhibits a more centralized regime but looser culture, and Japan with a more decentralized regime but tighter culture. They conclude that there is no one-size-fit-all strategy that can be used to combat COVID-19 on a global scale. This confirms that despite COVID-19 spreads worldwide, the strategies to defeat it cannot be designed outside of cultural values and political organisation.

In addition to checking the empirical evidence on behaviours, it is possible to delve into the determinants of compliance with the rules using experiments. Fischer et al. (2020) process the information provided by 3,102 individuals to show how to achieve enhanced adherence to health regulations without coercion. The participants were people residing in 77 different countries. They were asked to adhere to constraining behaviours, such as staying at home, keeping social distance, repeatedly washing hands and avoiding meeting seniors. These constraints restrict the personal freedom, but generate health benefits for both the individual and the entire population. This scenario can be modelled as a Chicken game (that motivates the players to cooperate, even when assuming the opponent does not). They find "that a cluster of short interventions, such as elaboration on possible consequences, induction of cognitive dissonance, addressing next of kin and similar others and receiving advice following severity judgements, improve individuals' health-preserving attitudes".

8.3 Behavioural Changes Due to COVID-19

The invisible threat of COVID-19 has generated fear and mistrust towards people and places that were traditionally considered safe. The isolation and distancing measures established to combat this threat have influenced elements such as work and consumption, with the consequent economic impact. The perception and form of consumption have also changed, as well as other more subjective issues, for example, the conception of the home, concern for health, trust in authority, prioritisation of the family. Citizens do not occupy their time in the same way, neither those who work not those who study. Responsibilities for training and obtaining financial resources have shifted from schools and usual places of work to the home. Teleworking and distance training have spread throughout the world. This new situation that implies staying at home longer, maybe accompanied by adverse effects, as coexistence problems, and a shift in some demands towards an increase in harmful habits. The following section describes how the COVID-19 pandemic has been able to influence these changes.

8.3.1 Consumption Patterns

Consumption habits have been modified in the world as a result of the pandemic and the subsequent mobility limitations imposed. Some essential products are still in demand, while those related to leisure have experienced notable changes, and prospects indicate declines in consumption. The products consumed have changed because needs are different, but also the way of consuming, much more biased towards buying online. Without being exhaustive, some examples can help raise awareness of the magnitude of the change. In October 2020, online traffic in the supermarket segment increased by 34.8% compared to the reference period in January 2020. Online visits in the tourism sector decreased by 43.7% during the same period, as fashion decreases 10.3%.[5] The UN World Tourism Organization estimates a loss of US$300–450 billion in international tourism receipts.

People are staying at home much more going forward, and telecommuting will presumably continue, which implies more home-linked consumption. Even when restrictions are relaxed, it is foreseeable that leisure, travel and restaurant consumption will not reach pre-pandemic levels immediately. On the other hand, consumers have less disposable income due to the effect of unemployment, which will reinforce the option of staying at home and will condition that many of the changes in consumption are lasting over time.

McKinsey & Company (2020a) conducts a study focused on 13 core countries, selected because of their economic significance and the impact that COVID-19 has had on their populations. Following the results, consumers in China, India, and Indonesia consistently report higher optimism than the rest of the world, while those in Europe and Japan remain less optimistic about their countries' economic conditions after COVID-19. Except for Italy, optimism has declined throughout European countries. China appears as the only exception to a global pattern of reducing holiday spending. McKinsey & Company (2020b) shows that some consumption trends that had been taking place in the past have been accelerated by the pandemic. For instance, online delivery has grown the same in eight weeks as in the previous ten years, or online entertainment, in five months the same as the previous seven years. Some changes are probably temporary as the reduction in international travel and increase in domestic tourism, but the reduction in discretionary spending, the trading down and price sensitivity seem more enduring behaviours. In terms of the shake-up of preferences, the reduction in on-the-go consumption and the trends of larger baskets and less frequency of purchase could be a temporary behaviour, while the preference for health and hygiene products seems a permanent trend.

---

[5] https://www.statista.com/statistics/1105486/coronavirus-traffic-impact-industry/.

Parady et al. (2020) distinguish three periods (before the spread, after the spread and before the emergency declaration, and after spread and after emergency declaration) for analysing the decrease in frequency for most activities in Japan. They observe a rebound in shopping activities after the emergency declaration, which would corroborate that the changes in consumption patterns would not be permanent. For activities like such as eating-out and leisure, the reductions in frequency persisted after the emergency declaration. Changes in these patterns cannot be just explained from the consumer side. They are also a result of changes in the supply side which in turn are imposed or suggested by governments. No closure obligations were imposed in Japan, but many eating-out and leisure establishments closed down or shortened their business in contrast to most shopping facilities which provided more essentials services.

In a study focused on 16 American cities, Yilmazkuday (2020) concludes that consumption carried out within the home (grocery, pharmacy, home maintenance) has increased by 56%, while that carried out outside the home (fuel, transportation, personal care services, restaurants) has decreased up to 51%. Online shopping has relatively increased up to 21%, while its expenditure share has relatively increased by up to 16% compared to the pre-COVID-19 period.

Using data on household financial transactions, Baker et al. (2020) illustrate the short-term responses of Americans' spending to the increase in COVID cases. They also analyse the responses to the policies implemented by municipal governments, such as confinement. Household spending was radically altered by these events across a wide range of categories, and the strength of the response depended in part on the severity of the outbreak in the state. Demographic characteristics such as age and family structure led to higher levels of heterogeneity in spending responses to COVID-19, while income did not. Furthermore, regardless of the political orientation, an increase in spending before the epidemic was observed and, at the same time, there were some differences in the political orientation in some categories, indicative of differential beliefs or risk exposure.

8.3.2 Unhealthy Consumption Habits

The confinement situation may have generated changes in unhealthy consumption habits, such as alcohol, tobacco or drugs. Vanderbruggen et al. (2020) use a web-based for Belgium where respondents reported consuming slightly more alcohol and smoking marginally more cigarettes than before the COVID-19 pandemic, while no significant changes in the consumption of cannabis were noted. The reasons for consuming more of the various substance were boredom, lack of social contacts, loss of daily structure, reward after a hard-working day, loneliness, and conviviality. Koopmann et al. (2020) confirm that during the COVID-19 pandemic the total revenue in alcoholic beverages in the German population increased significantly by 6.1% compared to the mean of corresponding weeks in the past year. As it remained unclear, whether this was due to stockpiling, or reflected real changes in alcohol drinking behaviour, they obtained responses via an anonymous online survey. A survey conducted for the Polish population (Szajnoga et al. 2020) found that the vast majority of respondents reduced the frequency of consumption of all types of alcohol. However, particular groups are more vulnerable to alcohol misuse: higher frequency of alcohol consumption lockdown was most often found in the group of men, people aged 18–24 years, inhabitants of big cities, and remote workers. Contrary to the previous references, Callinan et al. (2020) find a decrease in harmful alcohol consumption for Australia compared to the pre-pandemic period. This effect is observed in young adults in particular and explained by the closure of licensed premises, but there is no reason to assume that these decreases will not reverse when licensed premises re-open.

The modification of routines during confinement has also affected eating habits. Ruíz-Roso et al. (2020) describe how the COVID-19 pandemic has modified dietary trends of adolescents from Spain, Italy, Brazil, Colombia, and Chile. They conclude that "Due to confinement, it appears that families had more time to cook and improve eating habits by increasing legume, fruit, and vegetable intake, even though this did not increase the overall diet quality. Further, adolescents also exhibited a higher sweet food consumption, likely due to boredom and stress produced by COVID-19 confinement".

Attempts to project the longer-term impacts of the current pandemic depend on the extent to which changes in patterns become permanent once normality is restored. Not much is known about the long-term mental and physical health effects of lockdown and limitation of mobility, but the situation has led to adopt or reinforce unhealthy behaviours (physical activity decrease, sedentary behaviour increase, unhealthy nutritional habits). There is also a surge of addictive behaviours (implying substance use disorders) both new and relapse in this period. Depending on the addictive component of the behaviours which can be modelled even in rational terms (Becker and Murphy 1988), changes in current habits can pose a health problem in the long term.

8.3.3 Teleworking

Before the COVID-19 outbreak, just 15% of the employed in the EU had ever teleworked, since then, working from home has been the way to continue the activity for many workers. According to Eurofound (2020), almost 40% of those currently working in the EU began to telework full-time as a result of the pandemic. Telework is structurally more widespread in countries with larger shares of employment in knowledge and ICT-intensive services. The presence of teleworking will be conditioned by the rate of self-employment, flexibility, supervisory styles, and the organisation of work, which vary across countries. Espinoza and Reznikova (2020) find that while 30% of workers could telework across the OECD, the likelihood decreases for workers without tertiary education and with lower levels of numeracy and literacy. They also find that while an average of 56% of OECD workers in the top 20% of the income distribution can telework, the share stands at only 14% for those in the bottom 20%. López-Igual and Rodríguez-Modroño (2020) confirm that the most significant determinants of telework are still self-employment, a higher educational level, while other factors (age, living in urban areas, higher status, and better working conditions) are losing. The maintenance or extension of telework in the future will depend to a large extent on productivity under this work organisation. The evidence in the EU suggests that in normal times telework can sustain, or even enhance workers while enjoying a better work-life balance. Under the current exceptional circumstances, productivity and/or working conditions may be deteriorating due to, lack of childcare, unsuitable working spaces and ICT tools. The potential costs associated with teleworking such as loss of productivity, job quality, workers' work-life balance and mental health, may not affect family workers in the same way, depending on the distribution of household tasks that is established (Feng and Savani 2020). The reconciliation of teleworking with family life and dedication to children is one of the main difficulties encountered across countries, even in Portugal, where Tavares et al. (2020) check an easy and very quick adaptation to teleworking. Katsabian (2020) points out that telework has converted homes in hybrids spaces of work. Because of its hybrid nature, it reproduces in the labour market the gendered traditional roles within the family domain along with socioeconomic disparities among households concerning access to technology and technological skills.

8.3.4 Gender and Family Violence

At the same time that the value of the family is recognised in a circumstance of illness and risk of losing it, the "compulsory" coexistence for longer than usual can unleash situations of violence at home. The confinement during the pandemic can exacerbate the problem of domestic violence. Family members spend more time living together, economic instability makes tension and stress more acute, and isolation places victims in a more vulnerable position. The efforts to contain the virus are vital to protect global health, but expose women, children and adolescents to an increased risk of family violence. The rise in reports of domestic abuse and family violence have increased around the world since social isolation and quarantine measures came into force: 300% in China, 50% in Brazil, 30% in Cyprus, France, and New Zealand, 25% in the United Kingdom (UK), and 20% in Spain (Noman et al. 2021). Usher et al. (2020) also review the evidence in an intimate partner, women, and children violence due to isolation and quarantine in different countries, and highlights that France began commissioning hotels as shelters for those fleeing abuse. This strategy was followed also by the Italian government given the increasing number of people fleeing abusive situations. Boxall et al. (2020) provide evidence of onset of the frequency or severity of physical or sexual violence or coercive control for many women in Australia. Silverio-Murillo et al. (2020) show empirical evidence for Mexico City, during the lockdown: while official domestic violence crime reports decline, within-household violence continues during the pandemic. To reconcile this apparent contradiction, several causes are investigated, as the fact that confinement of victims with their perpetrators prevents reporting, the changes in bargaining power within the household, the alcohol consumption, or fear to be infected during reporting.

8.4 Discussion and Conclusions

Throughout history, humanity has faced numerous pandemics that have caused mortality figures even higher than those generated by COVID-19 that we suffer today. Although quarantine has been an effective way of controlling communicable disease outbreaks for centuries, it has always been debated, perceived as intrusive, and accompanied at all times and under all political regimes by a current of suspicion, mistrust and riots. When public health measures in the fight against pandemics imply restriction of freedoms and are not applied from a public power in which the citizens' trust, they may be frustrated. The acceptance and compliance of measures to stop the pandemic depend on the particular characteristics of the citizens, the values of the society in which they are inserted, the toughness and confidence in their effectiveness, as well as the trust in the governments that impose them.

The literature review that covers a large number of countries and cultures, has shown different determinants for explaining compliance and adherence with the measures that governments have taken to appease the virus have been found. The explanatory factors are studied both from the point of view of those who dictate the measures (governments) and those who receive them (citizenship). Concerning citizens, several characteristics appear to be explicative, including the perception of own vulnerability, fear of the virus, age, gender, size of the city of residence, level of education, social class, political leaning or moral and religious beliefs. Focusing on the characteristics and feelings towards those who dictate or enforce the rules, some circumstances seem relevant, such as being a feared authority, efforts on publicising information or trust in government. Besides, there are other determinants of context, such as the influence of the media, duration of confinements, opportunity to violate the rules, trust in the health system, power of peer pressure, institutional arrangements or cultural values.

How action measures are dictated to protect individuals from the pandemic is relevant to the way citizens respond. Imposing is not always more effective than informing. Planning, anticipating the duration of the measures and being transparent with the public can be more effective than it seems. There is no one-size-fit-all strategy that can be used to combat COVID-19 on a global scale, and the strategies to defeat it cannot be designed outside of cultural values, political organisation, or citizenship's characteristics.

The pandemic COVID-19 has generated fear and mistrust towards people and places that were traditionally considered safe. The isolation and distancing measures established to combat this threat have influenced elements such as work and consumption, with the consequent economic impact. The products consumed have changed because needs are different, with increases in products consumed within the home (grocery, pharmacy, home maintenance) and decrease in those carried out outside the home (fuel, transportation, personal care services, restaurants). The way of consuming is now much more biased towards buying online: online delivery has grown the same in eight weeks as in the previous ten years, or online entertainment, in five months the same as the previous seven years. Some changes are probably temporary as the reduction in international travel and increase in domestic tourism, but the reduction in discretionary spending, the trading down and price sensitivity seem more enduring behaviours. The new situation that forces people to stay longer at home, and the psychological effects created by confinement, may also lead to increases in the consumption of harmful goods, or damage eating habits, which has been proven to occur in some contexts.

Teleworking, spread throughout the world, has also been a way to reveal inequalities: it is the richest countries with the most advanced technologies that used more telework even before the pandemic, so they have been better able to adapt to the situation. The most significant determinants of telework are self-employment, a higher educational level, and non-manual occupations, which once again places the most disadvantaged workers in a worse situation by being forced to face-to-face work. The maintenance or extension of telework in the future will depend to a large extent on productivity under this work organisation. In normal times telework can sustain, or even enhance workers while enjoying a better work-life balance. Under the current exceptional circumstances, productivity and/or working conditions may be deteriorating due to, lack of childcare, unsuitable working spaces and ICT tools.

The circumstances that accompany periods of confinement during the pandemic can exacerbate the problem of domestic violence. Family members spend more time living together, economic instability makes tension and stress more acute, and isolation places victims in a more vulnerable position. There is evidence of a rise in reports of domestic abuse and family violence around the world since social isolation and quarantine measures came into force.

The fight against a pandemic represents a challenge for governments of uncommon magnitude. The new rules established by governments imply restriction of freedoms, and the acceptance of them depends a lot on the social, cultural, political and personal context. Distancing and staying at home have emerged as key elements in slowing the spread of the virus. Getting citizens to comply with these mandates can be difficult, and communication strategies, transparency of information and adaptation to the particular characteristics are essential for success. Although the problem is global, the countries are different, and for this reason it is not possible to find one-size-fit-all strategy that can be used to combat COVID-19 in any situation. This confirms that despite the virus spreads worldwide, the strategies to defeat it cannot be designed outside of cultural values and political organisation.

The scale, intensity, and speed of the interventions against the pandemic have diverged across territories. Many Asian countries promptly did extensive testing (not only on symptomatic), tracing and isolating (at institutions rather than at home), and surveillance systems were strengthened. These measures have been adopted much less quickly in Europe (Han et al. 2020). The use of masks was also much more widespread in Asia than in Europe and was adopted almost immediately in a massive way. Previous SARS and MERS epidemics have also prepared Asian health systems much better to fight a pandemic, while in Europe austerity policies have weakened public health infrastructure. Asian citizens are more predisposed to confront and cooperate with measures that restrict freedom than Europeans. The experience of past epidemics has made them aware of the convenience of renouncing individual freedoms for the benefit of the community. Experiences can certainly serve as learning for the future, as they have been for Asia, but it is difficult to know if this process of compliance by the population, preparation of the health system, and anticipation of governments will be applicable to other countries. Having gone through the same experience previously is not the only explanatory factor: economic, cultural and social issues will be decisive in the learning process.

The costs in terms of human lives, health and the economy are immeasurable quantitatively and qualitatively, but the pandemic has generated a new way of life based on distance and isolation that generates derived effects in many other areas. For this reason, the current challenges of governments are focused on dictating adequate measures and trying to enforce them, but future challenges are completely unpredictable.

References

A. Al-Hasan, D. Dobin Yim, J. Khuntia, Citizens' adherence to COVID-19 mitigation recommendations by the government: a 3-country comparative evaluation using web-based cross-sectional survey data. J. Med. Internet Res. **22**(8) (2020). https://doi.org/10.2196/20634

A. Ahmad, R. Krumkamp, R. Reintjes, Controlling SARS: a review on China's response compared with other SARS-affected countries. Trop. Med. Int. Health **14**(Suppl 1(s1)), 36–45 (2009). https://doi.org/10.1111/j.1365-3156.2008.02146.x

S.R. Baker, R.A. Farrokhnia, S. Meyer, M. Pagel, C. Yannelis, J. Pontiff, How does household spending respond to an epidemic? Consumption during the 2020 COVID-19 pandemic. The Review of Asset Pricing Studies **10**(4), 834–862 (2020)

G.S. Becker, K.M. Murphy, A theory of rational addiction. J. Polit. Econ. **96**(4), 675–700 (1988)

H. Boxall, A. Morgan, R. Brown, The prevalence of domestic violence among women during the COVID-19 pandemic. Australian Institute of Criminology. Statistical Bulletin, 28 July 2020

G. Briscese, N. Lacetera, M. Macis, M. Tonin, Expectations, reference points, and compliance with COVID-19 social distancing measures. NBER Working Paper No. 26916 (2020)

S. Callinan, K. Smit, Y. Mojica-Perez, S. D'Aquino, D. Moore, E. Kuntsche, Shifts in alcohol consumption during the COVID-19 pandemic: early indications from Australia. Soc. Study Addiction. Res. Rep. (2020). https://doi.org/10.1111/add.15275

H.F. Chan, M. Brumpton, A. Macintyre, J. Arapoc, D.A. Savage, A. Skali, et al., How confidence in health care systems affects mobility and compliance during the COVID-19 pandemic. PLoS ONE **15**(10), e0240644 (2020). https://doi.org/10.1371/journal.pone.0240644

C. Clark, A. Davila, M. Regis, S. Kraus, Predictors of COVID-19 voluntary compliance behaviors: an international investigation. Glob. Trans. **2**, 76–82 (2020). https://doi.org/10.1016/j.glt.2020.06.003

J.F. Daoust, Elderly people and responses to COVID-19 in 27 Countries. PLoS ONE **15**(7), e0235590 (2020). https://doi.org/10.1371/journal.pone.0235590

R. Espinoza, L. Reznikova, Who can log in? The importance of skills for the feasibility of teleworking arrangements across OECD countries. OECD Social, Employment and Migration Working Papers, No. 242 (2020)

Eurofound. Living, working and COVID-19 dataset, Dublin (2020). http://eurofound.link/covid19dataropean

Z. Feng, K. Savani, Covid-19 created a gender gap in perceived work productivity and job satisfaction: implications for dual-career parents working from home. Gender in Management: An International Journal **35**(7/8), 719–736 (2020). https://doi.org/10.1108/GM-07-2020-0202

J.S. Fernandez-Prados, A. Lozano-Díaz, J. Muyor Rodríguez, Factors explaining social resilience against COVID-19: the case of Spain. Eur. Soc. (2020). https://doi.org/10.1080/14616696.2020.1818113

I. Fischer, S. Avrashi, T. Oz, R. Fadul, K. Gutman, D. Rubenstein, G. Kroliczak, S. Goerg, A. Glöckner, The behavioural challenge of the COVID-19 pandemic: indirect measurements and personalized attitude changing treatments (IMPACT). R. Soc. Open Sci. **7**(8) (2020). https://royalsocietypublishing.org/doi/10.1098/rsos.201131

M. Hameleers, G.L. Toni, A. van der Meer, A. Brosius, Feeling "disinformed" lowers compliance with COVID-19 guidelines: evidence from the US, UK, Netherlands and Germany. Harv. Kennedy Sch. Misinformation Rev. **1**. Special Issue on COVID-19 and Misinformation (2020)

E. Han, M.M. Jin Tan, E. Turk, D. Sridhar, G.M. Leung, K. Shibuya et al., Lessons learnt from easing COVID-19 restrictions: an analysis of countries and regions in Asia Pacific and Europe. Lancet **396**, 1525–1534 (2020). https://doi.org/10.1016/S0140-6736(20)32007-9

C.A. Harper, L.P. Satchell, D. Fido, R.D. Latzman, Functional fear predicts public health compliance in the COVID-19 pandemic. Int. J. Ment. Health Addict. (2020). https://doi.org/10.1007/s11469-020-00281-5

T. Katsabian, The telework virus: how the COVID-19 pandemic has affected telework and exposed its implications for privacy and equality. https://doi.org/10.2139/ssrn.3684702

A. Koopmann, E. Georgiadou, F. Kiefer, T. Hillemacher, Did the general population in Germany drink more alcohol during the COVID-19 pandemic lockdown? Alcohol Alcohol. **55**(6), 698–699 (2020). https://doi.org/10.1093/alcalc/agaa058

D. Laborde, W. Martin, R. Vos, Poverty and food insecurity could grow dramatically as COVID-19 spreads, in *COVID-19 and Global Food Security*, ed. by J. Swinnen, J. McDermott. Part one: food security, poverty, and inequality. Chapter 2 (International Food Policy Research Institute (IFPRI), Washington, DC, 2020), pp. 16–19. https://doi.org/10.2499/p15738coll2.133762_02

P. López-Igual, P. Rodríguez-Modroño, Who is teleworking and where from? Exploring the main determinants of telework in Europe. Sustainability **12**(21), 87–97 (2020). https://doi.org/10.3390/su12218797

McKinsey & Company, Consumer sentiment and behavior continue to reflect the uncertainty of the COVID-19 crisis (2020a), https://www.mckinsey.com/business-functions/marketing-and-sales/our-insights/a-global-view-of-how-consumer-behavior-is-changing-amid-covid-19

McKinsey & Company, How COVID-19 is changing consumer behavior now and forever (2020b), https://www.mckinsey.com/industries/retail/our-insights/how-covid-19-is-changing-consumer-behavior-now-and-forever

D.M. Morens, G.K. Folkers, A.S. Fauci, What is a pandemic? J. Infect. Dis. **200**(7), 1018–1021 (2009). https://doi.org/10.1086/644537

A.H.M. Noman, M.D. Griffiths, S. Pervin, M.N. Ismail, The detrimental effects of the COVID-19 pandemic on domestic violence against women. J. Psychiatr. Res. **134**, 111–112 (2021). https://doi.org/10.1016/j.jpsychires.2020.12.057

G. Parady, A. Taniguchi, K. Takami, Travel behavior changes during the COVID-19 pandemic in Japan: analyzing the effects of risk perception and social influence on going-out self-restriction. Transp. Res. Interdiscip. Perspect. **7**, 100181 (2020). https://doi.org/10.1016/j.trip.2020.100181

S. Reicher, C. Stott, On order and disorder during the COVID-19 pandemic. British Journal of Social Psychology **59**, 694–702 (2020). https://doi.org/10.1111/bjso.12398

G. Roman Etxebarrieta, M. Álvarez-Rementería Álvarez, E. Pérez-Izaguirre, M. Dosil Santamaria, El papel de los medios de comunicación en situaciones de crisis sanitaria. La percepción de la población en torno al control y las normas sociales durante la pandemia del COVID-19. Revista Latina de Comunicación Social **78**, 437–456 (2020)

B. Van Rooij, A.L. de Bruijn, C. Reinders Folmer, E. Kooistra, E. Malouke, M. Brownlee, E. Olthuis, A. Fine, Compliance with COVID-19 mitigation measures in the United States (April 22, 2020). Amsterdam Law School Research Paper, 2020–2021. https://doi.org/10.2139/ssrn.3582626

M.B. Ruíz-Roso et al., Covid-19 confinement and changes of adolescent's dietary trends in Italy, Spain, Chile, Colombia and Brazil. Nutrients **12**(6) (2020). https://doi.org/10.3390/nu12061807

C.G. Sibley et al., Effects of the COVID-19 pandemic and nationwide lockdown on trust, attitudes toward government, and well-being. American Psychologist **75**(5), 618–630 (2020). https://doi.org/10.1037/amp0000662

A. Silverio-Murillo, J.R. Balmori de la Miyar, L. Hoehn-Velasco, Families under confinement: COVID-19, domestic violence, and alcohol consumption (2020). https://doi.org/10.2139/ssrn.3688384

A. Simonov, S.K. Sacher, J.P.H. Dubé, S. Biswas, The persuasive effect of FOX News: non-compliance with social distance during the COVID-19 pandemic. NBER Working Paper 27237 (2020). http://www.nber.org/papers/w27237

D. Szajnoga, M. Klimek-Tulwin, A. Piekut, COVID-19 lockdown leads to changes in alcohol consumption patterns. Results from the Polish national survey. J. Addict. Dis. (2020). https://doi.org/10.1080/10550887.2020.1848247

C. Tabernero, R. Castillo-Mayén, B. Luque, E. Cuadrado, Social values, self- and collective efficacy explaining behaviours in coping with Covid-19: self-interested consumption and physical distancing in the first 10 days of confinement in Spain. PLoS ONE **15**(9) (2020). https://doi.org/10.1371/journal.pone.0238682

F. Tavares, E. Santos, A. Diogo, V. Ratten, Teleworking in Portuguese communities during the COVID-19 pandemic. Journal of Enterprising Communities: People and Places in the Global Economy (2020). https://doi.org/10.1108/JEC-06-2020-0113

E. Tognotti, Lessons from the history of quarantine, from plague to Influenza A. Emerging Infection Diseases **19**(2), 254–259 (2013). https://doi.org/10.3201/eid1902.120312

K. Usher, N. Bhullar, J. Durkin, N. Gyamfi, D. Jackson, Family violence and COVID-19: Increased vulnerability and reduced options for support. International Journal of Mental Health Nursing **29**, 549–552 (2020). https://doi.org/10.1111/inm.12735

N. Vanderbruggen, F. Matthys, S. van Laere, D. Zeeuws, L. Santermans, S. van den Ameele, C.L. Crunelle, Self-reported alcohol, tobacco, and cannabis use during COVID-19 lockdown measures: results from a web-based survey. European Addiction Research **26**, 309–315 (2020). https://doi.org/10.1159/000510822

B. Yan, X. Zhang, L. Wu, H. Zhu, B. Chen, Why do countries respond differently to COVID-19? A comparative study of Sweden, China, France, and Japan. American Review of Public Administration **50**(6–7), 762–769 (2020). https://doi.org/10.1177/0275074020942445

H. Yilmazkuday, Changes in consumption amid COVID-19: Zip-code level evidence from the U.S. (2020), https://papers.ssrn.com/sol3/papers.cfm?abstract_id=3658518

Open Access This chapter is licensed under the terms of the Creative Commons Attribution 4.0 International License (http://creativecommons.org/licenses/by/4.0/), which permits use, sharing, adaptation, distribution and reproduction in any medium or format, as long as you give appropriate credit to the original author(s) and the source, provide a link to the Creative Commons license and indicate if changes were made.

The images or other third party material in this chapter are included in the chapter's Creative Commons license, unless indicated otherwise in a credit line to the material. If material is not included in the chapter's Creative Commons license and your intended use is not permitted by statutory regulation or exceeds the permitted use, you will need to obtain permission directly from the copyright holder.

Chapter 9
COVID-19 and Optimal Lockdown Strategies: The Effect of New and More Virulent Strains

Jonathan P. Caulkins, Dieter Grass, Gustav Feichtinger, Richard F. Hartl, Peter M. Kort, Alexia Prskawetz, Andrea Seidl, and Stefan Wrzaczek

Abstract Most nations have responded to the COVID-19 pandemic by locking down parts of their economies starting in early 2020 to reduce the infectious spread. The optimal timing of the beginning and end of the lockdown, together with its intensity, is determined by the tradeoff between economic losses and improved health outcomes. These choices can be modelled within the framework of an optimal control model that recognises the nonlinear dynamics of epidemic spread and the increased risks when infection rates surge beyond the healthcare system's capacity. Past work has shown that within such a framework very different strategies may be optimal

J. P. Caulkins (✉)
Heinz College, Carnegie Mellon University, 4800 Forbes Avenue, Pittsburgh, PA 15213, USA
e-mail: caulkins@andrew.cmu.edu

D. Grass · S. Wrzaczek
International Institute for Applied Systems Analysis (IIASA), Schlossplatz 1, 2361 Laxenburg, Austria
e-mail: dieter.grass@tuwien.ac.at

S. Wrzaczek
e-mail: wrzaczek@iiasa.ac.at

G. Feichtinger · A. Prskawetz
Institute of Statistics and Mathematical Methods in Economics, TU Wien, Wiedner Hauptstr. 8-10, 1040 Vienna, Austria
e-mail: gustav@eos.tuwien.ac.at

A. Prskawetz
e-mail: afp@econ.tuwien.ac.at

R. F. Hartl · A. Seidl
Department of Business Decisions and Analytics, University of Vienna, Oskar-Morgenstern-Platz 1, 1090 Vienna, Austria
e-mail: richard.hartl@univie.ac.at

A. Seidl
e-mail: andrea.seidl@univie.ac.at

P. M. Kort
Tilburg School of Economics and Management, Tilburg University, Warandelaan 2, 5037 AB Tilburg, The Netherlands
e-mail: Kort@uvt.nl

© The Author(s) 2022
M. C. Boado-Penas et al. (eds.), *Pandemics: Insurance and Social Protection*,
Springer Actuarial, https://doi.org/10.1007/978-3-030-78334-1_9

ranging from short to long and even multiple lockdowns, and small changes in the valuation on preventing a premature death may lead to quite different strategies becoming optimal. There even exist parameter constellations for which two or more very different strategies can be optimal. Here we revisit those crucial questions with revised parameters reflecting the greater infectivity of variants such as the "UK variant" of the SARS-CoV-2 virus and describe how the new variant may affect levels of mortality and other outcomes.

9.1 Introduction

To reduce social interactions and thereby also contagious transmission of the SARS-CoV-2 virus, most countries have implemented one or several lockdowns of non-essential parts of the economy. While lockdowns have succeeded to varying degrees in reducing new infections, the effects on the economy (Fernández-Villaverde and Jones 2020) can be substantial. The lockdowns can themselves harm health, either directly (e.g., when non-essential healthcare is deferred) or indirectly (unemployment and poverty can reduce life expectancy). The question therefore arises as to what is the ideal duration and intensity of lockdowns. If lockdowns are relaxed too soon, the epidemic may bounce back. If these measures are too severe or prolonged, needless economic hardship may result.

9.1.1 The Challenge of New Virus Variants

Such questions have become even more pressing with the discovery of new, mutated strains of the SARS-CoV-2 virus, notably one detected first in the UK and thus referred to as "UK variant". This strain appears to be much more virulent, in the sense of spreading more rapidly. In particular, the U.S. Centers for Disease Control (CDC) had previously advised that epidemiological models use as a base case assumption that the basic reproduction number (denoted by R_0) of the SARS-CoV-2 virus was 2.5, but the new variant is thought to be about 60% more contagious, suggesting a new R_0 of 4.0.

The new variant is not more lethal, so far as is understood at at the time of this writing; i.e., its infection fatality rate is not higher. So the consequences of COVID-19 spreading through the majority of the population are roughly the same, apart from greater mortality when cases are bunched up in time, swamping the healthcare system, but the severity of lockdown necessary to prevent such spread is greater and so more costly. In particular, lockdowns and other interventions that reduced social interaction by 60% would have been sufficient to stall the spread of the original virus, since $2.5 \times (1 - 60\%) = 1.0$, but with the new virus, those same interventions would have each infection leading to $4.0 \times (1 - 60\%) = 1.6$ new infections.

Since the course of the typical infection plays out over roughly two weeks, that would leave the number of new infections growing at a compound rate of about $1.6^2 - 1 = 150\%$ per month. If it will take another six to nine months to achieve herd immunity through mass vaccinations, that spread would be fast enough to infect essentially everyone in a country that currently has an average rate of infections. Hence, policies that were adequate or even optimal in the past may no longer be so today.

This paper explores how this greater virulence may or may not alter conclusions about what constitutes the ideal timing and duration of a lockdown. It also adjusts the time horizon until an effective vaccine has been widely deployed to $T = 1.5$ years, better reflecting the actual trajectory of vaccine development that has been observed.

9.1.2 Review of Past Findings

We begin here with a brief review of findings obtained from these lockdown models using the older, lower basic reproduction number of $R_0 = 2.5$. In Caulkins et al. (2020) we analysed an epidemiological model of the pandemic overlain with a simple optimal control model that considers the optimal starting and ending times of a lockdown that withdraws part of the population from the labour force. The objective function balances economic costs (lost output) and health costs (COVID-19 related deaths) while considering the limited capacity of intensive care units within the health care system. The number of deaths is modelled as being proportional to the number of infections plus an extra penalty for infections that happen when hospitals are overwhelmed.

Even that rather simple model produces some complex behaviour. For instance, sometimes starting a lockdown later might make it better to have a shorter, not a longer lockdown. Most interestingly, we found the formal mathematical version of the notion of "tipping points" that were popularised by Malcolm Gladwell's famous book of that name. In particular, for certain parameter values two very different strategies (e.g., long versus short lockdown) can be optimal when starting at the same initial condition, and slight deviations away from those starting points may make either type of solution optimal. In optimal control models, such tipping points with two or more alternative optimal solution trajectories have been called Skiba, Sethi-Skiba, DNS, and DNSS points (Grass et al. 2008).

Characterising such points is important for two reasons. First, it may help explain why different countries have pursued such different lockdown strategies despite having similar interests in balancing economic and health considerations. Secondly, these tipping points highlight the need to gather better information about the key parameters that cause these different lockdown strategies to be optimal.

Caulkins et al. (2021) extended Caulkins et al. (2020) by allowing for multiple lockdowns and also considering lockdowns of varying intensity, rather than treating a lockdown as an all or nothing binary choice. In addition, the economic modelling is richer in two respects. First, employment is represented by a state variable, and the policy maker's choices, or "control variable", adjusts that level of economic activity. Underemployment is costly of course (foregone economic activity), but so too is changes in that level; rapidly alternating between mild and severe lockdowns is more costly than maintaining an intermediate lockdown because change is disruptive to business. Furthermore, the adjustment costs are asymmetric to recognise that shutting down businesses may be easier than re-starting them.

Second, since the public's cooperation can wane when lockdowns are too intensive and long, Caulkins et al. (2021) include "lockdown fatigue" as an additional state variable which may undermine the efficiency of a lockdown.

Within this framework, the optimal lockdown strategies are quite diverse, ranging from long and forceful lockdowns to (a couple of) short and rather soft lockdowns. Again, the specific parameter values determine the optimal strategy. Similar to Caulkins et al. (2020), there are parameter constellations for which two very different strategies are both optimal. In addition, we also found triple Skiba points at which even three different strategies are optimal. The fact that such complex strategies result from rather stylised models hints at the complexity of designing lockdown strategies in practice. While our models cannot specify/recommend any single optimal lockdown strategy, our framework provides insight as to which are the most important parameters that drive the decision about the start, duration and intensity as well as the multiplicity of lockdowns.

Before investigating how the UK variant might affect these models and selected results we give a brief summary of related papers in the literature.

9.1.3   Review of Other Related Literature

Several papers discuss the balancing of health and economic interests (see Layard et al. 2020; Bloom et al. 2020; Scherbina 2020; Brodeur et al. 2020 for a careful evaluation). However only a minority of these papers have investigated the optimal timing, length and extent of the lockdown itself. These exceptions include e.g. Gonzalez-Eiras and Niepelt (2020) who start from a simple SIR model and investigate the optimal lockdown intensity and duration taking into account the tradeoff between health and economic consequences of the lockdown.

In Alvarez et al. (2020) the fraction of people going into lockdown is assumed to be the control variable. The model is derived with and without testing as a control variable where test availability implies that those who are recovered can be identified and are not subject to lockdown. It is shown that absence of testing will increase the economic costs of the lockdown and shorten its duration since the dynamics of the epidemiology imply that an increasing share of recovered will decrease the efficiency of the lockdown.

An optimal control model on reducing the transmission rate is presented in Abel and Panageas (2020) that also allows for positive vital rates (births and non-COVID deaths). They show that it is not optimal to eradicate the disease but to limit interactions until a cure or vaccination becomes available. In Acemoglu et al. (2020) a heterogeneous SIR model is applied that distinguishes between "young", "middle-aged" and "old". It is shown that a stricter lockdown on the old is particularly important. Compared to optimal uniform policies such targeted policies imply a considerable reduction in mortality but may also reduce economic damage since the young and middle-aged groups can be released from lockdown earlier. A similar argument is presented in Gershon et al. (2020) who show that if ICU beds are in short supply, partial quarantine of the most vulnerable group may be enough compared to a shut down of the whole economy. Aspri et al. (2021) consider a SEIRD model, where the population is divided into susceptibles, exposed but asymptomatic, infected, recovered and deceased. Similar to other papers they model the optimal tradeoff between reduction in fatalities and the loss in output. However, different to the literature so far, they assume that containment policies are piece-wise linear functions representing a more realistic policy modelling. Based on the specific value assumed for a statistical life they obtain multiple lockdowns as well as Skiba points. While the previous models apply numerical solution methods, Rachel (2020) presents an analytical model of COVID-19 lockdowns. By differentiating between the individual versus social optimal mitigation strategy it is shown that at the individual level too much social distancing will result in equilibrium relative to the social optimum. This result is explained by the fact that higher social distancing today will reduce infections and flatten out the curve, but raise infection rates later on. In contrast, a social planner considers the cumulative infection risk and not just the infection risk today. In Huberts and Thijssen (2020) a stochastic version of the SIR model is introduced. Based on a continuous-time Markov chain model the optimal timing of intervention and the option to end the intervention are studied. Federico and Ferrari (2021) present a model on the optimal lockdown policy where not only the transmission rate in the standard SIR model is stochastic, but also the time horizon is assumed to be stochastic. Within their framework they show that the optimal policy is first to let the epidemic evolve, followed by a pronounced containment policy and in the last phase to reduce the strength of the containment again. Similar to our model the limited capacity of health system is considered in Piguillem and Shi (2020). Testing is an important strategy to accrue welfare gains in their paper.

The body of this paper now proceeds in two parts. Section 9.2 explores a model in which the policy maker only gets to decide the start and end times of a lockdown, but the intensity of that lockdown is fixed. Section 9.3 then examines the more general situation when the lockdown intensity can be varied continuously over time, and there can even be more than one lockdown.

9.2 The Optimal Start and Length of a Lockdown

This section updates the model presented in Caulkins et al. (2020) to address the higher infectivity of the UK variant of the SARS-CoV-2 virus.

9.2.1 The Model

The epidemiological model we apply is based on an open-population SLIR model Kermack and McKendrick (1927) with a birth rate v and extra mortality for individuals who are infected (μ_I) above and beyond that for those who are susceptible or recovered (μ):

$$\dot{S}(t) = vN(t) - \beta \frac{S(t)\mathcal{I}(I(t), L(t))}{N(t)} - \mu S(t) \tag{9.1a}$$

$$\dot{L}(t) = \beta \frac{S(t)\mathcal{I}(I(t), L(t))}{N(t)} - (\mu + \varphi)L(t) \tag{9.1b}$$

$$\dot{I}(t) = \omega\varphi L(t) - (\alpha + \mu + \mu_I)I(t) \tag{9.1c}$$

$$\dot{R}(t) = (1 - \omega)\varphi L(t) + \alpha I(t) - \mu R(t) \tag{9.1d}$$

$$\beta := R_{\text{eff}}(t, \tau_1, \tau_2)\alpha \tag{9.1e}$$

$$\mathcal{I}(I, L) := I + fL \tag{9.1f}$$

$$N(t) := S(t) + L(t) + I(t) + R(t). \tag{9.1g}$$

The state variables $S(t)$, $L(t)$, $I(t)$ and $R(t)$ denote the number of individuals respectively who are susceptible to infection, have a latent (asymptomatic and pre-symptomatic) infection, are infected and symptomatic, and are recovered at time t. The term "recovered" is standard in the literature even though it is a bit of a misnomer because it includes not only those who have recovered from COVID-19 symptoms (i.e., passed through the I state), but also those who previously had asymptomatic infections (passed through the L state only). (CDC guidance is that about 40% of those who become infected remain asymptomatic.)

The parameter β is key to the epidemic dynamics. The term it modifies in Eq. (9.1a) counts potential interactions between those who are susceptible to becoming infected (those in the S state) and those are infected (those in the L and I states). The symbol $\mathcal{I}$ denotes the weighted sum of people in the I and L states, weighting by the (lower) relative likelihood of spreading the virus when in the L state. (As of this writing, the CDC recommends assuming this weighting parameter $f = 0.75$.) β is essentially a proportionality constant that converts social interactions into infections. Outside of the lockdown it has one (higher) value; during the lockdown its value is lower, e.g., because either the infected or susceptible person wears a mask, maintains social distance, interacts only virtually if one or both work from home, or the interaction simply does not occur because it has been banned by the lockdown.

Although lockdowns directly affect β, the effective reproduction number $R_{\text{eff}}(t, \tau_1, \tau_2)$ is more readily interpretable, so we describe the lockdown phases in terms of effects on $R_{\text{eff}}(t, \tau_1, \tau_2)$ and adjust β accordingly. (For a formal derivation of the relationship between the basic reproduction number R_0 and β, see Appendix 2 in Caulkins et al. (2020)). The start and end times of the lockdown are denoted by τ_1 and τ_2. They are chosen by the decision maker and—as we only allow one lockdown in this setup—define three periods: before, during and after that one lockdown.

Following CDC guidance, Caulkins et al. (2020) assumed that the basic reproduction number before the lockdown equals $R_0^1 = 2.5$. The lockdown was assumed to reduce that to $R_0^2 = 0.8$ and not bounce back fully, as some behavioural adjustments (such as not shaking hands) could be expected to continue even after people return to work. The extent to which those behavioural changes persisted depends on the duration of the lockdown. In particular, Caulkins et al. (2020) assumed that after the lockdown there exists a gap between the realised and potential value of $R_0^3 = 2.0$, with the potential value being reached only with increasing length of the lockdown.

Here, we assume that the reproduction numbers before and after the lockdown are 1.6 times larger (so $R_0^1 = 4.0$ and $R_0^3 = 3.2$) but continue to assume that during the lockdown $R_0^2 = 0.8$. I.e., we implicitly assume that the lockdown intensity is increased sufficiently to push the reproduction number appreciably below 1.0 despite the new variant's greater infectivity.

We model COVID-19 deaths by focusing on those who require hospitalisation and critical care. Some calculations (described in Caulkins et al. 2020) suggest that about $p = 2.31\%$ of people who develop symptoms will need critical care, and 45% of them will die prematurely as a result of COVID-19 even if they receive that care. The parameter p converts the 2.31% into a daily rate by multiplying by α, the reciprocal of the average duration of symptoms, which we take to be nine days. Likewise, the death rate per person-day spent in the I state by people who need and also receive critical care is $\mu_I = p\xi_1\alpha$.

In addition, there is an extra risk of death for people who need critical care but do not receive it because hospitals are overwhelmed. That term is proportional to $\max(\{0, pI - H_{max}\})$ where H_{max} is the number of critical care hospital beds available.[1] In the U.S., there are about 0.176 critical care beds per 1,000 people. Overall deaths are therefore represented by:

$$\xi_1 pI + \xi_2 \max(\{0, pI - H_{max}\}, \zeta),$$

where ξ_1 is the death rate from COVID-19 of infected people who need and receive critical care, and ξ_2 is the additional, incremental death rate when such individuals do not receive that care. One aim of the decision maker is to minimise these deaths.

It is of course very difficult to determine what value society should place on averting a premature death generally, or in the case of COVID-19 in particular. We represent that quantity by the parameter M, the cost per COVID-19 death, and consider a very wide range of values for that parameter.

The literature has traditionally used values for M ranging at least from 20 times GDP per capita (Alvarez et al. 2020) up to 150 times GDP per capita (Kniesner et al. 2012). Hammitt (2020) argues that lower values may be appropriate for COVID-19 deaths, so we consider a range from 10 to 150 times GDP per capita.

Economic activity is modelled as being proportional to the number of employed people raised to a power, as in a classic Cobb-Douglas model, with that exponent set

---

[1] In our numerical simulations we have replaced the max function that is not differentiable with a smooth function (see Caulkins et al. 2020, Fig. 1).

to $\sigma = 2/3$ (Acemoglu 2009). Since the time horizon is short, capital is presumed to be fixed and subsumed into the objective function coefficient K for economic activity. Susceptible, latent, and recovered individuals are eligible to work (symptomatic individuals are assumed to be either too sick to work, or are in quarantine). During a lockdown, only a proportion $\gamma(t)$ of those eligible to work are employed. We therefore assume that $\gamma(t) = 1.0$ before the lockdown, $\gamma(t) = 0.25$ during the lockdown, and after lockdown it only partially recovers. The longer the lockdown, the more jobs that are lost semi-permanently because firms go out of business. That recovery is modelled as decaying exponentially in the length of the lockdown with a time constant of 0.001 per day, so that if a lockdown ended after six months, 17% of jobs suspended during the lockdown would not reappear, at least until a vaccine became available.

These economic and mortality costs are summed up from time $t = 0$, when the virus arrives, until time $T = 1.5$ years, when a vaccine has been developed and widely deployed.

The objective function also includes a salvage value that reflects the reduction in economic activity at time T relative to what it was at time 0 (see Caulkins et al. 2020 for further discussion of the salvage value). The summary of the full model and the base case parameter values are given in Appendix 1 and Table 9.4 Appendix 3.

9.2.2 Results

COVID-19 spread very fast in early 2020, so lockdown initiation was often a rushed decision made so quickly that there was no time to build models or optimise them. Hence, we start, in Fig. 9.1, by considering the simpler problem of when to end a lockdown that has already started, answering that question for a wide range of start times. In particular, the left hand panel of Fig. 9.1 shows with the solid blue line how that optimal ending time (measured by the vertical axis, τ_2) varies as a function of the time the lockdown was started (given as the horizontal axis, τ_1). The gap between the blue line and the black line (45-degree line) indicates the duration of the lockdown.

For this model and these parameter values, if the lockdown starts promptly (so on the left side of that panel) the lockdown should be maintained almost until the time when the vaccine has been successfully deployed. That is assumed to happen in 1.5 years; since time is measured in days, that corresponds to 547.5 on the vertical axis. That the blue line starts out at a level of about 500 days indicates maintaining the lockdown until only a month or two before the vaccine has been successfully deployed would minimise total costs, including both health and economic costs. (Ending the lockdown before full deployment does not require an implausible degree of forecasting ability; predicting how long it will take to invent an effective vaccine is hard, but deployment takes approximately six months, so recognising when it is within a couple months of wrapping up is not that hard.)

So the first conclusion is, if a nation starts to lock down early, it should keep that lockdown in place more or less for the duration of the epidemic.

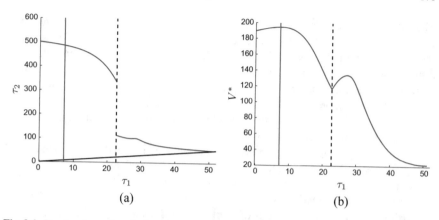

Fig. 9.1 Panel a shows the solutions for a fixed initial lockdown time τ_1 and optimally chosen time τ_2. In panel b the objective value is shown for the optimally chosen time τ_2. For $\tau_1 = 22.6$, which is indicated by the black dashed line, there exists a Skiba solution, i.e. there are two different solution paths which deliver the same objective value. The red vertical line denotes the optimally chosen τ_1. The parameter values are those of Table 9.4 in Appendix 3, with $M = 60,000$ and $R_0^2 = 0.8$

Now suppose the lockdown's initiation was delayed a bit, meaning we slide a little to the right along the horizontal axis of Fig. 9.1. Intuitively one might have expected that getting a late start would imply one should maintain the lockdown longer to compensate, but the opposite is true in this model. The fact that the blue line slopes downward implies that the later one starts the lockdown, the sooner it should end.

The second surprising result is that the blue line does not decline smoothly; it contains a discontinuous jump when the lockdown starts at $\tau_1 = 22.6$ days. As one delays the start of the lockdown from τ_1 from 0 up to 22.5 days, the ideal ending time τ_2 decreases smoothly from about 500 days (roughly a year and four months) down to a little less than a year. Then suddenly, when the lockdown starts just a little later, at $\tau_1 = 22.7$ days, it becomes optimal to end the lockdown fairly soon, at only $\tau_1 = 120$ days, or after about three months.

What has happened at that point is that the epidemic has had a chance to spread so widely in those first 22.7 days that it is just too hard to rein the epidemic in for it to be worthwhile. If a prolonged lockdown were going to spare most of the population from getting infected it would be worth the cost; but if the lockdown hasn't started until $\tau_1 = 22.7$ days, it is just too late for it to be wise to pursue that strategy. One should still lockdown, but only relatively briefly. That can "flatten the curve" a bit and avoid totally swamping the limited capacity of the healthcare system.

In simple words, if the lockdown starts too late, then one should abandon the "long lockdown" strategy that protects most people from infection, and instead employ a much more limited "curve flattening" strategy.

The discontinuity in the blue line shows that if the lockdown starts at just exactly $\tau_1 = 22.6$ days, then either the "long lockdown" or the "curve flattening" can be followed with equal results.

This equivalency is illustrated more directly in the right-hand panel of Fig. 9.1, which shows the so-called value function (V) versus the lockdown initiation time τ_1.

The value function indicates the performance achieved when the optimal strategy is followed. There is a kink in the value function right at $\tau_1 = 22.6$ days. To the left of that kink it is optimal to follow the "long lockdown" strategy, but the plunging value function shows that the "long lockdown" strategy performs less and less well as τ_1 increases. Likewise, to the right of $\tau_1 = 22.6$ days it is optimal to follow the "curve flattening" strategy, but as the lockdown start time decreases, approaching 22.6 from the right, the "curve flattening" strategy does less and less well. And right at τ_1 the two strategies' value functions cross.

The third surprising result pertains to where the blue line in the right-hand panel of Fig. 9.1 peaks. It is not ideal to start the lockdown immediately at $\tau_1 = 0$. Instead, the value function peaks at about $\tau_1 = 8$ days (a point in time indicated in the left hand panel by a vertical red line). The reason is that every day of lockdown is expensive, because people are out of work, but at the very beginning, when there are very, very few infected people, there are also very, very few new infections to be prevented. When the virus is very scarce, targeted approaches, such as testing and contact tracing, may be preferred to shutting down the entire economy.

Figure 9.1 illustrated two different strategies: a short "curve flattening" lockdown and a long lockdown that starts after a short delay. For other parameter values, two other strategies can be optimal: never locking down at all or a long lockdown that begins immediately.

That raises the question of under what conditions it is optimal to pursue each strategy. Figure 9.2 answers that question with respect to two key parameters: (1) The economic value placed on preventing a COVID-19 death, M, and (2) the epidemic's reproduction number during the lockdown, denoted by the parameter R_0^2. That figure, called a bifurcation diagram, shows for each combination of those two key parameters which strategy is optimal.

The base case values for those parameters were $R_0^2 = 0.8$, meaning the lockdown could still drive the reproduction number below the critical threshold of 1.0, and $M = 60,000$, meaning that the cost of a premature death is set at about 150 times GDP per capita. That point falls within the region labelled IIb, but for other values of R_0^2 and/or M different strategies may be optimal.

Not surprisingly, as M increases—meaning moving from left to right in Fig. 9.2— the optimal strategy changes to make greater and greater use of lockdowns. When M is very small, it may be optimal not to lockdown at all. When M is sufficiently large, then a long lockdown is best. For intermediate values of M, the "curve flattening" strategy may be best.

The verticality of the Skiba curve separating the regions where no lockdown vs. a short lockdown are optimal indicates that increasing the epidemic's reproductive number during the lockdown R_0^2 has little effect on the relative merits of not locking down versus using a short lockdown. That makes sense precisely because in neither of those strategies was the lockdown prolonged in any event. However, the Skiba curve separating regions where a short lockdown (Region I) and a long lockdown (Regions IIa and IIb) is preferred slopes up and to the right indicating that the larger R_0^2 is, the larger M must be in order to justify a long lockdown. That also makes sense. If the new variant's higher virulence sufficiently undermines the effectiveness of locking

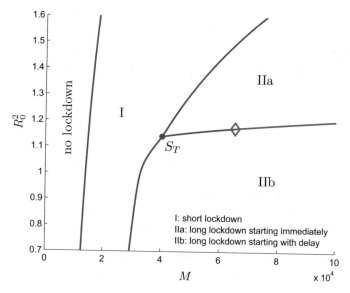

Fig. 9.2 Bifurcation diagram in the $R_0^2 - M$ space. The blue lines denote Skiba curves, separating (discontinuously) regions with different optimality regimes. At the red curve the regimes change continuously. The point S_T corresponds to a triple Skiba point, where three optimal solutions exist. At the red diamond the discontinuous Skiba solution changes into a continuous transition curve

down, then the value per life saved has to be greater to justify the imposition of a long lockdown.

One of the interesting features of this model is that two Skiba curves intersect, namely the curve separating Region I from Regions IIa or IIb and the curve separating Regions IIa and IIb. That intersection, which is denoted by the point S_T, is a triple Skiba point. If the parameters have exactly those values, then any of three distinct strategies can be optimal. It is akin to Snow Dome Mountain in Canada's Jasper National Park, where a drop of water could equally well flow west through the Columbia River system to the Pacific Ocean, east to Hudson's Bay and the Atlantic Ocean, or north via the Athabasca and McKenzie Rivers into the Arctic Ocean. Except that instead of being indifferent between flowing to different oceans, at this point a social planner is indifferent between starting a long lockdown immediately, starting a long lockdown after a short delay, and employing only a short lockdown.

9.3 The Optimal Lockdown Intensity

The previous section updated results from a model based on Caulkins et al. (2020) that sought to determine the optimal start and length of a lockdown when the intensity of that lockdown was given exogenously. We next present an extension of the model as given in Caulkins et al. (2021) that allows the intensity of the lockdown to vary continuously over time.

As in Caulkins et al. (2020), we define $\gamma(t)$ to be the share of potential workers who are employed at time t. However, now we model $\gamma(t)$ as a state variable that can be altered continuously via a control $u(t)$:

$$\dot{\gamma}(t) = u(t), \quad \gamma(0) = 1,$$

We set $\gamma(0) = 1.0$ because the planning horizon begins when COVID-19 first arrives, and so before there is any lockdown. We include a state constraint $\gamma(t) \leq 1$ for $0 \leq t \leq T$, since employment cannot exceed 100%. This formulation allows for multiple lockdowns, takes into account that employment takes time to adjust, and it recognises that changing employment levels induces adjustment costs, which we allow to be asymmetric, with it being harder to restart the economy than it is to shut it down.

Since public approval of lockdowns may wane the longer a lockdown lasts, we introduce a further state variable that models this "lockdown fatigue" $z(t)$:

$$\dot{z}(t) = \kappa_1 (1 - \gamma(t)) - \kappa_2 z(t),$$

where κ_1 governs the rate of accumulation of fatigue and κ_2 measures its rate of decay. Note that if the worst imaginable lockdown ($\gamma(t) = 0$) lasted forever then $z(t)$ would grow to its maximum possible value of $z_{max} = \kappa_1/\kappa_2$.

We use an epidemiological model based on an open-population SIR[2] model with a birth rate ν and extra mortality for individuals who are infected (μ_I) above and beyond that for those who are susceptible or recovered (μ). In addition, we allow a backflow of recovered individuals back into the susceptible state at a rate ϕ. How long immunity will last with SARS-CoV-2 virus is not known at the time of this writing, but immunity to other corona viruses often lasts 3–5 years, so we set ϕ to 0.001 per day in our base case, which corresponds to a mean duration of immunity of $1000/365 = 2.74$ years.

---

[2] Since the qualitative dynamics in Caulkins et al. (2020) did not change if we excluded the latent state, we opted for a more parsimonious model in our extensions Caulkins et al. (2021) and the parameters have been adapted accordingly.

The state dynamics in our extended model can then be written as

$$\dot{S}(t) = \nu N(t) - \beta(\gamma(t), z(t))\frac{S(t)I(t)}{N(t)} - \mu S(t) + \phi R(t)$$

$$\dot{I}(t) = \beta(\gamma(t), z(t)))\frac{S(t)I(t)}{N(t)} - (\alpha + \mu + \mu_I)I(t)$$

$$\dot{R}(t) = \alpha I(t) - \mu R(t) - \phi R(t)$$

$$\dot{\gamma}(t) = u(t), \quad \gamma(0) = 1$$

$$\dot{z}(t) = \kappa_1(1 - \gamma(t)) - \kappa_2 z(t), \quad z(0) = 0$$

$$\gamma(t) \leq 1, \quad 0 \leq t \leq T$$

where $N(t) = S(t) + I(t) + R(t)$ is the total population. As before, the factor $\beta(\gamma, z)$ captures the number of interactions and the likelihood that an interaction produces an infection. It is assumed to depend on both the intensity of the lockdown γ and the level of lockdown fatigue z in the following way:

$$\beta(\gamma, z) := \beta_1 + \beta_2\left(\gamma^\theta + f\frac{\kappa_2}{\kappa_1}z(1 - \gamma^\theta)\right)$$

This expression can be interpreted as follows. In the absence of lockdown fatigue, we might model $\beta(\gamma, 0)$ as some minimum level of infection risk β_1 that is produced just by essential activities plus an increment β_2 that is proportional to γ raised to an exponent $\theta > 1$. Having θ greater than 1 is consistent with locking down first the parts of the economy that generate the most infections per unit of economic activity (perhaps concerts and live sporting events) and shutting down last industries with high economic output per unit of social interaction (perhaps highly automate manufacturing and mining).

The term $(\kappa_2/\kappa_1)z$ is the lockdown fatigue expressed as a percentage of its maximum possible value. So if $f = 1$ and z reached its maximum value, then all of the potential benefits of locking down would be negated. Lockdown fatigue will not actually reach that maximum because the planning horizon is relatively short. Also, we choose a relatively small value of $f = 0.05$, so this lockdown fatigue has only a modest effect. Nonetheless, including this term at least acknowledges this human dimension of the public's response to lockdowns.

The objective function includes health costs (due to deaths from COVID-19), economic loss (due to locking down), and the adjustment costs of changing the employment level γ. We assume these adjustment costs to be quadratic in the control u and allow for them to be asymmetric with different constants for shutting down businesses c_l and reopening them c_r, with an extra penalty for reopening after an extended shut down so that

$$V_u(u, \gamma) := \begin{cases} c_l u^2 & u \leq 0 \\ c_r(z + 1)u^2 & u > 0. \end{cases}$$

The resulting optimal control model and the base case parameter values are summarised in Appedix 2 and Table 9.4 Appendix 3.

9.3.1   Results

9.3.1.1   The Effect of Increased Infectivity

Our main interest is in how a mutated strain that is more contagious alters what strategies are optimal. That is perhaps best captured in Fig. 9.3, which has two panels. The one on the left corresponds to the old reproduction number of $R_0 = 2.5$; the one on the right corresponds to the new, higher number of $R_0 = 4$. Both are similar to the right panel of Fig. 9.1 in that they show how the value function depends on the parameter M describing the cost per premature death.

This value function can be thought of as the "score" that a social planner "earns" when he or she follows the optimal strategy. Naturally in both panels the value function slopes down. The greater the penalty the social planner "pays" for each premature death, the lower the score. On the left side of each panel the value function slopes down steeply because there isn't much locking down so there are a lot of deaths; thus, a given increment in the cost per death gets "paid" many times. On the right side of each panel, the optimal strategy involves an extended lockdown, so there are fewer deaths and the same increment in the cost per death reduces the social planner's score by less.

There are, though, two noteworthy differences between the value functions across the two panels. First, the kink in the curve, indicating the point at which an extended

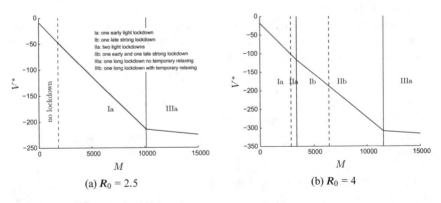

Fig. 9.3 Dependence of the value function on the social cost of a death M for the base case parameters in Table 9.4 Appendix 3 and lockdown fatigue $f = 0$. There are four main regimes (no lockdown, I, II, III) which differ by the duration, intensity, and number of lockdowns of the optimal solutions. For the value of M highlighted by solid vertical black lines two different solution paths are optimal. The dashed vertical lines denote continuous transitions from one regime to the other

lockdown becomes preferred, occurs at a larger value of M in the right-hand panel. That is because when the reproduction number is larger, it takes a more determined lockdown to pull off the extended lockdown strategy, making it more costly and less appealing unless the penalty per premature death is larger. The difference is not enormous though, with valuations equivalent to about ($M = 11, 560$) 32 times GDP per capita in the right panel and ($M = 10, 140$) 28 times GDP per capita in the left panel.

The second difference is that—at least with all other parameters at their base case values—increasing R_0 increased the number of different types of strategies that can be optimal. With $R_0 = 4$ there are five distinguishable types of lockdown strategies that can be optimal, not just two.

Here is how to interpret the labels of the four regions Ia, Ib, IIa, and IIb. The Roman numeral I or II refers to whether there are one or two lockdowns. The 'b' versus 'a' roughly indicates whether there is a substantial lockdown later in the planning horizon to prevent a rebound epidemic. (A rebound may be possible after an appreciable number of previously infected individuals have lost their immunity and returned to the susceptible state S via the backflow.)

Figure 9.4 shows example control trajectories for all five regions. The vertical axis is γ, the proportion of workers who are allowed to work, so any dip below 1.0 indicates a lockdown. If the social planner places a very low value on preventing COVID-19 deaths (e.g., $M = 1500$ in panel a), then there is only a small, short early lockdown which does little except to take a bit of the edge off the initial spike in infections. Such a small effort does not prevent many people from getting infected, but it shifts a few infections to later, when hospitals are less overwhelmed. When M is a little larger (specifically $M = 3200$ in panel c), then there is also a similarly small lockdown later, to take a bit of the edge off of the rebound epidemic. But in neither of those cases is there much locking down or much reduction in infections.

When M is still larger ($M = 5000$ in panel b) the later lockdown gets considerably larger—large enough to essentially prevent the rebound epidemic. Curiously, at this point the initial lockdown disappears, but it wasn't very big to begin with, so this qualitative change is not actually a very big difference substantively. When M increases further ($M = 11, 000$) the initial lockdown reappears, albeit as a very small blip.

Then rather abruptly when M crosses the Skiba curve separating type I and II strategies from type III strategies it becomes optimal to use a very large and sustained lockdown to reduce infections and deaths dramatically. Panel e shows the particular optimal lockdown trajectory when $M = 13, 000$, which is equivalent to valuing a premature death at 35 times GDP per capita. That sustained lockdown averts most of the infections and deaths, but at the considerable cost of almost 50% unemployment for about a year and a half.

Thus, when the lockdown intensity is allowed to vary continuously, many nuances emerge, but the overall character still boils down to an almost binary choice. If M is high enough, then use a sustained and forceful lockdown to largely preempt the epidemic despite massive levels of economic dislocation. Otherwise, lockdowns are too blunt and expensive to employ as the primary response to the epidemic. Thus, the model prescribes an almost all-or-nothing approach to economic lockdowns.

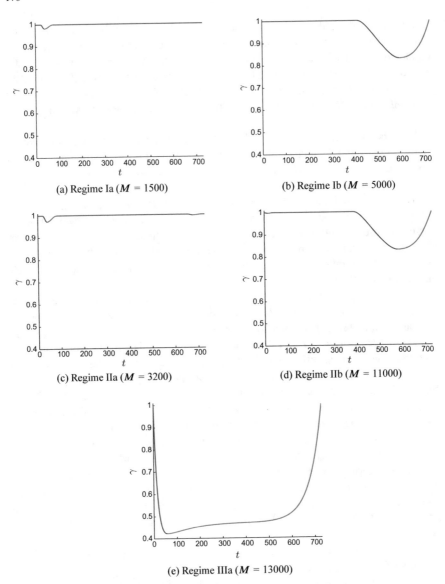

(a) Regime Ia ($M = 1500$)

(b) Regime Ib ($M = 5000$)

(c) Regime IIa ($M = 3200$)

(d) Regime IIb ($M = 11000$)

(e) Regime IIIa ($M = 13000$)

Fig. 9.4 Showing the time evolution of the optimal lockdown for the different regimes in Fig. 9.3b

For certain combinations of parameter values (e.g., Fig. 9.4 panels b and d corresponding to $M = 5,000$ and $M = 11,000$) it can be optimal to act fairly decisively against the rebound epidemic even if all one does in response to the first epidemic is a bit of curve flattening. It may seem odd to lock down more aggressively in response to the second, smaller epidemic, but the reason is eminently practical. When the reproduction number is high enough, it is very hard to prevent the epidemic from exploding if everyone is susceptible. But there is already an appreciable degree of herd immunity when the second, rebound epidemic threatens, so a less severe lockdown can be sufficient to preempt it.

9.3.1.2 Interpreting the Types of Lockdown Strategies that Can be Optimal

Table 9.1 summarises the nature and performance of each of the strategies in the right-hand panel of Fig. 9.3. Its columns merit some discussion. The lockdowns' start and end times are self-explanatory except to note that with strategies IIa and IIb, there are two separate lockdowns, so there are two separate start and end times. The intensity of the lockdown measures the amount of unemployment that the lockdown creates on a scale where 365 corresponds to no one in the population working for an entire year.

Table 9.1 shows that even when lockdown intensity and duration are allowed to vary continuously, there are basically only three sizes that emerge as optimal: very small (less than 1.04), modest (around 30–35, or the equivalent of the economy giving up one month of economic output), and large (around 360, or the equivalent of the economy giving up a full year of economic output).

The levels of deaths also fall into basically three levels. High (around 2.9% of the population) goes with small lockdowns. Medium-high deaths (around 2.4%) goes with modest lockdowns. Small deaths (around 0.2%) goes with large lockdowns. It would be nice to have a small number of deaths despite only imposing a small lockdown, but that just isn't possible.

In sum, there are basically three strategies: (1) Do very little locking down and suffer deaths both from the initial epidemic and also the rebound epidemic as people lose immunity, (2) Only do a bit of curve flattening during the first epidemic but use a modest sized lockdown later on to prevent the rebound epidemic and so have a medium-high number of deaths, or (3) Lockdown forcefully more or less throughout the entire planning horizon in order to avert most of the deaths altogether.

Figure 9.2 provides the corresponding information when $R_0 = 2.5$. It shows that when the virus is less contagious the large lockdown does not need to be quite as large (size of 257 or about 8.5 months of lost output, not a full year) in order to hold the number of deaths down to low levels. Perhaps surprisingly, the minimalist strategies (Ia) are less minimalist when $R_0 = 2.5$; when $R_0 = 4.0$ the epidemic is just so powerful that it is not even worth doing as much curve flattening as it is when $R_0 = 2.5$.

Table 9.1 $R_0 = 4$: Data characterising the optimal solutions for the different regimes of Fig. 9.4. The size of the lockdown is defined as $\int_{\tau_s}^{\tau_e} (1 - \gamma(t))dt$, where τ_s is the starting and τ_e the exit time of the lockdown

	Start time	End time	Duration	Size of lockdown	Deaths (%)	M
Lockdown: Ia	24.8	77.4	52.6	0.5	2.9	1500
Lockdown: Ib	412.8	730.0	317.2	32.8	2.4	5000
Lockdown: IIa						
1	21.4	81.3	59.9	0.8	–	–
2	659.5	713.8	54.2	0.1	–	–
Total	–	–	114.1	0.9	2.9	3200
Lockdown: IIb						
1	3.2	30.9	27.7	0.0	–	–
2	388.1	730.0	341.9	35.8	–	–
Total	–	–	369.6	35.8	2.3	11,000
Lockdown: IIIa	0.0	730.0	730.0	360.8	0.2	13,000

Table 9.2 $R_0 = 2.5$. The optimal solutions for $R_0 = 2.5$ evaluated at the same M values as for $R_0 = 4$

	Start time	End time	Duration	Size of lockdown	Deaths (%)	M
Lockdown: Ia	–	–	–	0	2.1	1500
Lockdown: Ib	35.2	145.8	110.6	4.8	2.0	5000
Lockdown: IIa	44.3	122.2	77.9	1.3	2.1	3200
Lockdown: IIb	0.0	730.0	730.0	257.6	0.2	11,000
Lockdown: IIIa	0.0	730.0	730.0	258.8	0.2	13,000

9.3.1.3 The Effects of Lockdown Fatigue

One feature of the current model is its recognition of lockdown fatigue. Recall that fatigue means that the infection-preventing benefits of an economic lockdown may be eroded over time by the public becoming less compliant, e.g., because the economic suffering produces pushback. The results above used parameter values that meant the power of that fatigue was fairly modest. In this subsection we explore how greater tendencies to fatigue can influence what strategy is optimal.

The tool again is a bifurcation diagram with the horizontal axis denoting M, the value the social planner places on preventing a premature death. (See Fig. 9.5.) Now, though, the vertical axis measures the strength of the fatigue effect, running from 0 (no effect) up to 1.0. The units of this fatigue effect are difficult to interpret, but roughly speaking, over the time horizons contemplated here, if $f = 1.0$ then when employing the sustained lockdown strategies, the lockdowns lose about half of their effectiveness by the time they are relaxed.

Figure 9.5 shows the results. When that fatigue parameter is small (lower parts of Fig. 9.5), the march across the various strategies with increasing M is the same as that depicted in Fig. 9.3. With large values (top of Fig. 9.5), there are two differences. First, Region Ib disappears but Region IIb remains, meaning if it is ever optimal to use a moderately strong lockdown to forestall a rebound epidemic, then one also does at least something in response to the first epidemic. Second, Region IIIa gives way to Region IIIb in which some degree of lockdown is maintained for an extended time, but it is relaxed somewhat between the first and rebound epidemics in order to let levels of fatigue dissipate somewhat.

The still more important lesson though pertains to the curve separating regions where some major lockdown is optimal (whether that is of type IIIa or IIIb) and regions where only small or moderate sized lockdowns are optimal (Regions Ia, Ib, IIa, or IIb). That boundary slopes upward and to the right, meaning that the greater the tendency of the public to fatigue, the higher the cost per premature death (M) has to be in order for a very strong and sustained lockdown to be optimal. That makes sense. If fatigue will undermine part of the effectiveness of a large lockdown, then the valuation of the lockdown's benefits has to be greater in order to justify its considerable costs.

This suggests that those advocating for very long lockdowns might want to think about whether there are ways of making that lockdown more palatable in order to minimise fatigue. For example, some Canadian provinces tempered their policies limiting social interaction to people within a household bubble so that people living alone were permitted to meet with up to two other people, to avoid the mental health harms of total isolation.

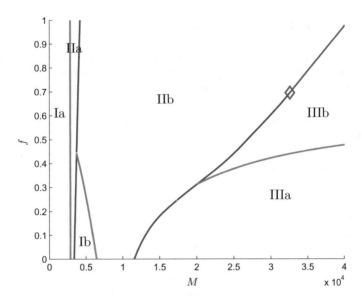

Fig. 9.5 This figure shows the different regions in the $M - f$ space. The green lines denote continuous transitions from region region I to II ($R_0 = 4$). The blue curve is a Skiba curve, where the transition from region II to III is discontinuous and at the Skiba curve two optimal solutions exist. The Skiba curve switch to a continuous transition curve (red) at the red diamond. In Regime IIIb the lockdown is relaxed in between and then tightened again, whereas in Regime IIIa the lockdown is steadily increased and then steadily relaxed

9.3.1.4 Illustrating Skiba Trajectories

One key finding here is that for certain sets of parameter values, two—or sometimes even three—very different strategies can produce exactly the same net value for the social planner. We close by illustrating this phenomenon in greater detail.

Returning to Fig. 9.3b, with the higher level of infectivity believed to pertain for the UK variant of the virus, as the valuation placed on preventing a premature death (M) increases, one crosses two Skiba thresholds, one at $M = 3395$ separating Regions IIa and Ib and another at $M = 11,560$ separating Regions IIb and IIIa. These thresholds are denoted in Fig. 9.3b by solid vertical lines. They can also be seen in Fig. 9.5 by moving left to right at the bottom level ($f = 0$).

Figure 9.6 shows the two alternate strategies, in terms of γ, the proportion of employees who are allowed to work. The left panel shows the two equally good strategies when $M = 3395$; the right-hand panel shows the two strategies that are equally good when $M = 11,560$. We have already discussed their nature. On the left side one is choosing between two very small lockdowns and one moderately large lockdown later. On the right side one is choosing between a pair of lockdowns (very small early and moderately large later) and one very deep and sustained lockdown.

Table 9.3 Data on costs for Skiba point at $M = 11,560$

	Uncontrolled	Flattening	long and sustained lockdown
Health costs	335.67	271.42	25.33
Economy	18.27	36.77	270.13
Adjustment costs	0	0.91	13.64
Total costs	353.94	309.10	309.10
Deaths (%)	2.9	2.3	0.2

The observation to stress for present purposes is just how different the trajectories are in each pairing. When one crosses a Skiba threshold, what is optimal can change quite radically. Likewise, when one is standing exactly at that Skiba threshold, one has two equally good options, but those options are radically different.

That means that when two people advocate very different lockdown strategies in response to COVID-19, one cannot presume that they have very different understandings of the science or very different value systems. They might actually share very similar or indeed even identical worldviews, but still favour radically different policies.

Table 9.3 illustrates how this can be so. Its first column summarises the outcomes (costs) when there is no control. Health costs are enormous because more or less everyone gets infected and 2.9% of the population dies; the numbers are on a scale such that 365 is one year's GDP, so the health cost of 335.7 is almost as bad as losing an entire year's economic output. There are also some economic losses from losing the productivity of those who die prematurely, producing a total cost of 353.9.

The second column shows that modest deployment of lockdowns only reduces health costs by 20%, to 271.4, whereas a severe and sustained lockdown reduces them by 92.5%, to 25.3. However, the severe and sustained lockdown multiplies costs of lost labor fifteenfold, to 270.1, and creates an additional cost equivalent to 13.6 days of output from forcing businesses to adjust to changing lockdown policies. Summing across all three types of costs produces the same total of 309.1 for both types of lockdown strategies.

Thus the two lockdown strategies produce the same aggregate performance (309.1), but with very different compositions. The moderate lockdown strategy creates smaller economic costs but only reduces health costs by 20%. The severe and sustained lockdown eliminates most of the healthcare costs but creates very large economic dislocation.

What is quite sobering is that either optimal policy only reduces total social cost by 13%, from 353.9 to 309.14. The COVID-19 pandemic is truly horrible; at least within this model, even responding to it optimally alleviates only a modest share of the suffering. Lockdowns can convert health harms to economic harms, but they cannot do much to reduce the total amount of harm.

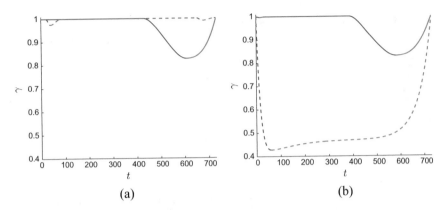

Fig. 9.6 Optimal time paths for the Skiba solutions for $M = 3395, 11560$ in Fig. 9.3b

9.4 Discussion

This paper investigated implications of the SARS-CoV-2 virus being more contagious than has previously been understood, e.g., because of a mutation or variant strain. In particular, it investigates implications for economic lockdown strategies within an optimal control model that balances health and economic considerations. A number of results were confirmed that had been obtained earlier with parameters reflecting the earlier understanding of the epidemic's reproduction number. In particular, we continue to find that:

- Very different lockdown policies can be optimal—ranging from very little to long and sustained lockdowns—depending on the value of parameters that are difficult to pin down, notably including the valuation placed on averting a premature death.
- For certain parameter constellations, the nature of the optimal policy can change radically even with quite small changes in these parameters.
- There are even situations in which two very different policies can both yield exactly the same aggregate performance, with one strategy's better performance at reducing deaths being exactly offset by its worse performance in other respects.
- As we have discussed previously in Caulkins et al. (2020), these results suggest a degree of humility is in order when advocating for one policy over another. Another person who favours a very different policy might actually share a very similar scientific understanding of the disease dynamics and even hold similar values, and yet still reasonably reach quite different conclusions.

There are, though, differences here. One is that a greater variety of strategies emerged as candidates. Some concerned how to address a potential rebound epidemic among people who were previously infected but then flowed back from the recovered to the susceptible state as their immunity wore off. For some parameter values, if the virus is sufficiently contagious and the time until a vaccine arrives long enough, it may be prohibitively difficult to substantially avoid the initial wave of infection, but nonetheless be desirable to use a moderately aggressive lockdown to avert a rebound epidemic, for two reasons. First, it is easier to deal with the rebound epidemic because there will still be some degree of herd immunity at that time, in contrast to the situation when the virus first arrives. Second, the time until a vaccine's arrival is obviously shorter when addressing a rebound as opposed to the initial epidemic, so any lockdowns do not need to be sustained as long.

Indeed, whereas in the past the multiplicity of strategies basically fell into two camps, either a fairly modest lockdown that served only to flatten the curve and a much more intensive and sustained lockdown that largely protected the population from infection, now there is a third category of strategies. It might be thought of as flattening the first wave and eliminating the second.

We also investigated more thoroughly than before the potential effects of lockdown fatigue. The primary results are perhaps as expected. The greater the tendency for fatigue to undermine the effectiveness of a lockdown, the higher one must value the benefits created by a lockdown in order for a large and sustained lockdown to be optimal. That suggests that those wishing to impose long and deep lockdowns might want to think about ways of reducing resistance to those measures.

In sum, when lockdown intensity is allowed to vary continuously, many nuances emerge, but the overall character still boils down to an almost binary choice, one made perhaps even more stark if the virus becomes more contagious. If the value placed on preventing a premature death is high enough, then a social planner should use a sustained and forceful lockdown to preempt the epidemic despite incurring massive economic dislocation. Otherwise, lockdowns are too blunt and expensive to employ as the primary response to the epidemic. Thus, the model prescribes an almost all-or-nothing approach to economic lockdowns.

That finding does not mean that modulated approaches to what might be termed social lockdowns do not have a role. It may be entirely possible for a government to ramp up or down when and where it requires masks and social distancing outside the workplace, or to do the same with travel restrictions and quarantines. Our model is looking only at economic lockdowns.

Here is one way to think about this conclusion. We credited policy makers with a degree of common sense that could have made intermediate levels of economic lockdowns appealing. In particular, we assumed that the benefits in terms of reduced infection were a concave function of the amount of the economy that is shut down. In plain language, we presumed policy makers would shut down first the economic activities that had the greatest ratio of infection risk to economic value (e.g., in-person concerts and other crowd gatherings) and shut down last those that produce a lot of economic value per unit of infection risk (e.g., mining and highly automated manufacturing). If the virus' behaviour were linear, that might be expected to favour lockdowns of intermediate intensity. However, the contagious spread of a virus is highly nonlinear, involving very powerful positive feedback loops that produce exponential growth. To speak informally, if the virus gets its nose into the tent and locking down is the only policy response, then the virus will rip through the population if the lockdown is anything other than very strong.

The good news is that policy makers do have other tools besides lockdowns. For example, rapid, intense testing and contact tracing might be able to hold down infections when there are relatively few people getting infected. But if the virus spreads beyond the ability of such targeted measures, and the only remaining tool is broad-based economic lockdowns, the analysis here suggests being decisive; waffling efforts may produce the worst of both worlds, with substantial economic losses and still high rates of infection.

We close with a caveat. Despite its apparent complexity, this model explored here is of course vastly simplified compared to the real world, and there is much that remains unknown and uncertain about optimal economic response to pandemic threats. We hope we have usefully provoked thinking and advanced understanding, but hope even more fervently that society will invest heavily in much more such analysis, so that we can all be better prepared the next time the world confronts a novel pandemic.

Acknowledgements The author Dieter Grass was supported for this research by the FWF Project P 31400-N32.

Appendix 1

The decision variables are τ_1 and τ_2, the times when the lockdown begins and ends, and the full model can be written as:

$$V(X_0, \tau_1, \tau_2) := \int_0^T \left(V_l(W(t), \tau_1, \tau_2) - V_h(I(t)) \right) dt$$
$$- (T + \Gamma) K W(0)^\sigma \gamma(0, \tau_1, \tau_2)^\sigma$$
$$+ \Gamma K W(T)^\sigma \gamma(T, \tau_1, \tau_2)^\sigma$$
$$V^*(X_0) := \max_{\tau_1, \tau_2} V(X_0, \tau_1, \tau_2), \quad X := (S, L, I, R), \quad W := S + L + R.$$

$$\text{s.t.} \quad \dot{X}(t) = \begin{cases} \text{SLIR}_1(X(t), \tau_1, \tau_2) & 0 \le t < \tau_1 \\ \text{SLIR}_2(X(t), \tau_1, \tau_2), & \tau_1 \le t \le \tau_2 \\ \text{SLIR}_3(X(t), \tau_1, \tau_2) & \tau_2 < t \le T \end{cases}$$

$$X(0) = X_0 \ge 0$$

$$\gamma(t, \tau_1, \tau_2) := \begin{cases} \gamma_1 & 0 \le t < \tau_1 \\ \gamma_2 & \tau_1 \le t \le \tau_2 \\ \gamma_3(\tau_1, \tau_2) := \gamma_2 + (\gamma_1 - \gamma_2)e^{\kappa_2(\tau_1 - \tau_2)} & \tau_2 < t \le T \end{cases}$$

$$R_{\text{eff}}(t, \tau_1, \tau_2) := \begin{cases} R_0^1 & 0 \le t < \tau_1 \\ R_0^2 & \tau_1 \le t \le \tau_2 \\ R_0^3(\tau_1, \tau_2) := \bar{R}_0^3 + (R_0^1 - \bar{R}_0^3)e^{\kappa_1(\tau_1 - \tau_2)} & \tau_2 < t \le T \end{cases}$$

with $R_0^2 \le \bar{R}_0^3 \le R_0^1$.

We specify the health care term and the economic (labor) term in the objective as

$$V_h(I) := M\,(\xi_1 pI + \xi_2 \max(\{0, pI - H_{\max}\}, \zeta))$$
$$V_l(W, \tau_1, \tau_2) := K\gamma(t, \tau_1, \tau_2)^\sigma W(t)^\sigma.$$

The derivation of the necessary optimality conditions can be found in the Appendix 1 in Caulkins et al. (2020). The Matlab toolbox OCMat is used for the numerical calculations (see http://orcos.tuwien.ac.at/research/ocmat_software).

Appendix 2

$$V(X_0, u(\cdot)) := \int_0^T \big(V_l(W(t), \gamma(t)) - V_h(I(t)) - V_u(u(t), z(t))\big)\mathrm{dt}$$
$$- (T + \Gamma)KW(0)^\sigma \gamma(0)^\sigma$$
$$+ \Gamma KW(T)^\sigma \gamma(T)^\sigma$$

$$V^*(X_0) := \max_{u(\cdot)} V(X_0, u(\cdot))$$

$$X(t) := (S(t), I(t), R(t), \gamma(t), z(t)), \quad W(t) := S(t) + R(t)$$
$$N(t) := S(t) + I(t) + R(t).$$

s.t.
$$\dot{S}(t) = \nu N(t) - \beta(\gamma(t), z(t))\frac{S(t)I(t)}{N(t)} - \mu S(t) + \phi R(t)$$

$$\dot{I}(t) = \beta(\gamma(t), z(t))\frac{S(t)I(t)}{N(t)} - (\alpha + \mu + \mu_I)I(t)$$

$$\dot{R}(t) = \alpha I(t) - \mu R(t) - \phi R(t)$$

$$\dot{\gamma}(t) = u(t), \quad \gamma(0) = 1$$

$$\dot{z}(t) = \kappa_1(1 - \gamma(t)) - \kappa_2 z(t), \quad z(0) = 0$$

$$\gamma(t) \le 1, \quad 0 \le t \le T$$

$$\beta(\gamma, z) := \beta_1 + \beta_2\left(\gamma^\theta + f\frac{\kappa_2}{\kappa_1}z(1 - \gamma^\theta)\right)$$

$$V_l(W, \gamma) := K\gamma^\sigma W^\sigma$$

$$V_h(I) := M\,(\xi_1 pI + \xi_2 \max_s(\{0, pI - H_{\max}\}, \zeta))$$

$$V_u(u, z) := \begin{cases} c_l u^2 & u \le 0 \\ c_r(z + 1)u^2 & u > 0 \end{cases}$$

Appendix 3

Table 9.4 Base case parameter values and initial state variable values. The $*$ denotes a free parameter and is specified in the figures

Variable	Model Sect. 9.2	Model Sect. 9.3	Description
α	1/9	1/15	Reciprocal of average duration of the infection
$R_0^{1,2,3}$	4, $*$, 3.2	–	Level of infection risk
$\gamma_0^{1,2,3}$	1, 0.25, 0.75	–	Level of lockdown intensity
β_1	–	0	Minimum level of infection risk
β_2	–	0.2	Increment in the level of infection risk
$H_{\max}$	1.76×10^{-4}	2×10^{-4}	Capacity of intensive care units
p	2.311×10^{-2}	2.25×10^{-2}	Probability that infected person needs critical care
M	$*$	$*$	Social cost of a premature death due to COVID-19
K	1	1	Coefficient on economic activity
Γ	365	365	Reflects time required to return to full employment
f	0.75	–	Likelihood of spreading the virus when in the L state
f	–	$*$	Impact of lockdown fatigue on infection risk
κ_1	2×10^{-3}	–	Rate of decay for lockdown intensity
κ_1	–	0.15	Rate of accumulation of fatigue
κ_2	10^{-3}	–	Rate of decay for reproduction number
κ_2	–	0.2	Rate of exponential decay of fatigue
ω	0.6	–	Proportion of infections that become symptomatic
φ	1/7.2	–	Reciprocal of average duration of the latency
σ	2/3	2/3	Labor elasticity in Cobb-Douglas production function
ν	0.01/365	0	Birth rate
μ	0.01/365	0	Death rate (not caused by COVID-19)
μ_I	13/10800	0	COVID-19 death rate
ζ	5000	5000	Parameter in the approximation of the max-function
ξ_1	0.05	0.03	Death rate of infected individual in critical care
ξ_2	0.55/9	0.55/15	Incremental death rate if IC capacity is exceeded
ϕ	–	0.001	Rate by which recovered get susceptible again
c_l	–	1000	Parameter in business shutting down costs
c_r	–	5000	Parameter in business reopening costs
θ	2	2	Exponent in the proportionality function $\beta(t)$
$S(0)$	0.999	0.999	Initial susceptible population
$L(0)$	0.001	–	Initial latent population
$I(0)$	0	0.001	Initial infected population
$R(0)$	0	0	Initial recovered population
$\gamma(0)$	–	1	Initial employment level
$z(0)$	–	0	Initial lockdown fatigue

References

A.B. Abel, S. Panageas, Optimal management of a pandemic in the short run and the long run. NBER Working Paper Series WP 27742 (2020)

D. Acemoglu, *Introduction to Modern Economic Growth* (Princeton University Press, Princeton, 2009)

D. Acemoglu, V. Chernozhukov, I. Werning, M.D. Whinston, Optimal targeted lockdowns in a multi-group SIR model. NBER Working Paper Series WP 27102 (2020)

F.E. Alvarez, D. Argente, F. Lippi, A simple planning problem for COVID-19 lockdown. Am. Econ. Rev. Insights (forthcoming) (2020)

A. Aspri, E. Beretta, A. Gandolfi, E. Wasmer, Mortality containment vs. economics opening: optimal policies in a SEIARD model. J. Math. Econ. (2021)

D.E. Bloom, M. Kuhn, K. Prettner, Modern infectious disease: Macroeconomic impacts and policy responses. J. Econ. Liter. (2020)

A. Brodeur, D.M. Gray, A. Islam, S. Bhuiyan, A literature review of the economics of COVID-19. IZA Discussion Paper No. 13411 (2020)

J.P. Caulkins, D. Grass, G. Feichtinger, R. Hartl, P.M. Kort, A. Prskawetz, A. Seidl, S. Wrzaczek, How long should the COVID-19 lockdown continue? PloS One **15**(12), e0243413 (2020)

J.P. Caulkins, D. Grass, G. Feichtinger, R.F. Hartl, P.M. Kort, A. Prskawetz, A. Seidl, S. Wrzaczek, The optimal lockdown intensity for COVID-19. J Math. Econ. (2021)

S. Federico, G. Ferrari, Taming the spread of an epidemic by lockdown policies. J. Math. Econ. (2021)

J. Fernández-Villaverde, C.I. Jones, Macroeconomic outcomes and COVID-19: A progress report. Nber working paper 28004 (2020)

D. Gershon, A. Lipton, H. Levine, Managing COVID-19 pandemic without destructing the economy. arXiv preprint arXiv:2004.10324 (2020). https://ui.adsabs.harvard.edu/abs/2020arXiv200410324G/abstract

M. Gonzalez-Eiras, D. Niepelt, On the optimal 'lockdown' during an epidemic. CEPR Discussion Paper 14612 (2020)

D. Grass, J.P. Caulkins, G. Feichtinger, G. Tragler, D.A. Behrens, *Optimal Control of Nonlinear Processes: With Applications in Drugs, Corruption, and Terror* (Springer, Berlin, 2008)

J.K. Hammitt, Valuing mortality risk in the time of COVID-19. SSRN Electron. J. (2020). https://doi.org/10.2139/ssrn.3615314

N. Huberts, J. Thijssen. Optimal timing of interventions during an epidemic. Available at SSRN 3607048, University of York (2020)

W.O. Kermack, A.G. McKendrick, A contribution to the mathematical theory of epidemics. Proc. Roy. Soc. London **115**(772), 700–721 (1927)

T.J. Kniesner, W.K. Viscusi, C. Woock, J.P. Ziliak, The value of a statistical life: evidence from panel data. Rev. Econ. Stat. **94**(1), 74–87 (2012). https://doi.org/10.1162/REST_a_00229

R. Layard, A. Clark, J.-E. De Neve, C. Krekel, D. Fancourt, N. Hey, G. O'Donnell, When to release the lockdown? A wellbeing framework for analysing costs and benefits. IZA Institute of Labor Economics IZA DP No. 13186 (2020)

F. Piguillem, L. Shi, Optimal COVID-19 quarantine and testing policies. Einaudi Institute for Economics and Finance EIEF Working Papers Series 20/04 (2020)

L. Rachel, *An Analytical Model of Covid-19 Lockdowns* (Technical report, Center for Macroeconomics, 2020)

A. Scherbina, *Determining the Optimal Duration of the COVID-19 Supression Policy: A Cost-Benefit Analysis* (Technical report, American Enterprise Institute, 2020)

Open Access This chapter is licensed under the terms of the Creative Commons Attribution 4.0 International License (http://creativecommons.org/licenses/by/4.0/), which permits use, sharing, adaptation, distribution and reproduction in any medium or format, as long as you give appropriate credit to the original author(s) and the source, provide a link to the Creative Commons license and indicate if changes were made.

The images or other third party material in this chapter are included in the chapter's Creative Commons license, unless indicated otherwise in a credit line to the material. If material is not included in the chapter's Creative Commons license and your intended use is not permitted by statutory regulation or exceeds the permitted use, you will need to obtain permission directly from the copyright holder.

Chapter 10
Diagnostic Tests and Procedures During the COVID-19 Pandemic

Sherry A. Dunbar and Yi-Wei Tang

Abstract Coronavirus disease 2019 (COVID-19) has brought a huge impact on global health and the economy. Early and accurate diagnosis of severe acute respiratory syndrome coronavirus 2 (SARS-CoV-2) infections is essential for clinical intervention and pandemic control. This book chapter addresses the evolving approach to the laboratory diagnosis of COVID-19 covering preanalytical, analytical, and postanalytical steps. The rapidly changing dynamics of the COVID-19 pandemic serve as an example which will be important for laboratories to plan for future pandemics. With the quick identification of the causative pathogen and availability of the genome sequence, it will be possible to develop and implement diagnostic tests within weeks of an outbreak. Laboratories will need to be flexible to continuously adapt to changing testing needs and burdens on the healthcare system, plan mitigation strategies for bottlenecks in testing and workflow due to limitations on resources and supplies, and prepare back-up plans now in order to be better prepared for future pandemics.

10.1 Introduction

In December 2019, a respiratory illness caused by a novel coronavirus was detected in Wuhan City, Hubei Province, China and rapidly evolved into the global pandemic known as COVID-19, see Lu et al. (2020a), Zhu et al. (2020). Coronaviruses are enveloped viruses comprised of a single stranded positive-sense RNA genome. Virions are spherical, with a spike glycoprotein (S) embedded in the envelope and additional structural proteins, including envelope (E), matrix (M), and nucleocapsid

S. A. Dunbar
Luminex Corporation, 12212 Technology Blvd, Austin, TX 78727, USA
e-mail: sdunbar@luminexcorp.com

Y.-W. Tang (✉)
Cepheid/Danaher Diagnostic Platform, Shanghai, China
e-mail: yi-wei.tang@cepheid.com

© The Author(s) 2022
M. C. Boado-Penas et al. (eds.), *Pandemics: Insurance and Social Protection*,
Springer Actuarial, https://doi.org/10.1007/978-3-030-78334-1_10

(N). This novel coronavirus, designated SARS-CoV-2, has been shown by genomic sequencing to be approximately 85-88% identical to bat SARS-like coronaviruses and 96.2% related to the bat coronavirus RaTG13, but distinct from SARS-CoV-1 with approximately 80% similarity, see Zhu et al. (2020), Lu et al. (2020c), Yan et al. (2020).

The binding of SARS-CoV-2 and entry into the host cell is mediated by the S protein where the S1 subunit contains the receptor binding domain (RBD) that binds to the peptidase domain of angiotensin-converting enzyme 2 (ACE 2), Cevik et al. (2020). The primary mechanism of transmission of SARS-CoV-2 is by infected respiratory droplets, through direct or indirect contact with nasal, conjunctival, and/or oral mucosa, Hui et al. (2020). The target host receptors are found mainly in the epithelium of the human respiratory tract in the oropharynx and upper airway; however, the conjunctiva and gastrointestinal tract are also susceptible.

In the respiratory tract, peak viral loads are observed at the time of symptom onset, typically in the first week of illness, with subsequent decline thereafter, Cevik et al. (2021). Quantitative reverse transcription polymerase chain reaction (qRT-PCR) can detect SARS-CoV-2 RNA in the upper respiratory tract for 17 days (on average) after symptom onset, although viral RNA detection may not equate to infectivity. Viral culture from PCR-positive upper respiratory tract samples is primarily negative beyond eight to nine days of illness, see Wölfel et al. (2020), Bullard et al. (2020).

Active replication and release of the virus in the cells of the lungs leads to non-specific symptoms such as fever, myalgia, headache, and other respiratory symptoms. A lymphocytic endotheliitis has been observed postmortem upon examination of lung, heart, kidney, and liver, indicating that the virus directly affects many organs, Varga et al. (2020). Clinical outcomes of infection are influenced by host factors such as older age, underlying medical conditions, and host-immune response, as well as the viral load.

By March 2021, more than 116 million cases of COVID-19 had been reported globally, leading to more than 2.5 million deaths, World Health Organization (2021). Individuals of all ages are at risk for infection and severe disease, but the most serious disease occurs in people aged 60 and older, residents of nursing homes and long-term care facilities, and those with immunosuppression and chronic medical conditions. In a study of more than 1.3 million cases in the U.S., a significantly higher percentage of hospitalisations, intensive care unit admissions, and deaths occurred in patients with chronic medical conditions and in individuals >70 years of age, Stokes et al. (2020), Guan et al. (2020), Wu et al. (2020).

As two main processes are thought to direct the pathogenesis of COVID-19, different therapeutic approaches may be employed in different stages of infection. Early in infection, the disease is driven by replication of the virus, whereas later in the course of infection, the disease is driven by an overactive immune/inflammatory response. Therefore, antiviral therapies should have the greatest effect early in the course of disease, while immunosuppressive and anti-inflammatory therapies are likely to be more beneficial in the later stages.

Several antiviral therapies continue to be investigated for the treatment of COVID-19. These drugs inhibit viral entry, viral membrane fusion and endocytosis, or the activity of SARS-CoV-2 protease (3CLpro) and RNA-dependent RNA polymerase, Sanders et al. (2020). Remdesivir, an antiviral agent that inhibits viral replication, is recommended for use in hospitalised patients, especially those who require supplemental oxygen, Beigel et al. (2020). Dexamethasone, a corticosteroid, has been found to improve survival in hospitalised patients who require supplemental oxygen, with the greatest effect observed in patients who require mechanical ventilation and is strongly recommended in this clinical setting, Horby et al. (2021).

In the earliest stages of infection and before the patient has mounted an effective immune response, anti-SARS-CoV-2 antibody-based therapies would have the greatest likelihood of having an impact. Preliminary data suggests that outpatients may benefit from receiving anti-SARS-CoV-2 monoclonal antibodies early in the course of infection, Ju et al. (2020), Wang et al. (2020). Several anti-SARS-CoV-2 monoclonal antibodies have been authorised for the treatment of outpatients with mild to moderate COVID-19, Chen et al. (2021), Weinreich et al. (2021).

Transmission of SARS-CoV-2 is thought to mainly occur through direct contact with an infected person or fomite, Cai et al. (2020). The second mode is via respiratory droplets transmitted by exhalation from an infectious person to others, generally within about six feet, Alsved et al. (2020). Less commonly, airborne transmission of small droplets and particles of SARS-CoV-2 can occur at distances greater than six feet, particularly in enclosed spaces, prolonged time of exposure, and inadequate ventilation or air handling, Li et al. (2007, 2020b), Lu and Yang (2020). The risk of SARS-CoV-2 transmission can be reduced by covering coughs and sneezes and maintaining a distance of at least six feet from others. When consistent distancing is not possible, face coverings may further reduce the spread of infectious droplets from individuals with SARS-CoV-2 infection to others. Frequent handwashing also effectively reduces the risk of infection, Centers for Disease Control and Prevention (2020b). Healthcare providers should follow institutional recommendations for infection control and appropriate use of personal protective equipment (PPE).

As more vaccines become available to prevent COVID-19, the pandemic may be alleviated through high vaccine effectiveness, effectiveness to new variants as they arise, and if a sufficient proportion becomes vaccinated to achieve herd immunity. As of February 2021, at least seven different vaccines across three platforms (inactivated virus, nucleic acid/mRNA, and recombinant viral vector-based) have become available with vulnerable populations at the highest priority for vaccination, World Health Organization (2020b). It is not currently known how long SARS-CoV-2 vaccine protective effect will last, whether they prevent asymptomatic infection or transmission, or whether they will prevent infection by all current or emergent strains of SARS-CoV-2. Clinical data continue to be collected and clinical trials for other SARS-CoV-2 vaccine candidates are ongoing. As of March 2021, in large, placebo-controlled trials, these vaccines were 80–95% effective in preventing SARS-CoV-2 infection or serious/severe COVID-19 disease after participants completed all doses. In this chapter, we describe the laboratory diagnosis of COVID-19 in terms of pre-analytical (specimen and biosafety), analytical (tests and platforms), and post-analytical (result interpretation) considerations.

10.2 Laboratory Diagnosis: Pre-analytical Issues

10.2.1 Specimen Types and Specimen Collection

Regardless of the sensitivity (ability to designate an individual with a disease as positive) and specificity (ability to designate an individual without a disease as negative) of available laboratory tests, the diagnosis of viral pneumonias, such as that caused by SARS-CoV-2, is dependent on collecting the correct specimen from the patient at the correct time. Within 5–6 days from onset of symptoms, patients with COVID-19 have demonstrated high viral loads in upper and lower respiratory tracts, Wölfel et al. (2020), Pan et al. (2020), Zou et al. (2020). As viral pneumonias do not typically result in the production of a purulent sputum, nasopharyngeal specimens, such as a nasopharyngeal swab are recommended, but nasal and oropharyngeal swabs are also acceptable specimen types, Zou et al. (2020), Kim et al. (2011), National Institutes of Health (2021), Centers for Disease Control and Prevention (2020c). However, nasopharyngeal specimens may not detect early infection and a lower respiratory tract specimen may be needed. Lower respiratory tract samples have a higher diagnostic yield than those from the upper tract but these specimens are usually not obtained because of potential risk of virus aerosolisation during sample collection, National Institutes of Health (2021). Therefore, bronchoalveolar lavage and sputum induction should only be performed after careful consideration of the risk of exposing staff to infectious aerosols. Endotracheal aspiration appears to carry a lower risk of aerosol generation than bronchoalveolar lavage (BAL), and some experts consider the sensitivity and specificity of endotracheal aspirates and BAL specimens comparable in detecting SARS-CoV-2. Repeated testing over time may increase the likelihood of detecting SARS-CoV-2 present in the nasopharynx. Upper respiratory specimens are collected using synthetic fiber swabs with thin plastic or wire shafts that have been designed for sampling the nasopharyngeal mucosa which are then placed into a transport tube for transportation to the laboratory. If both nasopharyngeal and oropharyngeal specimens are collected, it is recommended they be combined in a single transport media tube to maximise test sensitivity and limit use of testing resources, Centers for Disease Control and Prevention (2020c). Several studies have shown saliva as suitable specimen for SARS-CoV-2 testing, To et al. (2020), Wyllie et al. (2020).

Some tests that have received Emergency Use Authorization (EUA) from the U.S. Food and Drug Administration (FDA) may be performed on saliva specimens and some allow for self-collection of saliva which is then sent to a laboratory for testing. Self-collection of saliva samples eliminates direct interaction between healthcare workers and patients, which may present a risk for spread, increases the demand for supplies, and can be a bottleneck for testing workflows. Saliva (1–5 mL) is collected in a sterile, leakproof container and does not require preservative for transportation to the laboratory, which also facilitates self-collection by the patient.

Studies have also shown that a significant proportion of COVID-19 patients carry SARS-CoV-2 in the intestinal tract. A meta-analysis of 17 studies showed that SARS-

CoV-2 RNA was detected in 33.7% of specimens and 43.7% of patients, Wong et al. (2020). A subsequent report analysed the results from 79 studies and found that the mean duration of viral RNA shedding was 17.2 days in stool with a maximum of 126 days; however no live virus was detected beyond 9 days of illness, Cevik et al. (2021). Stool is an attractive specimen type since it can be self-collected and has the potential to improve case identification in the community.

10.2.2   Biosafety Considerations

Patients with confirmed or possible SARS-CoV-2 infection should wear a facemask when being evaluated medically and healthcare personnel should adhere to standard and transmission-based precautions when caring for patients with SARS-CoV-2 infection, Centers for Disease Control and Prevention (2021). Precautions should be taken in handling specimens that are suspected or confirmed for SARS-CoV-2. All laboratories should identify and mitigate risks depending on the procedures they perform and the associated hazards, including the competency level of the individuals performing the procedures and the facility, equipment, and resources available, Centers for Disease Control and Prevention (2020d). Laboratory personnel should follow standard precautions when handling clinical specimens which may contain potentially infectious materials and follow routine laboratory practices and procedures for decontamination of work surfaces and management of laboratory waste.

Processing of respiratory specimens should be done in a class II biological safety cabinet (BSL-2), including procedures using automated instruments and analyzers, molecular analysis of extracted nucleic acid preparations, packaging of specimens for transport to other diagnostic laboratories for additional testing, and procedures using inactivated specimens (i.e., such as specimens in nucleic acid extraction buffer), Centers for Disease Control and Prevention (2020d), Chu et al. (2020). Lysis buffer for nucleic acid extraction should contain a guanidinium-based inactivating agent as well as a nondenaturing detergent. Buffers that are used with most commercial extraction platforms contain guanidium and detergents and are able to inactivate viable coronavirus, Blow et al. (2004), Kumar et al. (2015), Welch et al. (2020). Self-enclosed sample-to-answer systems which integrate nucleic acid extraction, amplification, and detection such as ID NOW (Abbott, San Diego, CA) Nie et al. (2014), Wang et al. (2018), cobas Liat (Roche Molecular Systems, Pleasanton, CA), ARIES® (Luminex Corporation, Austin, TX), and GeneXpert (Cepheid, Sunnyvale, CA) Ling et al. (2018), meet local regulatory requirements for SARS-CoV-2 testing. Once the specimen in viral transport medium is added into the cartridge in a BSL-2 biosafety cabinet, the cartridge is sealed. Many of the sealed, random access testing devices are suitable for point-of-care testing for hospitals and clinics without biosafety cabinets; however, staff collecting the specimen should use appropriate PPE and avoid spills of transport solution during specimen transfer to the cartridge. If any spills occur, decontamination should be performed as appropriate.

10.3 Laboratory Diagnosis: Analytical Issues

Traditional testing methodologies for diagnosis of respiratory viral infections, including cell culture, antigen- and antibody-based immunoassays, and nucleic acids tests
have all been applied to the detection of SARS-CoV-2, Carter et al. (2020). However,
diagnosis of acute infection with SARS-CoV-2 should be performed using nucleic
acid amplification tests (NAAT) with a sample collected from the upper respiratory
tract, National Institutes of Health (2021). Real-time reverse transcription-PCR (RT-
PCR) assays are preferred, with immunoassay methods being used as supplementary
tests and for epidemiological purposes and to detect past infection, Tang et al. (2020).
Table 10.1 shows the various laboratory diagnostic platforms available for SARS-
CoV-2 detection. These testing methods and platforms, as well as specific use cases,
are described in the following sections.

10.3.1 Non-molecular Methods

Viral isolation in cell culture is important for characterisation of SARS-CoV-2, to
recover isolates and strains, to identify neutralizing antibodies, and to support development of therapeutic agents and vaccines. However, cell culture for human coronaviruses is not routinely performed for diagnostic purposes and is not recommended
for diagnosis of SARS-CoV-2. In addition to biosafety concerns, viral culture generally has a long turnaround time, is labor-intensive, and requires specific expertise
to interpret the results, Loeffelholz and Tang (2020).

Viral antigen immunoassays are less sensitive than RT-PCR-based tests but
demonstrate similar high specificity. Antigen tests for SARS-CoV-2 perform best
early in the course of symptomatic infection when the viral load is at its highest
but there is concern that due to variability of viral loads in COVID-19 patients,
antigen detection may miss cases due to low virus burden or sampling variability,
Tang et al. (2020). The advantages of antigen-based tests are primarily low cost and
fast turnaround time. Antigen tests may be used for screening purposes, to exclude
SARS-CoV-2 infection in asymptomatic persons, or to determine whether a previously infected person is still contagious. Based on available data, the U.S. Centers for
Disease Control and Prevention (CDC) has developed an antigen testing algorithm
for these specific use cases, Centers for Disease Control and Prevention (2020e).

Serologic or antibody tests measure the host response to infection and can detect
recent or past infection and therefore is an indirect measure of infection that should be
used retrospectively, or in combination with other tests, Zhang et al. (2020). Several
serologic assays have received EUAs from the U.S. FDA for detection of antibodies
that bind to SARS-CoV-2 antigens, U.S. Food and Drug Administration (2021a).
These tests are primarily recommended for epidemiology and public health use, to
estimate the proportion of the population exposed to SARS-CoV-2, or clinically, as a
supplementary test for patients who are strongly suspected of having SARS-CoV-2

Table 10.1 Current lab diagnostic platforms: antigen, serology, culture, and different molecular methods

Method	Characteristics	Test time	Application	References
Antigen EIA	Rapid, poor sensitivity, some are CLIAwaived	<30 min	Diagnosis (detection)	Chen et al. (2016), Lau et al. (2004), Sastre et al. (2011)
Antigen IFA	Good sensitivity and specificity, subjective interpretation	1–4 h	Diagnosis (detection)	Liu et al. (2005), Sizun et al. (1998)
Cell culture	Gold standard, pure culture for further research and development, time consuming	1–7 days	Diagnosis (detection, differentiation, typing and characterization) and research	Hamre et al. (1967), Ksiazek et al. (2003), Tyrrell and Bynoe (1965), Zaki et al. (2012)
Serology	Retrospective, cross-reaction	2–8 h	Infection confirmation, epidemiology and research, vaccine evaluation	Chan et al. (2009), Shao et al. (2007), Peiris et al. (2003), Zhang et al. (2020)
NAAT, monoplex, pan-HCoV	High sensitivity with universal coverage of all species of HCoV	1–8 h	Diagnosis (detection), discovery and research	Zlateva et al. (2013), Kuypers et al. (2007), Woo et al. (2005)
NAAT, monoplex, specific-HCoV	High sensitivity and specificity for special species, potential quantification	1–8 h	Diagnosis (detection, differentiation, and limited typing) and research	Chan et al. (2015), Dare et al. (2007)
NAAT, multiplex	High sensitivity and specificity, covering other pathogens, FilmArray RP EZ is CLIA-waived	1–8 h	Diagnosis (detection, differentiation, and limited typing) and research	Babady et al. (2012, 2018), Gaunt et al. (2010), Tang et al. (2016)
NAAT, POCT	Rapid and safe, good sensitivity and specificity, some are CLIAwaived	15-30 min	Diagnosis (detection and limited differentiation) and research	Beal et al. (2020), Kozel and Burnham-Marusich (2017)

Modified from Loeffelholz and Tang (2020) and updated in March 2021
Abbreviations: *EIA* enzyme; *IFA* immunofluorescent assay; *NAAT* nucleic acid amplification test; *CLIA* Clinical Laboratory Improvement Act

infection but have tested negative by NAAT or antigen-based tests, Kucirka et al. (2020), Centers for Disease Control and Prevention (2020f). The rapid point-of-care immunoassays are generally lateral flow assays, but high-throughput automated versions are also available for population-level screening. Serologic assays may detect total antibody, IgM, IgG, IgA, or in various combinations although assays that detect IgG and total antibodies may have higher specificity to detect past infection, National Institutes of Health (2021).

Serologic tests could play a role in treatment for COVID-19 by helping to identify individuals who have developed an immune response to SARS-CoV-2 and may donate convalescent plasma. At the time of writing, it is not yet known if the presence of antibodies conveys an immunity to prevent or reduce the severity of re-infection, nor how long antibodies persist after infection, or the duration for which immunity lasts. As more vaccines become available and more individuals become vaccinated, serologic tests may be used to differentiate natural infection from vaccine-induced antibody responses to the SARS-CoV-2 spike protein antigen. Because the SARS-CoV-2 nucleocapsid protein is not part of the current vaccines, serologic tests that detect antibodies to the nucleocapsid protein can be used to distinguish natural infection from vaccine-induced antibody responses.

10.3.2 Molecular Methods

Commonly used molecular detection methods for SARS-CoV-2 include RT-PCR/real-time RT-PCR, next generation sequencing (NGS), isothermal amplification, Clustered Regularly Interspaced Short Palindromic Repeats (CRISPR), and some tests are available for use at the point-of-care (POC or POCT). As of March 5, 2021, there are 220 molecular diagnostic tests that have received EUA for detection of SARS-CoV-2 nucleic acid, U.S. Food and Drug Administration (2021b). Molecular platforms that are capable of automating sample extraction, amplification, and detection in a closed system with rapid turnaround time are particularly attractive as they are considered to be moderately complex and can be deployed in more laboratories than high complexity methods. Several of the current molecular methods are available for POCT and some are even authorised for home use. Table 10.2 lists the rapid, sample-to-answer molecular diagnostic tests that have received EUA for detection of SARS-CoV-2 nucleic acid as of March 5, 2021.

A real-time RT-PCR method is recommended for molecular testing, Tang et al. (2020), Loeffelholz and Tang (2020). A major advantage of real-time RT-PCR assays is that amplification and detection are done simultaneously in a closed system to minimise the risk of false-positive results associated with amplicon contamination. Several RT-PCR protocols targeting multiple genes/regions of SARS-CoV-2 were developed and quickly published which allowed laboratories to develop and validate their own tests until commercial assays became available, see Corman et al. (2020), Centers for Disease Control and Prevention (2020g, h). Corman et al. (2020) targeted the E gene as a screening test, followed by the RdRp gene as a confirmatory test, while

Table 10.2 Current lab diagnostic platforms: antigen, serology, culture, and different molecular methods

Assay	Specimen type[a]	Gene target(s)[a]	Technology	Limit of detection (LOD)[a]	Time to result (min)	Manufacturer	Notes
ID NOW COVID-19 assay	Nasal swab, nasopharyngeal swab, throat swab	RdRp	Isothermal amplification, based on NEAR	125 genome equivalents/ml	13	Abbott Diagnostics, Scarborough, ME, USA	POCT with CLIA Certificate of Waiver
Alinity m SARS-CoV-2	Nasal, nasopharyngeal and oropharyngeal swab, bronchoalveolar lavage	RdRp, N	Real-time RT-PCR, moderate complexity	100 copies/ml		Abbott Molecular, Des Plaines, IL, USA	Pooled samples (up to 5), high complexity only
BD SARS-CoV-2 Reagents for BD MAX System	Nasopharyngeal, nasal, midturbinate, and oropharyngeal swab, nasopharyngeal wash/aspirate or nasal aspirate	N1, N2	Real-time RT-PCR	640 GC/ml	120	Becton, Dickinson & Company, Sparks, MD, USA	
BioGX SARSCoV-2 Reagents for BD MAX System	Nasopharyngeal, nasal, midturbinate, and oropharyngeal swab, nasopharyngealwash/aspirate or nasal aspirate	N1, N2	Real-time RT-PCR	40 GE/ml	120	Becton, Dickinson & Company, Sparks, MD, USA	
BD SARS-CoV-2/Flu for BD MAX System	Nasopharyngeal and anterior nasal swab	N1, N2	Real-time RT-PCR	700 GTC/ml	120	Becton, Dickinson & Company, Sparks, MD, USA	
BioFire COVID-19 Test	Nasopharyngeal swab	ORF 1ab, ORF8	Real-time RT-PCR, moderate complexity	330 GC/ml	50	BioFire Defense, Salt Lake City, UT, USA	Pooled samples (up to 8), high complexity only
BioFire Respiratory panel 2.1	Nasopharyngeal swab	S, M	Real-time RT-PCR	160 copies/ml	45	BioFire Diagnostics, Salt Lake City, UT, USA	

(continued)

Table 10.2 (continued)

Assay	Specimen type	Gene target(s)[a]	Technology	Limit of detection (LOD)[a]	Time to result (min)	Manufacturer	Notes
BioFire Respiratory panel 2.1-EZ	Nasopharyngeal swab	S, M	Real-time RT-PCR	500 copies/ml	45	BioFire, Salt Lake City, UT, USA	POCT with CLIA Certificate of Waiver
Xpert Xpress SARS-CoV-2	Nasopharyngeal, oropharyngeal, nasal, midturbinate swab, nasal wash/aspirate	E, N2	Real-time RT-PCR	250 copies/ml (0.0200 PFU/ml)	45	Cepheid, Sunnyvale, CA, USA	POCT with CLIA Certificate of Waiver (nasopharyngeal, nasal, and midturbinate swab only)
Xpert Xpress SARS-CoV-2/Flu/RSV	Nasopharyngeal, oropharyngeal, nasal, midturbinate swab, nasal wash/aspirate	E, N2	Real-time RT-PCR	131 copies/ml	45	Cepheid, Sunnyvale, CA, USA	POCT with CLIA Certificate of Waiver (nasopharyngeal and nasal swab only)
Xpert Omni SARS-CoV-2	Nasopharyngeal, oropharyngeal, nasal, midturbinate swab, nasal wash/aspirate	E, N2	Real-time RT-PCR	400 copies/ml	45	Cepheid, Sunnyvale, CA, USA	
Cue COVID-19 Test	Nasal swab	N	Isothermal amplification	GR: 1300 cps/ml Infectious virus: 0.0100 PFU/mL	20	Cue Health Inc., San Diego, CA, USA	POCT with CLIA Certificate of Waiver; home and over the counter use
MobileDetect Bio BCC19 Test Kit	Nasopharyngeal, oropharyngeal, mid-turbinate, nasal swab	N, E	Isothermal amplification, RTLAMP	75,000 copies/ml	30	Detectachem Inc., Sugar Land, TX	
Simplexa™ COVID-19 Direct kit	Nasal swab, nasopharyngeal swab, nasal wash/aspirate, and bronchoalveolar lavage	ORF1a, S	Real-time RT-PCR	500 (NPS, NW/A); 242 (NS); 1208 (BAL) copies/ml	60	DiaSorin Molecular LLC, Cypress, CA, USA	

(continued)

Table 10.2 (continued)

Assay	Specimen type	Gene target(s)[a]	Technology	Limit of detection (LOD)[a]	Time to result (min)	Manufacturer	Notes
ePlex SARS-CoV-2 Test	Nasopharyngeal swab		RT-PCR	750 and 1000 copies/ml		GenMark Diagnostics, Inc., Carlsbad, CA, USA	
ePlex Respiratory Pathogen Panel 2	Nasopharyngeal swab		RT-PCR	250 copies/ml		GenMark Diagnostics, Inc., Carlsbad, CA, USA	
Lucira COVID-19 All-In-One Test Kit	Nasal swab	N	Isothermal amplification, RT-LAMP	900 copies/ml	30	Lucira Health, Inc., Emeryville, CA, USA	POCT with CLIA Certificate of Waiver; prescribed for home use
ARIES®	Nasopharyngeal swab	ORF1ab, N	Real-time RT-PCR	333 copies/ml	120	Luminex Corporation, Austin, TX, USA	
Accula SARSCov-2 Test	Nasal Swab	N	RT-PCR	200 GE/reaction	30	Mesa Biotech Inc., San Diego, CA, USA	POCT with CLIA Certificate of Waiver
NeuMoDx SARS-CoV-2 Assay	Nasal, nasopharyngeal, oropharyngeal swab, bronchoalveolar lavage, saliva (using saliva collection kit)	Nsp2, N	Real-time RT-PCR	150 copies/ml		NeuMoDx Molecular, Inc.	
QIAstat-Dx Respiratory SARS-CoV-2 Panel	Nasopharyngeal swab	E, ORF1b, RdRp	Real-time RT-PCR	500 copies/ml	60	Qiagen, Hilden, Germany	

(continued)

Table 10.2 (continued)

Assay	Specimen type	Gene target(s)[a]	Technology	Limit of detection (LOD)[a]	Time to result (min)	Manufacturer	Notes
Solana SARSCoV-2 Assay	Nasopharyngeal, nasal swab	pp1ab	Isothermal amplification, RTHDA	11,600 copies/ml		Quidel Corporation, San Diego, CA USA	
cobas SARSCoV-2 & Influenza A/B	Nasal, nasopharyngeal swab	ORF1ab, E	Real-time RT-PCR			Roche Molecular Systems, Inc., Pleasonton, CA, USA	
cobas SARSCoV-2 & Influenza A/B for cobas Liat	Nasal, nasopharyngeal swab	ORF1ab, N	Real-time RT-PCR	12 copies/ml		Roche Molecular Systems, Inc., Pleasonton, CA, USA	
cobas SARSCoV-2	Nasal, nasopharyngeal, oropharyngeal swab	ORF1ab, E	Real-time RT-PCR	0.009 TCID50/ml		Roche Molecular Systems, Inc., Pleasonton, CA, USA	Pooled samples (up to 6), high complexity only
T2SARS-CoV-2 Panel	Nasal, midturbinate, nasopharyngeal, oropharyngeal swab, bronchoalveolar lavage		RT-PCR	2000 GE/ml	<120	T2 Biosystems, Inc., Lexington, MA, USA	
Visby Medical COVID-19	Nasopharyngeal, nasal, mid-turbinate swabs	N	RT-PCR	1112 copies/ml	30	Visby Medical, Inc., San Jose, CA, USA	
Visby Medical COVID-19 Point of Care Test	Nasopharyngeal, anterior nasal, mid-turbinate swabs	N1	RT-PCR	435 copies/swab	30	Visby Medical, Inc., San Jose, CA, USA	POCT with CLIA Certificate of Waiver

[a]SARS-CoV-2 target(s) only. *GR* Genomic RNA; Modified from U.S. Food and Drug Administration (2021b)

the U.S. CDC assay targets two nucleocapsid targets, N1 and N3, Centers for Disease Control and Prevention (2020g, h). The University of Hong Kong used the N gene as the screening assay with an Orf1ab assay for confirmation, To et al. (2020), Chan et al. (2020), National Institute for Viral Disease Control and Prevention (2020).

Targeting multiple genes can help avoid cross-reactivity with endemic coronaviruses as well as mitigate false negative results due to genetic drift of SARS-CoV-2 and variation in emerging SARS-CoV-2 strains. Shortly thereafter, commercial manufacturers began the release of a multitude of real-time RT-PCR assays for SARS-CoV-2 which have received EUAs from the U.S. FDA. These tests are available in a variety of formats to accommodate POC testing, testing single or few specimens at a time, high-throughput fully automated systems that are typically performed in large reference laboratories, and everything in between. Furthermore, assays which can be performed in a closed device within an automated, sample-to-answer system are extremely useful in a pandemic setting as they require very minimal handling, which reduces the risk of exposure for staff and risk of incorrect results due to contamination or user error. As of March 5, 2021, more than a dozen rapid, sample-to-answer tests using real-time RT-PCR were available for emergency use (Table 10.2).

Metagenomic next-generation sequencing methods and random amplification deep-sequencing methods played a key role in the identification of SARS CoV-2, Chen et al. (2020), Zhou et al. (2020). Next-generation sequencing methods will continue to be important to identify mutations in SARS-CoV-2 and for epidemiological assessment of new variants but are generally not practical for diagnostics. In one technique, amplicon-based sequencing and metagenomics sequencing are both used to identify SARS-CoV-2 and to assess the background microbiome of infected individuals, Carter et al. (2020), Moore et al. (2020). This method allows for identification of SARS-CoV-2 and other pathogens that may be contributing to secondary infections in COVID-19 patients and has potential for contact tracing, epidemiology, and to check for mutations, sequence divergence, and viral evolution.

Isothermal nucleic acid amplification allows amplification at a constant temperature, eliminating the need for a thermal cycler and thus are well-suited for low-resource settings, field applications, and POCT. Isothermal detection techniques are rapid with minimal sample preparation requirements, the results are typically available in minutes, and tend to be highly sensitive, detecting down to hundreds or fewer copies/ml, Khan et al. (2020). Data generated by isothermal amplification can be provided in a variety of formats, such as fluorescence, change in pH, colorimeteric, or luminometric, making them easy to use and widely accessible in various locations. Several isothermal techniques have been applied to SARS-CoV-2 diagnostics with several assays commercially available for laboratory, POCT, and even home use.

One of the most well-known commercial isothermal SARS-CoV-2 assays is the ID NOW COVID-19 POCT assay (Abbott Diagnostics, Scarborough, ME, USA) which is based on NEAR (nicking enzyme amplification reaction) technology. NEAR uses two primers, a nicking enzyme, and a DNA polymerase to amplify short 20-30 nucleotide products 108- to 1010-fold in less than 10 min, Van Ness et al. (2003). The ID NOW COVID-19 test amplifies a target in the RdRp gene with a 5–13-min reaction time. Available reports on performance as compared to RT-PCR published in July and August 2020 have been mixed, with >90% positive agreement observed in some studies, Rhoads et al. (2020), but only ~55–74% in others, see Basu et al. (2020), Smithgall et al. (2020). This is possibly due to different sample types used (dry nasal swab vs. nasopharyngeal swab in viral transport media) and/or low viral loads in some samples.

RT-LAMP (loop-mediated isothermal amplification) technology uses a strand-displacing DNA polymerase with 4 to 6 primers in a single step amplification reaction to rapidly amplify target sequences with high specificity, Notomi et al. (2000), Wong et al. (2018). RT-LAMP has been developed for SARS-CoV-2 detection in multiple research applications using fluorescence and colorimetric detection methods and was shown to be able to detect SARS-CoV-2 nucleic acid in standards and in a variety of contrived respiratory specimens down to 3–1000 copies per reaction in about 30–40 min, see Khan et al. (2020), Lamb et al. (2020), Lu et al. (2020d), Yu et al. (2020). The Lucira COVID-19 All-In-One Test kit (Lucira Health, Emeryville, CA) is available for POCT and for at home testing (by prescription) and uses RT-LAMP with a colorimetric (pH change) readout to detect down to 900 copies/ml in 30 min, U.S. Food and Drug Administration (2021b).

Helicase-dependent amplification (HDA) is an isothermal amplification chemistry that relies on the complementary strand displacing ability of DNA helicase, Dunbar and Das (2019), Vincent et al. (2004). As with other isothermal amplification methods, HDA assays are rapid, sensitive, inexpensive, and can be performed without sophisticated instrumentation, Huang et al. (2013). The Solana SARS-CoV-2 assay uses RT-HDA targeting the pp1ab region for detection of SARS-CoV-2 RNA through fluorescent detection and demonstrated a limit of detection (LoD) of 11,600 copies/ml.

Clustered Regularly Interspaced Short Palindromic Repeats (CRISPR) are a family of nucleic acid sequences found in prokaryotes which are recognised and cleaved by a set of bacterial enzymes (e.g., Cas9, Cas12, and Cas13), Carter et al. (2020). Specific enzymes within the Cas12 Rusk (2019) and Cas13 Freije et al. (2019) families can be programmed to target and cut viral RNA sequences. CRISPR-based methods use the specificity of the amplification primers and the guide RNA reporter detection, enhancing specificity to the single nucleotide level, Khan et al. (2020). Two CRISPR-based assays have been developed which have received EUA for detection of SARS-CoV-2 RNA in respiratory specimens, U.S. Food and Drug Administration (2021b). Both assays are considered high complexity tests and require nucleic acid extraction. The Sherlock CRISPR SARS-CoV-2 assay (Sherlock Biosciences, Boston, MA, USA) uses RT-LAMP followed by Cas13 cleavage of an activated CRISPR complex within the amplified SARS-CoV-2 ORF1ab and N targets. Fluorophore-labelled reporter RNA sequences are released, and the fluorescence is measured on a plate reader. The test takes approximately 1 h (post-extraction) and has a reported LoD of 6750 copies/ml. The second assay, the SARS-CoV-2 DETECTR Reagent Kit (Mammoth Biosciences, San Francisco, CA, USA) relies on isothermal RT-LAMP amplification of a SARS-CoV-2 N gene target, followed by Cas12 cleavage of reporter RNA, resulting in a fluorescent readout. This assay has a reported LoD of 20,000 copies/ml and takes approximately 45 min (post-extraction) to complete.

10.3.3 Point-of-Care and Home Sample Collection and Testing

As of March 5, 2021, eleven of the molecular EUA tests are Clinical Laboratory Improvement Act (CLIA)-waived and available for use as POCTs and two of the isothermal amplification assays are available for at home testing either over the counter (Cue COVID-19 Test for Home and Over The Counter (OTC) Use) or by prescription (Lucira COVID-19 All-In-One Test Kit). Additionally, dozens of devices for home collection of nasal swab or saliva specimens are available for SARS-CoV-2 molecular testing, U.S. Food and Drug Administration (2021b). POCTs, specimen self-collection, and at-home testing helps alleviate the burden on the healthcare system and laboratory to accommodate the demand for testing under constrained resources. Self-collection of specimens also eliminates direct interaction between healthcare workers and patients, which may reduce the risk of exposure and spread.

10.3.4 Assay Selection

Which diagnostic test or tests that should be implemented for SARS-CoV-2 detection depends on a variety of factors, including the size of the population being served and the prevalence of COVID-19 within that population, the capability of the testing laboratory (i.e., high or moderate complexity testing), the required turnaround time for results, and the availability of resources (reagents, consumables, and labor). Having timely and accurate SARS-CoV-2 test results is crucial for reducing COVID-19 transmission and reducing the associated public health, economic, and social effects, World Health Organization (2020a). Multiple strategies may be required to meet the demand for testing and deliver results in a timely manner, such as a combination of high-throughput tests for high sample volumes and rapid sample-to-answer systems to use between runs as more samples arrive. Supply chain issues with commercial manufacturers may require implementation of multiple testing platforms to mitigate backorder issues and have a sufficient supply of tests. Because availability of COVID-19 diagnostic testing may be limited, the Infectious Diseases Society of America (IDSA) developed a four-tiered testing algorithm to help clinicians prioritise COVID-19 diagnostic testing, Infectious Diseases Society of America (2020). These recommendations are summarised in Table 10.3.

During the early phase of the COVID-19 pandemic, the immediate need for diagnostic tests led to the rapid design and development of hundreds of molecular and immunoassay tests by many test manufacturers, Vandenberg et al. (2021). Most, but not all, of these tests are standalone SARS-CoV-2 diagnostic assays targeting one or more genes or proteins of the virus. As we move past one year since the virus was first identified, and anticipate the need for long-term, routine testing, many manufacturers have or are in the process of incorporating the SARS-CoV-2 target into multi-analyte respiratory panels. This may better accommodate seasonal testing

Table 10.3 IDSA four-tiered approach to COVID-19 diagnostic testing

Tier level	Population
1	• Critically ill patients receiving ICU level care with unexplained viral pneumoniae or respiratory failure, regardless of travel history or close contact with suspected or confirmed COVID-1 patients
	• Any person, including health care workers, with fever or signs/symptoms of a lower respiratory tract illness and close contact with a laboratory-confirmed COVID-19 patient within 14 days of symptom onset (including all residents at a long-term care facility that has a laboratory-confirmed COVID-19 case)
	• Any person, including health care workers, with fever or signs/symptoms of a lower respiratory tract illness and a history of travel within 14 days of symptom onset to geographical regions where sustained community transmission has been identified; (iv) Individuals with fever or signs/symptoms of a lower respiratory tract illness who are also immunosuppressed (including patients with HIV), elderly, or have underlying chronic health conditions
	• Individuals with fever or signs/symptoms of a lower respiratory tract illness who are critical to pandemic response, including health care workers, public health officials and other essential leaders
2	• Hospitalized (non-ICU) patients and long-term care residents with unexplained fever and signs/symptoms of a lower respiratory tract illness. The number of confirmed COVID-19 cases in the community should be considered. As testing becomes more widely available, routine testing of hospitalized patients may be important for infection prevention and management of discharge
3	• Patients in outpatient settings who meet the criteria for influenza testing. This includes individuals with co-morbid conditions including diabetes, COPD, congestive heart failure, age >50, immunocompromised hosts among others. Given limited available data, testing of pregnant women and symptomatic children with similar risk factors for complications is encouraged. The number of confirmed COVID-19 cases in the community should be considered
4	• For community surveillance as directed by public health and/or infectious diseases authorities

Modified from Lu et al. (2020b) and updated in March 2021

when there is overlap with other pathogens, such as influenza and RSV, and identify SARS-CoV-2 as the causative agent in future waves or regional outbreaks.

10.3.5 Pooled Screen Testing

A pooled specimen testing strategy to expand diagnostic or screening testing capacity can be useful to preserve resources in settings where the prevalence is low, Centers for Disease Control and Prevention (2020a). A 10:1 pooled test strategy on-site at an airport of China was pursued, resulting in increased test throughput, limited use of reagents, and increased testing efficiency without loss of sensitivity. This testing approach has the potential to reduce the need for contact tracing when the results are delivered first time, Li et al. (2020a). When a pooled test result is negative, all specimens in the pool can be presumed negative and further testing is not required. But, when the pool test result is positive or indeterminant, all specimens in the pool must be retested individually. This strategy can help preserve testing resources, reduce the time to result, and lower the overall testing cost. Three of the EUA SARS-CoV-2 molecular tests have been approved for pooled sample testing (5–8 specimens/pool) in high complexity laboratories (Table 10.2). Pooled testing is discussed in detail in the following chapter, "Pooled testing in the COVID-19 Pandemic" by Matthew Aldridge and David Ellis.

10.3.6   Viral Load Testing

Several studies have shown a correlation between SARS-CoV-2 viral load, days from symptom onset, and COVID-19 disease severity, suggesting that viral load might be used for risk stratification of COVID-19 patients, Cevik et al. (2021), Pan et al. (2020), To et al. (2020), Fajnzylber et al. (2020). Real-time RT-PCR cycle threshold (Ct) values represent the number of amplification cycles required for the target amplicon to exceed a threshold level and are thus inversely related to viral load. While Ct values can provide an indirect measure of the viral load in the sample, it is also influenced by the amplification efficiency of the specific assay, the quality of the specimen, and the sample matrix, Bustin and Mueller (2005). Ct cutoff values are established by the test manufacturer during validation and are then verified by the implementing laboratory for their specific laboratory setting. The current real-time RT-PCR assays for SARS-CoV-2 are qualitative but some have advocated for the reporting of the Ct and reference ranges (low, medium, high) with the result, Tom and Mina (2020), and some states in the U.S. are requiring laboratories to report the Ct values, Florida Department of Health (2020). However, it's unclear how Ct values should be applied in clinical settings with no standardisation across platforms, nor clinical studies validating use of Ct to guide management of COVID-19 cases, American Association for Clinical Chemistry (2020).

10.4   Laboratory Diagnosis: Post-analytical Issues

In the post-analytical stage, test results should be carefully interpreted using both molecular and serological findings. Test result interpretation is summarised in Table 10.4.

Table 10.4 General molecular and serology test result interpretation in COVID-19

RNA	IgM	IgG	Interpretation
+	–	–	Patient in the two-week period prior to immune response
+	+	–	Patient in early infection
+	–	+	Patient in mid to late infection; confirmation if IgG titer in convalescence is 4 times higher than acute phase
+	+	+	Patient in active infection with decent immune response
–	+	–	Patient has active infection with a false-negative RNA assay
–	–	+	Patient with previous infection; virus has been cleared
–	+	+	Patient with recent infection and in convalescence; virus has been cleared; active infection with false-negative RNA assay

Modified from Lu et al. (2020b) and updated in March 2021

10.5   Interpretation of Serology Results

As previously mentioned, many serologic assays have received emergency authorisation for the detection of antibodies (IgM, IgG, Total) produced during SARS-CoV-2 infection, U.S. Food and Drug Administration (2021b). These tests are available in a variety of formats, from simple lateral flow immunoassay to ELISA, chemiluminescent immunoassays, and even T-cell receptor beta NGS to detect an adaptive T-cell immune response to SARS-CoV-2. Serological test results, in combination with SARS-CoV-2 RNA test results, can be helpful in determining at what stage in the course of infection the patient is in. For example, a positive RNA result with no antibody present would indicate an early infection, prior to development of an immune response. Positive RNA with positive IgM, IgG, or both can differentiate an early, mid to late, or active infection with a high immune response, respectively. Negative RNA with positive antibody can indicate a recent or previous infection, or a false negative RNA test result.

Several studies have shown that most individuals produce antibodies by day 5-8 of symptom onset and that an immune response can also be measured in asymptomatic individuals with positive or negative RNA test results, Zhang et al. (2020), Hung et al. (2020). Therefore, serological tests can be valuable for confirming the diagnosis of COVID-19 and will play an important role in the epidemiology of COVID-19 and determining the immune status of asymptomatic patients, but are unlikely to be useful for screening or diagnosis of early infections, Tang et al. (2020), Zhang et al. (2020). However, a combination of RT-PCR and serology could be implemented for case finding and contact tracing to expedite early diagnosis, isolation for infection control, and treatment, Hung et al. (2020). As more individuals are vaccinated for COVID-19, serological assays that measure antibody responses to multiple antigens, such as S/RBD and N, will be useful for differentiating past infection from vaccine response.

10.5.1 Interpretation of Molecular Results

Molecular assays for detection of SARS-CoV-2 RNA are the most used and most reliable test for COVID-19 diagnosis. Detection is unlikely early in infection, before symptom onset, but likely detected in the first 4 weeks after symptom onset, Sethuraman et al. (2020). However, viral RNA but may persist longer in lower respiratory tract samples and may continue to be shed in stool. It is important to recognise that a positive RNA test result does not indicate viability or infectivity of SARS-CoV-2. False-negative test results may occur primarily due to inappropriate timing of sample collection and/or inadequate sampling technique. False-positive results are typically due to technical errors and or contamination. Most of the molecular assays have a specificity of 100% because the primer design is specific to the SARS-CoV-2 genome sequence but should be monitored by in silico analysis as new variants are identified and sequenced.

A systematic review of the literature found that 89% of nasopharyngeal samples were RT-PCR positive at 0–4 days post-symptom onset while 81% were positive at 0–4 days post-hospitalisation, but dropped to 54% at 10–14 days post-symptoms and 45% at 10–14 days post-admission, Mallett et al. (2020). Intermittent false negative results occurred when the level of virus is close to the limit of detection of the assay. Several studies have investigated the relationship between the viral load and pathogenesis, disease progression, and mortality, Fajnzylber et al. (2020), Pujadas et al. (2020). Some published reports support the conversion of qualitative RT-PCR testing to quantitative viral load measurements, to assist with early risk stratification in COVID-19; however, more work is needed to assess the correlation of viral load with other disease biomarkers and clinical features to possibly develop algorithms to predict infectivity and risk.

10.5.2 Tests Beyond Detection and Diagnosis

In addition to disease diagnosis, laboratory testing and the measurement of appropriate biomarkers play a critical role in managing patients with COVID-19. Multiplicity of pathologic features can be used to characterise severe disease in patients with COVID-19. These include the cytokine release syndrome, downregulation of adaptive cellular immunity, increased thrombotic risk, lung and acute kidney dysfunction, and cardiomyocyte injury. Several types of biomarkers have been described for monitoring progression, prognostication, prediction of treatment response, and risk stratification, Weidmann et al. (2021). One good example is to use biomarkers to identify patients who potentially respond to a particular therapy, such as IL-6 for tocilizumab, Harwood et al. (2021).

Improving patient outcomes will require earlier detection of these issues, targeted treatments, and appropriate triage of patients, particularly those who are susceptible to the most severe course of this disease. Monitoring patients with resolution of

COVID-19 pneumonia may also be important in terms of when they should be discharged from the hospital. Two consecutive negative RT-PCR tests to cease self-quarantine/return to work has been suggested, but this has not been recommended by the U.S. CDC, Tang et al. (2020). Random-access, integrated devices available at the point of care with scalable capacities will facilitate the rapid diagnosis and monitoring of SARS-CoV-2 infection status and control the spread of the virus as test of infectivity/isolation, Lu et al. (2020b).

10.6 Concluding Remarks

Researchers are focusing on developing rapid and efficient methods for laboratory diagnosis and monitoring of SARS-CoV2 infections. In the preanalytical stage, collecting the proper respiratory tract specimen at the right time from the right anatomic site is essential for a prompt and accurate molecular diagnosis of COVID-19. Appropriate measures are required to keep laboratory staff safe while producing reliable test results. In the analytic stage, while real-time RT-PCR assays remain the test of choice for the etiologic diagnosis of SARS-CoV-2 infection, several new platforms using isothermal amplification and CRISPR detection are gradually becoming available. In the postanalytical stage, testing results should be carefully interpreted using both molecular and serological findings. Finally, in addition to disease diagnosis, laboratory testing and the measurement of appropriate biomarkers play a critical role in managing patients with COVID-19, including monitoring disease progression, prognostication, prediction of treatment response, and risk stratification.

In preparation for future pandemics, the rapid identification, isolation, and genomic sequencing of the causative pathogen is critical for fast development and implementation of diagnostic tests in the clinical laboratory, as well as for the development of vaccines, particularly recombinant and nucleic acid-based vaccines. In addition to diagnostic tests, tests for epidemiology and surveillance and immune monitoring are critically important for following the progression of a pandemic in real-time and determining the efficacy of public health measures to prevent the spread and bring the pandemic under control. Understanding the immune response and the variation in immune response to the pathogen is also vital to help develop appropriate treatments and therapeutic agents to alleviate the symptoms, pathology, and long-term effects of the disease.

References

M. Alsved et al., Exhaled respiratory particles during singing and talking. Aerosol Sci. Technol. **54**(11), 1245–1248 (2020)

American Association for Clinical Chemistry, SARS-CoV-2 cycle threshold: a metric that matters (or not). Clinical Laboratory News, December 3, 2020 (2020). https://www.aacc.org/cln/cln-stat/2020/december/3/sars-cov-2-cycle-threshold-a-metric-that-matters-or-not. Accessed 9 March 2021

N. Babady et al., Comparison of the Luminex xTAG RVP Fast assay and the Idaho Technology FilmArray RP assay for detection of respiratory viruses in pediatric patients at a cancer hospital. J. Clin. Microbiol. **50**(7), 2282–2288 (2012)

N. Babady et al., Multicenter evaluation of the ePlex respiratory pathogen panel for the detection of viral and bacterial respiratory tract pathogens in nasopharyngeal swabs. J. Clin. Microbiol. **56**(2) (2018)

A. Basu et al., Performance of Abbott ID Now COVID-19 rapid nucleic acid amplification test using nasopharyngeal swabs transported in viral transport media and dry nasal swabs in a New York City Academic Institution. J. Clin. Microbiol. **58**(8), e01136-20 (2020)

S. Beal et al., Performance and impact of a CLIA-waived, point-of-care respiratory PCR panel in a pediatric clinic. Pediatr. Infect. Dis. J. **39**(3), 188–191 (2020)

J. Beigel et al., Remdesivir for the treatment of Covid-19—final report. N. Engl. J. Med. **383**(19), 1813–1826 (2020)

J. Blow et al., Virus inactivation by nucleic acid extraction reagents. J. Virol. Methods **119**(2), 195–198 (2004)

J. Bullard et al., Predicting infectious SARS-CoV-2 from diagnostic samples. Clin. Infect. Dis. (2020)

S. Bustin, R. Mueller, Real-time reverse transcription PCR (qRT-PCR) and its potential use in clinical diagnosis. Clin. Sci. (Lond.) **109**(4), 365–379 (2005)

J. Cai et al., Indirect virus transmission in cluster of COVID-19 cases, Wenzhou, China, 2020. Emerg. Infect. Dis. **26**(6), 1343–1345 (2020)

L. Carter et al., Assay techniques and test development for COVID-19 diagnosis. ACS Cent. Sci. **6**(5), 591–605 (2020)

Centers for Disease Control and Prevention, interim guidance for use of pooling procedures in SARS-CoV-2 diagnostic, screening, and surveillance testing (2020a), https://www.cdc.gov/coronavirus/2019-ncov/lab/pooling-procedures.html. Accessed 9 March 2021

Centers for Disease Control and Prevention, Covid-19: how to protect yourself & others (2020b), https://www.cdc.gov/coronavirus/2019-ncov/prevent-getting-sick/prevention.html. Accessed 4 March 2021

Centers for Disease Control and Prevention, Interim guidelines for collecting, handling, and testing clinical specimens from persons for coronavirus disease 2019 (COVID-19) (2020c), https://www.cdc.gov/coronavirus/2019-ncov/lab/guidelines-clinical-specimens.html. Accessed 1 March 2021

Centers for Disease Control and Prevention, Interim laboratory biosafety guidelines for handling and processing specimens associated with coronavirus disease 2019 (covid-19) (2020d), https://www.cdc.gov/coronavirus/2019-nCoV/lab/lab-biosafety-guidelines.html. Accessed 5 March 2021

Centers for Disease Control and Prevention, Interim guidance for antigen testing for SARS-CoV-2 (2020e), https://www.cdc.gov/coronavirus/2019-ncov/lab/resources/antigen-tests-guidelines.html. Accessed 7 March 2021

Centers for Disease Control and Prevention, Interim guidelines for COVID-19 antibody testing (2020f), https://www.cdc.gov/coronavirus/2019-ncov/lab/resources/antibody-tests-guidelines.html. Accessed 7 March 2021

Centers for Disease Control and Prevention, Research use only 2019-novel coronavirus (2019-nCoV) real-time RT-PCR primers and probes (2020g), https://www.cdc.gov/coronavirus/2019-ncov/lab/rt-pcr-panel-primer-probes.html. Accessed 8 March 2021

Centers for Disease Control and Prevention, DC 2019-nCoV real-time RT-PCR diagnostic panel instructions for use (2020h), https://www.fda.gov/media/134922/download. Accessed 8 March 2021

Centers for Disease Control and Prevention, Interim infection prevention and control recommendations for healthcare personnel during the coronavirus disease 2019 (covid-19) pandemic (2021), https://www.cdc.gov/coronavirus/2019-ncov/hcp/infection-control-recommendations.html. Accessed 4 March 2021

M. Cevik et al., Virology, transmission, and pathogenesis of SARS-CoV-2. BMJ **371**, 3862 (2020)

M. Cevik et al., SARS-CoV-2, SARS-CoV, and MERS-CoV viral load dynamics, duration of viral shedding, and infectiousness: a systematic review and meta-analysis. Lancet Microbe **2**(1), e13–e22 (2021)

C. Chan et al., Examination of seroprevalence of coronavirus HKU1 infection with S protein-based ELISA and neutralization assay against viral spike pseudotyped virus. J. Clin. Virol. **45**(1), 54–60 (2009)

J. Chan et al., Development and evaluation of novel real-time reverse transcription-PCR assays with locked nucleic acid probes targeting leader sequences of human-pathogenic coronaviruses. J. Clin. Microbiol. **53**(8), 2722–2726 (2015)

J. Chan et al., Genomic characterization of the 2019 novel human-pathogenic coronavirus isolated from a patient with atypical pneumonia after visiting Wuhan. Emerg. Microbes Infect. **9**(1), 221–236 (2020)

L. Chen et al., RNA based mNGS approach identifies a novel human coronavirus from two individual pneumonia cases in 2019 Wuhan outbreak. Emerg. Microbes Infect. **9**(1), 313–319 (2020)

P. Chen et al., SARS-CoV-2 neutralizing antibody LY-CoV555 in outpatients with Covid-19. N. Engl. J. Med. **384**(3), 229–237 (2021)

Y. Chen et al., A highly specific rapid antigen detection assay for on-site diagnosis of MERS. J. Infect. **73**(1), 82–84 (2016)

D. Chu et al., Molecular diagnosis of a novel coronavirus (2019-nCoV) causing an outbreak of pneumonia. Clin. Chem. **66**(4), 549–555 (2020)

V. Corman et al., Detection of 2019 novel coronavirus (2019-nCoV) by real-time RT-PCR. Euro Surveill. **25**(3), 2000045 (2020)

R. Dare et al., Human coronavirus infections in rural Thailand: a comprehensive study using real-time reverse-transcription polymerase chain reaction assays. J. Infect. Dis. **196**(9), 1321–1328 (2007)

S. Dunbar, S. Das, Amplification chemistries in clinical virology. J. Clin. Virol. **115**, 18–31 (2019)

J. Fajnzylber et al., SARS-CoV-2 viral load is associated with increased disease severity and mortality. Nat. Commun. **11**(1), 5493 (2020)

Florida Department of Health, Mandatory reporting of COVID-19 laboratory test results: reporting of cycle threshold values (2020). 9 March 2021

C. Freije et al., Programmable inhibition and detection of RNA viruses using Cas13. Mol. Cell **76**(5), 826–837.e11 (2019)

E. Gaunt et al., Epidemiology and clinical presentations of the four human coronaviruses 229E, HKU1, NL63, and OC43 detected over 3 years using a novel multiplex real-time PCR method. J. Clin. Microbiol. **48**(8), 2940–2947 (2010)

W. Guan et al., Clinical characteristics of coronavirus disease 2019 in China. N. Engl. J. Med. **382**(18), 1708–1720 (2020)

D. Hamre, D. Kindig, J. Mann, Growth and intracellular development of a new respiratory virus. J. Virol. **1**(4), 810–816 (1967)

R. Harwood et al., A national consensus management pathway for pediatric inflammatory multisystem syndrome temporally associated with COVID-19 (PIMS-TS): results of a national Delphi process. Lancet Child Adolesc. Health **5**(2), 133–141 (2021)

P. Horby et al., Dexamethasone in hospitalized patients with Covid-19. N. Engl. J. Med. **384**(8), 693–704 (2021)

S. Huang et al., Low cost extraction and isothermal amplification of DNA for infectious diarrhea diagnosis. PLoS ONE **8**(3), e60059 (2013)

K. Hui et al., Tropism, replication competence, and innate immune responses of the coronavirus SARS-CoV-2 in human respiratory tract and conjunctiva: an analysis in ex-vivo and in-vitro cultures. Lancet Respir. Med. **8**(7), 687–695 (2020)

I. Hung et al., SARS-CoV-2 shedding and seroconversion among passengers quarantined after disembarking a cruise ship: a case series. Lancet Infect. Dis. **20**(9), 1051–1060 (2020)

Infectious Diseases Society of America, COVID-19 prioritization of diagnostic testing (2020), https://www.idsociety.org/globalassets/idsa/public-health/covid-19-prioritization-of-dx-testing.pdf. Accessed 17 March 2021

B. Ju et al., Human neutralizing antibodies elicited by SARS-CoV-2 infection. Nature **584**(7819), 115–119 (2020)

P. Khan, L. Aufdembrink, A. Engelhart, Isothermal SARS-CoV-2 diagnostics: tools for enabling distributed pandemic testing as a means of supporting safe reopenings. ACS Synth. Biol. **9**(11), 2861–2880 (2020)

C. Kim et al., Comparison of nasopharyngeal and oropharyngeal swabs for the diagnosis of eight respiratory viruses by real-time reverse transcription-PCR assays. PLoS ONE **6**(6), e21610 (2011)

T. Kozel, A. Burnham-Marusich, Point-of-care testing for infectious diseases: past, present, and future. J. Clin. Microbiol. **55**(8), 2313–2320 (2017)

T. Ksiazek et al., A novel coronavirus associated with severe acute respiratory syndrome. N. Engl. J. Med. **348**(20), 1953–1966 (2003)

L. Kucirka et al., Variation in false-negative rate of reverse transcriptase polymerase chain reaction-based SARS-CoV-2 tests by time since exposure. Ann. Intern. Med. **173**(4), 262–267 (2020)

M. Kumar et al., Inactivation and safety testing of Middle East Respiratory Syndrome Coronavirus. J. Virol. Methods **223**, 13–18 (2015)

J. Kuypers et al., Clinical disease in children associated with newly described coronavirus subtypes. Arch. Virol. **119**(1), e70-76 (2007)

L. Lamb et al., Rapid detection of novel coronavirus (COVID-19) by reverse transcription-loop-mediated isothermal amplification (2020), https://www.medrxiv.org/content/10.1101/2020.02.19.20025155v1

S. Lau et al., Detection of severe acute respiratory syndrome (SARS) coronavirus nucleocapsid protein in SARS patients by enzyme-linked immunosorbent assay. J. Clin. Microbiol. **42**(7), 2884–2889 (2004)

H. Li et al., Real-time screening of specimen pools for coronavirus disease 2019 (COVID-19) infection at Sanya Airport, Hainan Island, China. Clin. Infect. Dis. (2020a)

Y. Li et al., Role of ventilation in airborne transmission of infectious agents in the built environment—a multidisciplinary systematic review. Indoor Air **17**(1), 2–18 (2007)

Y. Li et al., Evidence for probable aerosol transmission of SARS-CoV-2 in a poorly ventilated restaurant (2020b), https://www.medrxiv.org/content/10.1101/2020.04.16.20067728v1

L. Ling et al., Parallel validation of three molecular devices for simultaneous detection and identification of influenza A and B and respiratory syncytial viruses. J. Clin. Microbiol. **56**(3)(2018)

I. Liu et al., Immunofluorescence assay for detection of the nucleocapsid antigen of the severe acute respiratory syndrome (SARS)-associated coronavirus in cells derived from throat wash samples of patients with SARS. J. Clin. Microbiol. **43**(5), 2444–2448 (2005)

M. Loeffelholz, Y.-W. Tang, Laboratory diagnosis of emerging human coronavirus infections—the state of the art. Emerg. Microbes Infect. **9**(1), 747–756 (2020)

H. Lu, C. Stratton, Y. Tang, Outbreak of pneumonia of unknown etiology in Wuhan, China: the mystery and the miracle. J. Med. Virol. **92**(4), 401–402 (2020a)

H. Lu, C. Stratton, Y. Tang, An evolving approach to the laboratory assessment of COVID-19. J. Med. Virol. **92**(10), 1812–1817 (2020b)

J. Lu, Z. Yang, COVID-19 outbreak associated with air conditioning in restaurant, Guangzhou, China, 2020. Emerg. Infect. Dis. **26**(11), 2791–2793 (2020)

R. Lu et al., Genomic characterisation and epidemiology of 2019 novel coronavirus: implications for virus origins and receptor binding. Lancet **395**(10224), 565–574 (2020c)

R. Lu et al., Development of a novel reverse transcription loop-mediated isothermal amplification method for rapid detection of SARS-CoV-2. Virol. Sin. **35**(3), 344–347 (2020d)

S. Mallett et al., At what times during infection is SARS-CoV-2 detectable and no longer detectable using RT-PCR-based tests? A systematic review of individual participant data. BMC Med. **18**(1), 346 (2020)

S. Moore et al., Amplicon based MinION sequencing of SARS-CoV-2 and metagenomic characterisation of nasopharyngeal swabs from patients with COVID-19 (2020), https://www.medrxiv.org/content/10.1101/2020.03.05.20032011v1

National Institute for Viral Disease Control and Prevention, Specific primers and probes for detection 2019 novel coronavirus (2020), https://ivdc.chinacdc.cn/kyjz/202001/t20200121_211337.html. Accessed 8 March 2021

National Institutes of Health, NIH COVID-19 treatment guidelines. Testing for SARS-CoV-2 infection (2021), https://www.covid19treatmentguidelines.nih.gov/overview/sars-cov-2-testing/. Accessed 1 March 2021

S. Nie et al., Evaluation of Alere i Influenza A& B for rapid detection of influenza viruses A and B. J. Clin. Microbiol. **52**(9) (2014)

T. Notomi et al., Loop-mediated isothermal amplification of DNA. Nucleic Acids Res. **28**(12), e63 (2000)

Y. Pan et al., Viral load of SARS-CoV-2 in clinical samples. Lancet Infect. Dis. **20**(4), 411–412 (2020)

J. Peiris et al., Coronavirus as a possible cause of severe acute respiratory syndrome. Lancet **361**(9366), 1319–1325 (2003)

E. Pujadas et al., SARS-CoV-2 viral load predicts COVID-19 mortality. Lancet Respir. Med. **8**(9), e70 (2020)

D. Rhoads et al., Comparison of Abbott ID Now, DiaSorin Simplexa, and CDC FDA emergency use authorization methods for the detection of SARS-CoV-2 from nasopharyngeal and nasal swabs from individuals diagnosed with COVID-19. J. Clin. Microbiol. **58**(8), e00760-20 (2020)

N. Rusk, Spotlight on Cas12. Nat. Methods **16**(3), 215 (2019)

J. Sanders et al., Pharmacologic treatments for coronavirus disease 2019 (COVID-19): a review. JAMA **323**(18), 1824–1836 (2020)

P. Sastre et al., Differentiation between human coronaviruses NL63 and 229E using a novel double-antibody sandwich enzyme-linked immunosorbent assay based on specific monoclonal antibodies. Clin. Vaccine Immunol. **18**(1), 113–118 (2011)

N. Sethuraman, S. Jeremiah, A. Ryo, Interpreting diagnostic tests for SARS-CoV-2. JAMA **323**(22), 2249–2251 (2020)

X. Shao et al., Seroepidemiology of group I human coronaviruses in children. J. Clin. Virol. **40**(3), 207–213 (2007)

J. Sizun, N. Arbour, P. Talbot, Comparison of immunofluorescence with monoclonal antibodies and RT-PCR for the detection of human coronaviruses 229E and OC43 in cell culture. J. Virol. Methods **72**(2), 145–152 (1998)

M. Smithgall et al., Comparison of Cepheid Xpert Xpress and Abbott ID Now to Roche cobas for the rapid detection of SARS-CoV-2. J. Clin. Virol. **128**, 104428 (2020)

E. Stokes et al., Coronavirus disease 2019 case surveillance—United States, January 22–May 30, 2020. MMWR Morb. Mortal. Wkly. Rep. **2020**(69), 759–765 (2020)

Y. Tang et al., Clinical evaluation of the Luminex NxTAG respiratory pathogen panel. J. Clin. Microbiol. **54**(7), 1912–1914 (2016)

Y. Tang et al., Laboratory diagnosis of COVID-19: current issues and challenges. J. Clin. Microbiol. **58**(6) (2020)

K. To et al., Consistent detection of 2019 novel coronavirus in Saliva. Clin. Infect. Dis. **71**(15), 841–843 (2020)

M. Tom, M. Mina, To interpret the SARS-CoV-2 test, consider the cycle threshold value. Clin Infect Dis **71**(16), 2252–2254 (2020)

D. Tyrrell, M. Bynoe, Cultivation of a novel type of common-cold virus in organ cultures. Br. Med. J. **1**(5448), 1467–1470 (1965)

U.S. Food and Drug Administration, EUA authorized serology test performance (2021a), https://www.fda.gov/medical-devices/coronavirus-disease-2019-covid-19-emergency-use-authorizations-medical-devices/eua-authorized-serology-test-performance. Accessed 7 March 2021

U.S. Food and Drug Administration, In vitro diagnostics EUAs (2021b), https://www.fda.gov/medical-devices/coronavirus-disease-2019-covid-19-emergency-use-authorizations-medical-devices/vitro-diagnostics-euas#imft3. Accessed 7 March 2021

J. Van Ness, L. Van Ness, D. Galas, Isothermal reactions for the amplification of oligonucleotides. Proc. Natl. Acad. Sci. **100**(8), 4504–4509 (2003)

O. Vandenberg et al., Considerations for diagnostic COVID-19 tests. Nat. Rev. Microbiol. **19**(3), 171–183 (2021)

Z. Varga et al., Endothelial cell infection and endotheliitis in COVID-19. Lancet **395**(10234), 1417–1418 (2020)

M. Vincent, Y. Xu, H. Kong, Helicase-dependent isothermal DNA amplification. EMBO Rep. **5**(8), 795–800 (2004)

C. Wang et al., A human monoclonal antibody blocking SARS-CoV-2 infection. Nat. Commun. **11**(1), 2251 (2020)

H. Wang, J. Deng, Y.-W. Tang, Profile of the Alere i Influenza A & B assay: a pioneering molecular point-of-care test. Expert Rev. Mol. Diagn. **18**(5)(2018)

M. Weidmann, K. Ofori, A. Rai, Laboratory biomarkers in the management of patients with COVID-19. Am. J. Clin. Pathol. **155**(3), 333–342 (2021)

D. Weinreich et al., REGN-COV2, a neutralizing antibody cocktail, in outpatients with Covid-19. N. Engl. J. Med. **384**(3), 238–251 (2021)

S. Welch et al., Analysis of inactivation of SARS-CoV-2 by specimen transport media, nucleic acid extraction reagents, detergents, and fixatives. J. Clin. Microbiol. **58**(11) (2020)

M. Wong et al., Detection of SARS-CoV-2 RNA in fecal specimens of patients with confirmed COVID-19: a meta-analysis. J. Infect. **81**(2), e31–e38 (2020)

Y. Wong et al., Loop-mediated isothermal amplification (LAMP): a versatile technique for detection of micro-organisms. J. Appl. Microbiol. **124**(3), 626–643 (2018)

P. Woo et al., Characterization and complete genome sequence of a novel coronavirus, coronavirus HKU1, from patients with pneumonia. J. Virol. **79**(2), 884–895 (2005)

World Health Organization, Laboratory testing strategy recommendations for COVID-19 (2020a), https://apps.who.int/iris/bitstream/handle/10665/331509/WHO-COVID-19-lab_testing-2020.1-eng.pdf. Accessed 21 March 2021

World Health Organization, COVID-19 vaccines (2020b), https://www.who.int/emergencies/diseases/novel-coronavirus-2019/covid-19-vaccines. Accessed 1 March 2021

World Health Organization, Coronavirus disease (COVID-2019) situation reports, 2020 (2021), https://www.who.int/emergencies/diseases/novel-coronavirus-2019/situation-reports/. Accessed 4 March 2021

C. Wu et al., Risk factors associated with acute respiratory distress syndrome and death in patients with coronavirus disease 2019 pneumonia in Wuhan, China. JAMA Intern. Med. **180**(7), 934–943 (2020)

A. Wyllie et al., Saliva or nasopharyngeal swab specimens for detection of SARS-CoV-2. N. Engl. J. Med. **383**(13), 1283–1286 (2020)

R. Wölfel et al., Virological assessment of hospitalized patients with COVID-2019. Nature **581**(7809), 465–469 (2020)

R. Yan et al., Structural basis for the recognition of SARS-CoV-2 by full-length human ACE2. Lancet **367**(6485), 1444–1448 (2020)

L. Yu et al., Rapid detection of COVID-19 coronavirus using a reverse transcriptional loop-mediated isothermal amplification (RT-LAMP) diagnostic platform. Clin. Chem. **66**(7), 975–977 (2020)

A. Zaki et al., Isolation of a novel coronavirus from a man with pneumonia in Saudi Arabia. N. Engl. J. Med. **367**(19), 1814–1820 (2012)

W. Zhang et al., Molecular and serological investigation of 2019-nCoV infected patients: implication of multiple shedding routes. Emerg. Microbes Infect. **9**(1), 386–389 (2020)

P. Zhou et al., A pneumonia outbreak associated with a new coronavirus of probable bat origin. Nature **579**(7798), 270–273 (2020)

N. Zhu et al., A novel coronavirus from patients with pneumonia in China, 2019. N. Engl. J. Med. **382**(8), 727–733 (2020)

K. Zlateva et al., No novel coronaviruses identified in a large collection of human nasopharyngeal specimens using family-wide CODEHOP-based primers. Arch. Virol. **158**(1), 251–255 (2013)

L. Zou et al., SARS-CoV-2 viral load in upper respiratory specimens of infected patients. N. Engl. J. Med. **382**(12), 1177–1179 (2020)

Open Access This chapter is licensed under the terms of the Creative Commons Attribution 4.0 International License (http://creativecommons.org/licenses/by/4.0/), which permits use, sharing, adaptation, distribution and reproduction in any medium or format, as long as you give appropriate credit to the original author(s) and the source, provide a link to the Creative Commons license and indicate if changes were made.

The images or other third party material in this chapter are included in the chapter's Creative Commons license, unless indicated otherwise in a credit line to the material. If material is not included in the chapter's Creative Commons license and your intended use is not permitted by statutory regulation or exceeds the permitted use, you will need to obtain permission directly from the copyright holder.

Chapter 11
Pooled Testing and Its Applications in the COVID-19 Pandemic

Matthew Aldridge and David Ellis

Abstract When testing for a disease such as COVID-19, the standard method is individual testing: we take a sample from each individual and test these samples separately. An alternative is pooled testing (or 'group testing'), where samples are mixed together in different pools, and those pooled samples are tested. When the prevalence of the disease is low and the accuracy of the test is fairly high, pooled testing strategies can be more efficient than individual testing. In this chapter, we discuss the mathematics of pooled testing and its uses during pandemics, in particular the COVID-19 pandemic. We analyse some one- and two-stage pooling strategies under perfect and imperfect tests, and consider the practical issues in the application of such protocols.

11.1 Introduction

When testing for a disease such as COVID-19, the standard method is *individual testing*: we take a sample from each individual and test these samples separately. Under the convenient mathematical model of perfect testing, a sample from an infected individual always gives a positive result, while a sample from a noninfected individual always gives a negative result. For N individuals, this requires N tests, and we can accurately classify all the individuals as infected or noninfected. The infected

M. Aldridge
School of Mathematics, University of Leeds, LS2 9JT Leeds, UK
e-mail: m.aldridge@leeds.ac.uk

D. Ellis (✉)
School of Mathematics, University of Bristol, The Fry Building, Woodland Road, Bristol BS8 1UG, UK
e-mail: david.ellis@bristol.ac.uk

© The Author(s) 2022
M. C. Boado-Penas et al. (eds.), *Pandemics: Insurance and Social Protection*,
Springer Actuarial, https://doi.org/10.1007/978-3-030-78334-1_11

individuals can be advised to self-isolate and their contacts can be traced, while the noninfected individuals are reassured that they are free of the disease.

An alternative to individual testing is *pooled testing*, also called *group testing*. Instead of testing individual samples, we can instead pool samples together and test that pooled sample. Again under the convenient model of perfect testing, a pool consisting entirely of uninfected samples gives a negative result, while a pool containing one or more infected samples gives a positive result. Thus a negative result demonstrates that every individual in the pool is noninfected, while a positive result requires further information to work out which individuals in the pools are infected.

As we shall see in this chapter, when the prevalence of a disease is low enough and the accuracy of a test is high enough, pooled testing can accurately classify individuals as infected or noninfected in fewer than N tests. This can be more efficient—and often *much* more efficient—than individual testing.

This chapter is structured as follows. In the remainder of this section, we introduce background material. In Sect. 11.2, we analyse some algorithms for pooled testing under an idealised model of perfect tests. In Sect. 11.3, we adapt this analysis to more realistic models of testing with errors. In Sect. 11.4, we discuss some practical issues/problems with the application of pooled testing for COVID-19. In Sect. 11.5, we survey some uses of pooled testing during the pandemic, so far. In Sect. 11.6, we conclude, and give some of our own views on potential applications of pooled testing for COVID-19.

11.1.1 Testing for COVID-19

As well as discussing the general theory of pooled testing, much of this chapter concerns applications of pooled testing in the COVID-19 pandemic, so we proceed to give some background on the existing tests for detecting current SARS-CoV-2 infection.

In the real world, testing is not perfect. We distinguish between two types of test errors:

- *False positive test errors*, where a sample (individual or pool) that does not contain any infection wrongly gives a positive result. The probability that an infection-free sample correctly gives a negative result is called the *specificity*.
- *False negative test errors*, where a sample (individual or pool) that does contain infection wrongly gives a negative result. The probability that an infected sample correctly gives a positive result is called the *sensitivity*.

The most commonly used test for current SARS-CoV-2 infection is the RT-PCR test (reverse transcription polymerase chain reaction test, or just 'PCR test' for short). A PCR test for SARS-CoV-2 infection typically works as follows. First, a swab is taken from the nose or upper throat of the individual to be tested. The swab is then sent to a laboratory, where material from the swab analysed to find out whether it contains genetic material from the SARS-CoV-2 virus. We refer the reader to the previous chapter of this book, by Dunbar and Tang, for more details. The process typically takes from four to six hours from the receipt of the swab until the output of the result, depending on the laboratory (Mahase 2020).

The PCR test is very highly specific, with specificity estimated at ranging from 97.4 to 99.98% (Skitrall et al. 2021), meaning that false positive test errors are extremely rare. (The tests used in the UK's ONS Coronavirus Infection Survey, for example, have a specificity of more than 99.92%; see Office for National Statistics (2020).) On the other hand, the PCR test is only moderately sensitive, with sensitivity in the range 70–90% being typical (Böger et al. 2021; Woloshin et al. 2020). The sensitivity depends on the laboratory protocol being used, and can be affected by shortages of reagent or improper procedures. Another significant source of insensitivity is improperly taken swabs; this can depend on the level of training of the person taking the swabs, so sensitivity can be lower in community settings than in healthcare settings (Watson and Whiting 2020). Sensitivity for a given individual also depends on the viral load and the how long after illness onset the swab was taken. The mathematical lesson from all this is that a negative test does not definitively rule out the individual being infected.

Another test for SARS-CoV-2 infection is the RT-LAMP (reverse transcription loop-mediated isothermal reaction) test. Pooled testing can certainly be used with RT-LAMP tests; however, they are not yet widely available, so we will focus our attention here on PCR tests when discussing COVID-related applications of pooled testing.

A third test is the lateral flow test, which at the time of writing is being used in the UK for mass-testing in certain areas (such as areas where there is high prevalence, or where certain new variants of SARS-CoV-2 have been detected), and for the regular screening of secondary-school pupils (with pupils being asked to self-test twice per week, at home, from the week beginning 15 March 2021). Lateral flow testing kits are cheap, easily portable, require little training to use, and produce a result in around 30 min; on the flip side, they have much lower sensitivity than PCR tests. For example, in a large pilot study in the city of Liverpool, the sensitivity of community-based lateral flow testing was estimated at 48.9% (95% CI: 33.7–64.2%) (García-Fiñana et al. 2020). We believe that pooled testing is unlikely to be compatible with the lateral flow testing programme in the UK, in view of the low sensitivity of the test, the level of training of those administering the tests, and the premium placed on rapid turnaround time.

Much of our analysis in this chapter is applicable to testing for other pandemic diseases, such as pandemic influenza, sometimes with adjustments to the assumed sensitivities and specificities of the tests being used. In particular, the mathematical models used are independent of the disease in question, though the assumptions may be more or less appropriate in the case of other diseases. For example, if very accurate and rapid tests are available for a certain pandemic disease, then pooled testing algorithms with more than two sequential stages may be well worth considering, as they can yield even greater resource-savings than pooled testing algorithms with one or two stages.

11.1.2 Stages of a Pooled Testing Algorithm

Pooled testing was first proposed in 1943 by Dorfman (1943) for the detection of cases of syphilis in those called up for US army service during the Second World War. (The textbook of Du and Hwang (2020, Chap. 1) gives more information about the early history of pooled testing.) *Dorfman's algorithm* is perhaps the simplest of all pooled testing algorithms, and has also been the most widely-used one in disease control, both prior to and during the COVID-19 pandemic. It proceeds as follows. We assume for the moment that tests are perfectly accurate, with 100% sensitivity and 100% specificity.

Suppose we have N individuals, and we wish to identify who among those N individuals is infected.

1. We choose a pool size s, and we divide the N individuals into N/s disjoint groups of size s each. (We assume, for simplicity, that N is an exact multiple of s.) We take a sample from each of the N individuals, and then, for each of the N/s groups, we pool the samples from that group into a single pooled sample. We then run a test on each of the N/s pooled samples.
2. a. If a pool tests negative, we know all the individuals in the corresponding group are noninfected.
   b. If a pool tests positive, we then follow up by individually test all the individuals in the corresponding group. These individual tests discover which of the samples in the pool were infected or noninfected.

At the end of this process, under our perfect testing model, we have correctly classified all the individuals as infected or noninfected. This is illustrated in Fig. 11.1, in the case $N = 15$ and $s = 5$.

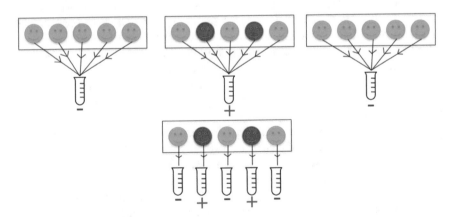

Fig. 11.1 Schematic illustration of the use of Dorfman's algorithm (under perfect testing) to identify all the infected individuals in a group of 15, using pools of size 5. In the above case, there are two infected individuals, and only eight tests are required to identify them. If the two infected individuals had been in different pools at the first step, then 13 tests would have been required

We shall see later that:

- Under perfect testing, if the prevalence is lower than 30%, then Dorfman's algorithm uses fewer than N tests on average, so is more efficient than individual testing. (See Sect. 11.2.3.)
- Under perfect testing, the optimal pool size s is easy to calculate, and is approximately $s = 1/\sqrt{p}$, where p is the *prevalence* of the disease. (See Sect. 11.2.1 for the formal definition of prevalence.)
- Even under imperfect noisy testing, Dorfman's algorithm can be more efficient than individual testing for sufficiently low prevalence. (See Sect. 11.3.2.)
- For even lower prevalence and higher accuracy, other pooled testing algorithms cannot only outperform individual testing but outperform Dorfman's algorithm as well. (See Sects. 11.2.4, 11.2.5, and 11.3.2.)

Note that individual testing is a *one-stage* or *nonadaptive* algorithm, in that all the tests are designed in advance and can be carried out in parallel. Meanwhile, Dorfman's algorithm is a *two-stage* algorithm: the first stage of pooled tests is designed in advance and carried out in parallel, but then the results must be analysed before designing and carrying out the follow-up individual tests in a second stage. There are also pooled testing algorithms with more than two stages. There is typically a tradeoff between the number of stages and the efficiency of the algorithm—more stages allows one to use fewer tests, but more stages take more laboratory time.

It is estimated that approximately 70% (95% CI: 52–90%) of the transmission of COVID-19 typically takes place either before symptom onset or in the first 48 h after symptom onset (see He et al. (2020), and the very slight correction in Ashcroft et al. (2020)). Hence, a fast turnaround time is an important factor to consider when choosing which protocol to use for case detection. If the tests were to have a very rapid processing time, it might be possible to use algorithms with many stages. However, as stated above, PCR tests for COVID-19 typically have a processing-time of four to six hours. It is likely that many laboratories worldwide will be able to perform two sequential stages in a 24-h period (for logistical reasons, a turnaround time of less than 24 h from swabbing to result announcement is often hard to achieve anyway), but adding more sequential stages may increase turnaround time too much. Moreover, laboratories under pressure may struggle to keep track of samples over more than two sequential stages. For these reasons, we focus our attention in this chapter on pooled testing algorithms with at most two sequential stages. For the state of the art in *fully adaptive* algorithms, with no limit on the number of stages, see Aldridge (2019), as surveyed in Aldridge (2019, Sect. 5.5).

For a more comprehensive (but pre-pandemic) surveys of the mathematics of pooled testing, we refer the reader to Aldridge et al. (2019), Du and Hwang (2020).

In this chapter, we only consider pooled testing where each test-result is simply either 'positive' or 'negative'. A different form of pooled testing is where an attempt is made to measure how much viral RNA is present in each pooled sample, and to make use of this information; this is known as *quantitative pooled testing*, or *quantitative group testing*, and it is a special case of the well-studied 'compressed sensing' problem. For a comprehensive introduction to the mathematics of compressed sensing, with emphasis on algorithms, the reader is referred to Foucart and Rauhut (2013). For an account of a protocol for using quantitative pooled testing for COVID-19 case identification, the reader is referred to Ghosh et al. (2020).

11.1.3 Who and Why to Test

There are two different potential applications of pooled testing for a pandemic diseases such as COVID-19. The first application, which we have discussed so far, is for *case identification*, where it is desired to identify which members of a group are infected, for the purposes of infection-control. There is also a second application, for *surveillance*, where the goal is only to estimate the infection prevalence, without necessarily identifying which individuals are those infected.

In this chapter, we focus mainly on the first application, for case identification, as we believe this is where the most useful applications of pooled testing for COVID-19 are most likely to be found. Briefly, we believe that in the UK, for example, the utility of pooled testing for surveillance on a national scale may be limited in the medium term, because, first, the UK already has a well-developed and extensive national surveillance programmes based on individual testing using random population sampling, such as the ONS Coronavirus Infection Survey, and second, using pooled testing for surveillance only yields large efficiency gains over individual testing when prevalence is lower than it has often been in the UK since the start of the pandemic. Pooled testing for surveillance does, however, still have some potential utility—for example, if prevalence in the UK becomes sufficiently low and it is desired to reduce the resource requirements of the ONS Infection Survey while still monitoring the prevalence of infection. It is also quite possible that pooled testing could be useful for the surveillance of new variants of the coronavirus, which is important in view of the risks posed by the latter to the effectiveness of vaccination programmes. We return to these issues in Sect. 11.6.

We also draw a distinction between testing symptomatic people, among whom the prevalence is likely to be high, and testing asymptomatic people, where the prevalence is likely to be lower. As we shall argue in Sect. 11.6, we believe that pooled testing for case identification is most likely to be useful for the screening of asymptomatic people—and possibly for the testing of contacts of confirmed cases, provided the prevalence of infection among the group to be tested is thought to be sufficiently low. On the other hand, we believe that pooled testing is unlikely to be useful for the testing of symptomatic people, because the prevalence of COVID-19 infection among those presenting symptoms is usually sufficiently high that the resource savings of pooled testing would be modest compared to individual testing, and are arguably outweighed by the down-sides of pooled testing, such as increased turnaround time compared to individual testing.

11.2 Pooled Testing Algorithms for Perfect Tests

11.2.1 Outline and Model

In this section, we look at some algorithms for pooled testing for a disease, and assess their performance under the mathematically convenient model of perfect test results. (Later, in Sect. 11.3, we look at the performance of these algorithms in the more realistic model of tests that are highly-but-imperfectly specific and moderately sensitive.)

Unsurprisingly, a key quantity in our model is the *prevalence* of the disease, denoted by p: this is the fraction of individuals in the population in question, who are infected with the disease, at the time when testing is being done. Equivalently, it is the probability that an individual selected at random from the population, is infected.

We assume that N individuals are being tested, and that these individuals are drawn from a large population. Each member of the population is assumed to be infected with probability p (where p is the prevalence, as defined above), independently of all other members of the population. (This independence assumption is not quite realistic in many settings, since some of the individuals being tested will often be contacts of one another, so clustering can occur. However, as we shall see, clustering actually makes pooled testing algorithms more efficient than if clustering is not present.)

Assuming that tests are perfect, we usually aim to correctly classify all N individuals as infected or noninfected. We often summarise the performance of algorithms through the *expected tests per individual*. If an algorithm uses a (possibly random) number of tests T to classify N individuals as either 'infected' or 'non-infected', then the expected tests per individual is $(\mathbb{E}T)/N$, where $\mathbb{E}T$ denotes the expectation (or mean, or average) of the random variable T. Clearly, it is desirable for the expected tests per individual to be as small as possible. Note that individual testing clearly has $(\mathbb{E}T)/N = N/N = 1$. The expected tests per individual is useful for comparing how much better (or worse) an algorithm is than individual testing.

A standard information-theoretic bound called the *counting bound* (see, for example, Aldridge (2019), Aldridge et al. (2019), Baldassini et al. (2013)) states that, for any successful pooled testing procedure, the expected tests per individual satisfies the lower bound

$$\frac{\mathbb{E}T}{N} \geq H(p). \tag{11.1}$$

Here, p is the prevalence of the disease, and $H(p)$ is the *binary entropy* function, defined by

$$H(p) = p \log_2 \frac{1}{p} + (1 - p) \log_2 \frac{1}{1 - p}.$$

The bound (11.1) immediately implies that, when the prevalence is high enough, pooled testing cannot significantly outperform individual testing (under the model of perfect tests). For example, when the prevalence of infection in the population being tested is 20% ($p = 0.2$), we have $H(0.2) \approx 0.72$, and therefore no pooled testing algorithm can use less than 72% of the number of tests per individual required by individual testing, at this prevalence level (under the model of perfect tests). In fact, a deeper mathematical result of Fischer et al. (1999) states that (under the model of perfect tests), individual testing is optimal among all pooled testing algorithms whenever the prevalence is at most 38.2% ($p \leq 0.382$).

It is useful to consider, for different pooled testing algorithms, how close their 'expected tests per individual' is to the counting bound (11.1), at different prevalence levels, under the model of perfect tests, and this is what we shall do in this section.

In this section, we study the following four classes of pooled testing algorithms:

- individual testing;
- Dorfman's algorithm, where each individual's sample appears in exactly one pool, in the first stage;
- grid-based designs, where each individual's sample appears in exactly two pools, in the first stage;
- (r, s)-regular designs, where each individual's sample appears in exactly r pools in the first stage, each pool containing s samples.

Figure 11.2 shows the performance of these algorithms (with the grid and (r, s)-regular methods algorithms in what we will later call their 'conservative two-stage' variants—see Sect. 11.2.4.1). The top subfigure shows the expected tests per item, and indicates the potential benefit of using pooled testing (compared to individual testing) when the prevalence is below 20%.

However, this top subfigure does not sufficiently convey the difference between the pooling algorithms at very low prevalence. The bottom subfigure shows the *rate*, defined by

$$\text{rate} = \frac{H(p)}{\mathbb{E}T/N} = \frac{H(p)N}{\mathbb{E}T}.$$

The rate measures how close an algorithm gets to the counting bound (11.1)—higher rates are better. (We remark that the rate can also be interpreted information-theoretically, as the number of bits of information learned per test. See for example Aldridge et al. (2019), Baldassini et al. (2013).) The rate illustrates better the comparison between the different pooling methods, and shows the advantage of the (r, s)-regular design at very low prevalences (e.g., below 2%).

11.2.2 Individual Testing

Clearly, the individual testing of N individuals requires $T = N$ tests, yielding an expected number of tests per individual equal to $T/N = N/N = 1$.

An obvious advantage of individual testing is that one does not need an estimate of the prevalence. It is a one-stage algorithm, so has the fastest possible turnaround time.

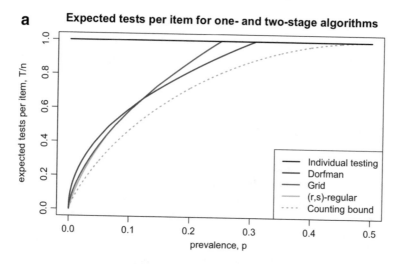

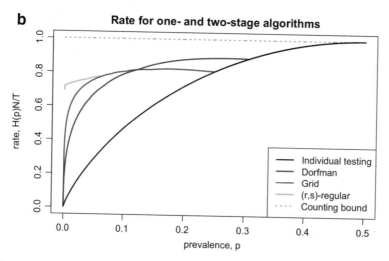

Fig. 11.2 Performance of one- and two-stage group testing algorithms under perfect testing, as measured by **a** the expected tests per item, and **b** the rate. 'Grid' and '(r, s)-regular' refer to the conservative two-stage variants. At all but the lowest prevalences, the (r, s) regular design has optimal parameter $r = 1$, so is equivalent to Dorfman's algorithm, or $r = 2$, and is equivalent to the grid algorithm

It is also the simplest testing algorithm to implement. However, we will see that other algorithms yield huge resource savings over individual testing, at low prevalence-levels, and are therefore well worth considering, when prevalence is fairly low and testing capacity is constrained.

11.2.3 Dorfman's Algorithm

In Sect. 11.1.2, we introduced Dorfman's original two-stage algorithm. In this section, we discuss it in more detail.

Suppose we receive samples from N individuals in a large population where the prevalence of infection is p, and we run Dorfman's algorithm on these N individuals, using pools of size $s \geq 2$, where N is assumed to be a multiple of s. Clearly, one test for each pool is always used at the first pooled testing stage. A pool will also require an extra s individual tests in the second stage if the pooled test was positive. The pooled test will be positive unless all s items are noninfected, so the probability it will test positive is $1 - (1 - p)^s$. So for each of the N/s pools we definitely have 1 pooled test, then with probability $1 - (1 - p)^s$ another s individual tests are required, giving an expected number of tests as

$$\mathbb{E}T = \frac{N}{s}\left(1 + s\left(1 - (1 - p)^s\right)\right) = \left(\frac{1}{s} + 1 - (1 - p)^s\right)N.$$

Hence the expected tests per individual is

$$\frac{\mathbb{E}T}{N} = \frac{1}{s} + 1 - (1 - p)^s.$$

It is easy to check that $s = 2$ is never the best choice of pool size s, but that $s = 3$ improves on individual testing for $p < 1 - (1/3)^{1/3} = 0.307$. That is, Dorfman's algorithm improves on individual testing for prevalences below roughly 30%.

If the prevalence p is known accurately beforehand, we should choose the pool size s so as to minimise the quantity $1/s + 1 - (1 - p)^s$, and thereby minimise the expected number of tests required. For fixed p, the function $s \mapsto 1/s + 1 - (1 - p)^s$ is very well-behaved, and it is very easy to numerically find the integer s that minimises it. But it is useful also to note that when p is small, we may use the approximation

$$\frac{1}{s} + 1 - (1 - p)^s \approx \frac{1}{s} + 1 - (1 - ps) = \frac{1}{s} + ps,$$

which is minimised over the reals at $s = 1/\sqrt{p}$. This gives an expected tests per individual of approximately $2\sqrt{p}$. More formally, choosing $s = \lfloor 1/\sqrt{p} \rfloor$ yields that the expected number of tests per individual is $1/s + 1 - (1 - p)^s = 2\sqrt{p} + O(p)$,

Table 11.1 Resource requirements of using Dorfman's algorithm to test N individuals, at different prevalence levels, with tests of perfect sensitivity and specificity

Prevalence p (%)	Dorfman's algorithm	
	Optimal pool size s	Expected tests per individual
5	5	0.43
2	8	0.27
1	11	0.20
0.5	15	0.14
0.2	23	0.088
0.1	32	0.063

where the error term $O(p)$ is small compared to $\sqrt{p}$ when p is small. Hence, when p is small, a good estimate for p is known beforehand, and the pool-size s is chosen sensibly, our model predicts that Dorfman's algorithm uses, on average, approximately $2\sqrt{p}N$ tests to identify all the infected individuals among a population of size N.

Table 11.1 shows the predicted resource requirements (under the simple model above, i.e., perfect tests), of using Dorfman's algorithm to test N individuals, at different prevalence levels, when the prevalence level is known accurately beforehand, and the pool-size s is chosen optimally.

There are shortcomings in this analysis. We mention three here. Firstly, tests used in the real world do not have perfect sensitivity or perfect specificity; we will refine the model to take this into account, in the following section.

Secondly, an accurate estimate for the prevalence may not be known beforehand, so it may not be possible to choose the pool size s optimally beforehand. We return to this issue in Sect. 11.4.

Thirdly, the independence assumption may fail because infections of different individuals being tested by the laboratory are unlikely to be truly independent. But if the overall prevalence remains the same, then Dorfman's algorithm will not perform any worse than the above analysis predicts. In fact, it is advantageous in terms of resource requirements if there is a 'clustering' of infected individuals in the same pool. In terms of resource use, the worst case for Dorfman's algorithm is when infected individuals are spread between many pool, and the best-case is when infected individuals are concentrated in few pools.

11.2.4 Grid Algorithms

In Dorfman's algorithm, each individual was tested once in the first stage. In a family of algorithms called *grid algorithms*, each item is tested twice in the first stage. One variant attempts to classify samples as infected or noninfected just from this single stage, while other variants follow up with a second stage of individual testing.

11.2.4.1 Variants of Grid Algorithms

The grid algorithms always begin as follows. Suppose we have N individuals to test. We split these into N/s^2 groups each of s^2. (We assume for simplicity that N is a multiple of s^2.) Let us concentrate on a single group. Each of the s^2 individuals is swabbed, and the sample from each swab is divided in two, so it can later be a part of two different sample pools. We now picture the s^2 individuals as laid out on a $s \times s$ grid. In the first stage we conduct $2s$ pooled tests: we make one pool from each of the s rows of the grid, and one pool from each of the s columns. A PCR test is run on each of these sample pools.

We assume, again, that tests are perfectly accurate. What can we learn from these results of these pooled tests?

Case 0: If none of the s^2 individuals are infected, then all $2s$ tests will be negative. We can confidently state that all the samples are noninfected.

Case 1: If exactly one out of the s^2 individuals is infected, then exactly one row test and one column test will be positive. We can confidently state that the individual at the intersection of that row and column in the grid is infected and that all the other individuals are noninfected.

Case 2: If two or more of the s^2 individuals are infected, then the test results may be ambiguous. If we are lucky, it could be that all the infected individuals are in the same row or the same column of the grid, in which case they can be identified with complete confidence—we call this **Case 2A**—but more often we cannot be certain exactly which individuals are infected—we call this **Case 2B**.

What should one do after receiving the pooled test results? Here, we briefly look at three possible choices, leading to three different variants of the grid algorithm:

- *One-stage grid algorithm*: In the one-stage variant, we do not perform any follow-up tests. Any individual that was in at least one negative-testing pool is confidently declared to be noninfected. In Cases 0, 1 or 2A, the remaining individuals (i.e., those who appeared in two positive-testing pools) can be confidently declared to be infected, whereas in Case 2B, the remaining individuals are declared to have unclear results. Running just one single stage has the benefit that a laboratory can quickly process the results, but the downside of sometimes producing inconclusive results.

- *Standard two-stage grid algorithm*: In this variant, we can run a second stage of individual testing, to clear up ambiguous results, if there are any. That is, in Cases 0, 1 or 2A, the algorithm is exactly the same as the one-stage grid algorithm above, but in Case 2B, all individuals who were not in at least one negative-testing pool, are given an individual test in the second stage; those who test negative are then declared negative, and those who test positive are declared positive.

Table 11.2 Resource requirements of using two variants of the grid algorithm at different prevalence levels, with tests of perfect sensitivity and specificity. In the one-stage variant, parameters are chosen such that each grid of s^2 individuals is correctly classified with probability at least 0.99. A dash—means that the method is worse than individual testing

Prevalence p (%)	One-stage grid algorithm		Conservative two-stage grid algorithm	
	Optimal s	Tests per ind.	Optimal s	Exp. tests per ind.
5	–	–	9	0.38
2	–	–	16	0.21
1	3	0.67	25	0.14
0.5	5	0.40	38	0.086
0.2	8	0.25	68	0.047
0.1	11	0.18	106	0.030

- *Conservative two-stage grid algorithm*: Alternatively, we can run a second stage where all individuals that appeared in two positive-testing pools are given an individual test—even in Case 1 or Case 2A, where the infected individuals can be logically determined after the first stage. This has the advantage that every infected individual can be definitely confirmed as infected by the 'gold standard' of an individual (non-pooled) PCR test, which might be personally reassuring for the individual or their employer, beneficial when tests are imperfect, or required by regulators. Two-stage algorithms where infection must be confirmed with an individual test are often known in the literature as *trivial two-stage algorithms*, but in this chapter we will use the term *conservative two-stage algorithm*, as it is more descriptive. Conservative two-stage algorithms can be easier to analyse mathematically, than the standard two-stage ones.)

In real-life situations, when using the one-stage variant of the grid algorithm, one must make a decision on what to do with individuals with an inconclusive result. One option would be to inform all those individuals they should self-isolate as a precaution; another option would be to individually re-test each of them at a later date (effectively running a two-stage algorithm with a delayed second stage); another option would be to restart the testing from scratch with different grids; a more reckless option would be to inform all the individuals being tested that the test results were inconclusive and that they should continue their lives as if they had tested negative. Which option the laboratory or the regulatory authorities choose may depend on the impact of letting an infection go undetected: if the individual in question is a school pupil or a member of the general community in a mass-testing programme, the impact is likely to be much less than if the individual is a healthcare worker working with highly vulnerable patients or a resident-facing social care worker, for example.

In the next subsection, we give a brief analysis of the conservative two-stage variant of the grid algorithm, under our convenient assumption that the tests are perfectly accurate. We summarise these results in Table 11.2.

11.2.4.2   An Analysis of the Conservative Two-Stage Grid Algorithm

Recall that in the conservative two-stage variant of the grid algorithm, every individual in two positive pooled tests receives an individual test. Our analysis here is similar to that in Aldridge (2020), Broder and Kumar (2020).

For each of the N/s^2 grids of s^2 individuals, there are $2s$ pooled tests in the first stage. An individual can then receive an individual test for one of two reasons. One is that the individual is infected, which occurs with probability p. The second is that the individual is uninfected (which occurs with probability $q = 1 - p$), and that further, its row contains an infected individual (this happens with probability $1 - q^{s-1}$), and its column also contains an infected individual (this also happens with probability $1 - q^{s-1}$).

Hence, using the linearity of expectation, the expected number of tests is

$$\mathbb{E}T = \frac{N}{s^2}\big(2s + s^2(p + q(1 - q^{s-1})^2)\big) = N\left(\frac{2}{s} + p + q(1 - q^{s-1})^2\right),$$

and the expected tests per individual is $(\mathbb{E}T)/N = \frac{2}{s} + p + q(1 - q^{s-1})^2$.

As before, given the prevalence p, this is a well-behaved function of s, so it is easy to numerically choose the optimal s. Results were shown in Table 11.2 above, and far outperform the one-stage variant when a second stage is available. Comparing with Table 11.1, we see that the conservative two-stage grid algorithm also outperforms Dorfman's algorithm for the values of p under consideration, though not by more than a factor of approximately two, for these values of p. We note, however, that the optimal choice of s (the pool-size in the conservative two-stage grid algorithm) for $p = 0.1\%$, is 106, and large pool-sizes do have practical down-sides (usually requiring automation, reducing sensitivity, and posing regulatory problems—see Sect. 11.4).

From a more mathematical perspective, a similar approximation to that for Dorfman's algorithm shows that, for small p, the optimal s is of the order $1/p^{2/3}$; this yields an expected number of tests per individual of the order $p^{2/3}$. For p sufficiently small, this is an improvement on Dorfman's algorithm (where the expected number of tests per individual is of the order $p^{1/2}$).

11.2.5   Pooling Algorithms Based on (r, s)-Regular Designs

The (r, s)-regular designs are a family of algorithms that generalise the algorithms we have seen so far. These algorithms have a first stage where each individual's sample appears in exactly r different pools, and each pool contains samples from exactly s individuals. The second stage (if there is a second stage) consists of individual testing—as with the grid algorithm discussed in the previous section, in the 'standard' variant, an individual is given an individual test at the second stage only when their infection status cannot be determined from the first stage, whereas in the 'trivial' variant, an individual is given an individual test at the second stage whenever all the pools they appear in test positive at the first stage.

Individual testing is a special case of an (r, s)-regular design, with $r = 1$, $s = 1$ and a single stage. Dorfman's algorithm is a special case with $r = 1$, $s > 1$, with two stages. The grid algorithms we discussed above are special cases with $r = 2$.

There are a number of ways to construct a pooling design that is (r, s) regular.

- *Randomly*: Given a number of individuals N, a testing procedure that tests each individual in r pools with each pool consisting of s samples can be chosen uniformly at random from all such procedures. This is easy to do computationally, and convenient for proving mathematical statements. However, the random choice means that rare bad designs are possible, and the lack of structure can make it awkward to carry out in a laboratory setting.

- *Hypercube*: This method generalises the grid algorithm to higher dimensions. It is required that $s = a^{r-1}$ for some positive integer a. Assume that N is a multiple of a^r. We split the N individuals into groups of size a^r, and we focus our attention on just one group. Imagine that those a^r individuals are placed on an r-dimensional $a \times a \times \cdots \times a$ hypercube. Each pool corresponds to an $(r - 1)$-dimensional slice of this hypercube, containing $a^{r-1} = s$ individuals. Note that each individual is sampled in r pools, one for each of the r slice directions. Taking $r = 2$, we obtain the grid algorithm. The structure of the hypercube can be convenient for implementation, although automation is usually required for pooling the samples, and for $r \geq 3$, the conditions give a somewhat restricted set of possibilities for s.

- *Code-based*: A classical construction of Kautz and Singleton shows how to construct an (r, s)-regular design from an error-correcting linear code with appropriate parameters. We point readers to Aldridge et al. (2019, Sect. 5.7) or Kautz and Singleton's original work (Kautz and Singleton 1964) for further details. The extra structure often gives good performance when N is small, although for some values of r and s it is not possible to find a code with appropriate parameters.

We note that, by counting the number of times a sample appears in a pool in two different ways, the number of pooled tests T_1 used by the first stage of an (r, s)-regular design satisfies $Nr = T_1 s$, and therefore the number of tests per individual used by the first-stage of an (r, s)-regular design is $T_1/N = r/s$.

As stated above, an (r, s)-regular pooled testing algorithm can be used in the form of a one-stage, a 'standard' two-stage, or a conservative two-stage algorithm, just as with the grid algorithm.

We briefly present a summary of an analysis of the conservative two-stage variant, following Aldridge (2020), Broder and Kumar (2020). With this variant, any individual whose r stage-one pooled tests are all positive receives and individual test in the second stage. For large N, the expected tests per individual in the random (r, s)-design (described above) satisfies the following with high probability:

Table 11.3 Resource requirements of using an (r, s)-regular design in a conservative two-stage algorithm to test N individuals, at different prevalence levels, with tests of perfect sensitivity and specificity

Prevalence p (%)	Conservative two-stage (r, s)-regular algorithm	
	Optimal parameters (r, s)	Expected tests per individual
5	(3, 13)	0.37
2	(4, 31)	0.19
1	(5, 63)	0.11
0.5	(7, 147)	0.063
0.2	(8, 351)	0.029
0.1	(9, 700)	0.016

$$\frac{\mathbb{E}T}{N} \sim \frac{r}{s} + p + q(1 - q^{s-1})^r, \tag{11.2}$$

where p is the prevalence and $q = 1 - p$. Here, r/s is the number of tests per individual in the first stage. An individual requires retesting in the second stage either if it is infected, with probability p, or if it is noninfected, with probability q, but all r of its tests are positive, each of which happens with probability $1 - q^{s-1}$. Thus if the results of the tests containing a given noninfected individual were independent, then (11.2) would hold exactly. It turns out that a randomly sampled (r, s)-design satisfies this independence condition for most individuals (with high probability); in fact, with positive probability, it satisfies the condition for all individuals.

Further, a lower bound is given in Aldridge (2020), which shows that, among all conservative two-stage algorithms, the random (r, s)-regular design is extremely close to optimal for all $p < 0.3$.

Table 11.3 shows the performance of the (r, s)-regular design according to (11.2) with an optimal choice of r and s. We note that the (r, s)-regular algorithm outperforms individual testing, Dorfman's algorithm, and the grid algorithms for all values of p in the table.

Table 11.3 shows results with the mathematically optimal choice of (r, s), but as the prevalence gets small, these parameter choices can get quite large. This could be unwieldy or even infeasible for a laboratory to carry out, and large values of s (the pool-size) provoke worries about sensitivity (with imperfectly sensitive tests). However, typically r can be reduced somewhat and s reduced quite a lot with only a marginal reduction in performance. For example, at $p = 0.5\%$, the optimal choice is $r = 7, s = 147$, giving an expected tests per individual of 0.063. But reducing the parameters to the much more manageable $r = 3, s = 62$ still gives an expected tests per individual of 0.072, which is only slightly worse. The practically best choice of parameters will depend on a laboratory's capability for carrying out complicated procedures, and worries about the impact of dilution on test sensitivity (see Sect. 11.4).

11.3 Pooled Testing Algorithms for Imperfect Tests

11.3.1 The Model

In this section, we refine the model of the previous section to take into account the fact that the tests we are dealing with do not always give the correct answer. Recall that the PCR test has very high sensitivity, typically higher than 99%, meaning false positive test results are extremely rare, and has moderate sensitivity, typically between 70 and 90%, meaning that false negative results are not uncommon.

Here, we use a very simple model for such tests. We assume that each test on a pool containing at least one infected sample has a fixed probability u of correctly returning a positive result, and that each test on a pool of entirely noninfected samples has a fixed probability v of correctly returning a negative result, independently of the outcomes of all other tests (including of tests on overlapping pools), and independently of the size of the pool (that is, with no 'dilution' effect).

Whether or not this is a realistic model will depend upon the main sources of false negatives and of false positives, and therefore on the precise protocol being used and the practical situation. If the main source of insensitivity or nonspecificity is a shortage of reagents, faulty equipment, or faulty lab-procedures, then it is probably quite realistic. If individuals are frequently swabbed incorrectly (as can happen when individuals are asked to self-swab), then incorrectly taken swabs will be an important source of insensitivity, and in this case, unless individuals are re-swabbed at each successive stage of a group-testing algorithm and there are no overlapping pools at any single step, the independence assumption will not be valid. Moreover, the assumption that dilution (where a small number of positive samples are diluted by a large number of negative samples), does not affect sensitivity, is likely to be fairly realistic with pool-sizes of 10 or less and with typical viral loads, but will be less realistic with pool-sizes of 100 or more. See Sect. 11.4 for a further discussion of this issue.

In the rest of this section, we look at some results regarding two-stage and one-stage algorithms, under this model for noisy tests.

11.3.2 Analysis of Individual Testing and Dorfman's Algorithm

For a given algorithm, there are (at least) three things we want to know: First, how many tests do we expect to use? Second, how many false negative declarations do we expect to make? Third, how many false positive declarations do we expect to make? A useful quantity for comparing algorithms is the *expected number of tests per isolated individual* (ETI): that is, the expected number of tests used, divided by the expected number of infected individuals correctly discovered and instructed to isolate. Since the isolation of infected individuals is the main public-health goal of a screening programme, ETI is a good measure of how much benefit we are getting per test used (though it does not take into account turnaround time).

Let us start with individual testing. To test N individuals, this requires exactly N tests. There are pN infected individuals on average, and we find each one if its test correctly gives a positive result, which happens with probability u. So on average we correctly find upN infected individuals but falsely miss $(1-u)pN$ of them; so the ETI is up. Similarly, also on average, of the $(1-p)N$ noninfected individuals, we correctly identify $v(1-p)N$ of them, but falsely declare $(1-v)(1-p)N$ of them to be infected.

Now consider using Dorfman's algorithm to test N individuals with pools of size s, with N a multiple of s, and suppose we use the protocol of declaring an individual to be infected only if both their pooled test and their individual test are positive. (If a pool tests positive in the first stage but all the corresponding individual tests in the second stage are negative, the pooled test is assumed to be a false positive.)

First, an individual will make it through to the second stage if either the pool is infected, and correctly gives a positive result; or if the pool is noninfected, but incorrectly gives a positive result. Thus the expected tests per individual is

$$\frac{\mathbb{E}T}{N} = \frac{1}{s} + u(1-q^s) + (1-v)q^s,$$

where $q = 1 - p$. Here, $1/s$ represents the requirements of the first stage (i.e., the pooled tests), $u(1-q^s)$ represents the requirements of the second stage in the case of a true positive pool result, and $(1-v)q^s$ represents the requirements of the second stage in the case of a false positive pool result. For small p, it turns out that an essentially optimal choice for minimising this quantity is $s = \lfloor 1/\sqrt{up} \rfloor$; this can be shown in a similar way to in the previous section. This yields an expected number of tests per individual which is approximately $2\sqrt{up} + (1-v)$, compared to 1 for individual testing. For $p = 0.02$, $u = 0.8$, $v = 0.995$, we get an improvement from 1 test per individual to 0.25 tests per individual.

Second, the expected number of infected individuals found is $u^2 pN$, as there are pN infected individuals on average, and they are found if both their pooled test and their individual test are correctly positive. The other (on average) $(1-u^2)pN$ infected individuals get false negative declarations. Compared to individual testing, where the total expected number of false negatives is simply $(1-u)pN$, we have

$$\frac{(1-u^2)pN}{(1-u)pN} = 1 + u \le 2,$$

and therefore the expected number of false negatives under Dorfman's algorithm can never be more than twice that when individual testing is used.

Third, a noninfected individual is falsely declared infected if both their pooled test is positive—either due to a false positive test or the presence of an infected individual and a true positive test—and their individual test is a false positive. This event has probability

Table 11.4 Expected resource requirements and impact of using individual testing or Dorfman's algorithm to test N individuals, at different prevalence levels, with tests of sensitivity 80% and specificity 99.5%

Prevalence p (%)	Individual testing			Dorfman's algorithm			
	$\mathbb{E}$ # tests	$\mathbb{E}$ # false neg	$\mathbb{E}$ # false pos	s	$\mathbb{E}$ # tests	$\mathbb{E}$ # false neg	$\mathbb{E}$ # false pos
5	N	$0.01N$	$0.005N$	6	$0.38N$	$0.02N$	$0.0009N$
2	N	$0.004N$	$0.005N$	9	$0.25N$	$0.007N$	$0.0006N$
1	N	$0.002N$	$0.005N$	12	$0.18N$	$0.004N$	$0.0004N$
0.5	N	$0.001N$	$0.005N$	17	$0.13N$	$0.002N$	$0.0003N$
0.2	N	$0.0004N$	$0.005N$	26	$0.08N$	$0.0007N$	$0.0002N$
0.1	N	$0.0002N$	$0.005N$	36	$0.06N$	$0.0004N$	$0.0002N$

$$\left((1 - q^{s-1})u + q^{s-1}(1 - v)\right)(1 - v).$$

Hence, the expected number of false positives is

$$\left((1 - q^{s-1})u + q^{s-1}(1 - v)\right)(1 - v)qN.$$

Again, for small p with $s = \lfloor 1/\sqrt{up}\rfloor$, this is approximately $(\sqrt{pu} + 1 - v)(1 - v)qN$. Compared to the expected number of false positives under individual testing, which is $(1 - v)qN$, Dorfman gives an improvement by a factor of $\sqrt{up} + 1 - v$. For $p = 0.02$, $u = 0.8$, $v = 0.995$, Dorfman gives an expected number of false positives which is approximately 0.12 times its value under individual testing. Thus Dorfman produces far fewer false positives than individual testing, a feature which is common to many other pooled testing algorithms.

Under the assumption that the sensitivity u of each test is 0.8 and the specificity v of each test is 0.995, Table 11.4 summarises, for different prevalences, the expected number of tests, false negatives, and false positives for individual testing and for Dorfman's algorithm. Note that, compared to individual testing, Dorfman's algorithm dramatically decreases the number of tests required and the number of false positives, but roughly doubles the number of false negatives.

The ETI of Dorfman's algorithm (used in the way above), is $(1/s + u(1 - q^s) + (1 - v)q^s)/(u^2p)$; if p is small and we choose $s = \lfloor 1/\sqrt{up}\rfloor$, as described above, then this is approximately $(2\sqrt{up} + 1 - v)/(u^2p)$.

11.3.3 One-Stage Testing

We now briefly consider general one-stage (nonadaptive) pooling algorithms, under our simple model of imperfect tests. Here, we see the results of the tests, and we must try to come up with a 'best guess', from those results, as to which individuals were infected. The precise meaning of 'best guess' depends (for example) on the down-sides of missing infected cases (false negatives), and of false positives. This kind of problem is known as an *inference problem*.

We suppose the pooling design is chosen according to a *pooling matrix* $A = (a_{ti}) \in \{0, 1\}^{T \times N}$, a matrix of zeros and ones, where $a_{ti} = 1$ if the sample from individual i is included in the tth pooled test, and $a_{ti} = 0$ otherwise. The T rows of the matrix A represent the T pooled tests, and the N columns represent the N individuals being tested.

Some further notation is useful. Let $\mathbf{x} = (x_i) \in \{0, 1\}^N$ be the vector of zeros and ones where $x_i = 1$ if individual i is infected, and $x_i = 0$ if individual i is uninfected; the vector $\mathbf{x}$ represents which individuals are truly infected and which are not, and it is what we really want to guess; we refer to it as the 'infection vector'.

We write $\mathbf{y} = (y_t) \in \{0, 1\}^T$ for the actual outcomes of the tests, i.e., $y_t = 1$ if the tth pool tests positive, and $y_t = 0$ otherwise. Finally, we write $\tilde{\mathbf{y}} = \tilde{\mathbf{y}}(A, \mathbf{x})$ for what the T outcomes of the pooled tests on the infection vector $\mathbf{x}$ would be, under perfect testing, i.e., $\tilde{y}_t = 1$ if the tth pool would test positive under perfect testing, and $\tilde{y}_t = 0$ otherwise. Explicitly, $\tilde{y}_t = 1$ if $(A\mathbf{x})_t \geq 1$, and $\tilde{y}_t = 0$ if $(A\mathbf{x})_t = 0$.

Given $\mathbf{y}$ and A, we must come up with an estimate $\hat{\mathbf{x}}$ for $\mathbf{x}$. If we are only interested in estimating the most likely set of infected individuals (i.e., we do not want to err on the side of caution when we report to the individuals whether they are infected or not), then it makes sense to report a maximum a posteriori (MAP) estimate (though there are other reasonable alternatives, e.g. minimising the expected number of false positives plus false negatives). It can be shown using the standard techniques that the MAP choice for $\hat{\mathbf{x}}$ is one where $\tilde{\mathbf{y}}(A, \hat{\mathbf{x}})$ is chosen to minimize the 'penalty function'

$$f(\hat{\mathbf{x}}) = a \times \#\{i : \hat{x}_i = 1\} + b \times \#\{t : y_t = 1, \tilde{y}_t = 0\} + c \times \#\{t : y_t = 0, \tilde{y}_t = 1\},$$

for some constants $a, b, c \geq 0$—these constants depend on the assumed prevalence, and the assumed sensitivity/specificity of the (pooled) tests. As a simple example, the P-BEST algorithm (see Sect. 11.5.2) minimises the penalty function with $a = 0$ and $b = c = 1$. We note that, if the estimated cost of declaring an infected person uninfected is much greater than the estimated cost of declaring an uninfected person infected, then the estimate with minimum expected cost may be different from the MAP estimate, and the penalty function to be minimized, should be changed, but Bayesian analysis can still be applied. We do not here get into the question of how to efficiently perform such an analysis.

11.4 Practical Challenges for Pooled Testing

The practical challenges and the downsides of implementing a pooled testing algorithm for COVID-19 testing—either as part of a national or local testing programme or within an autonomous institution or company—depend to some extent on the algorithm in question. A cost-benefit analysis is of course desirable, in each setting where pooled testing may be considered.

Small benefit at high prevalence. As mentioned earlier, when prevalence is greater than 38.2%, no pooled testing algorithm can outperform individual testing (Fischer et al. 1999) (under the assumption of perfect tests). Even when prevalence is significantly less than 38.2%, it will often be judged that the down-sides of pooled testing outweigh the advantage of resource-savings. For example, pooled testing has been approved in India only for use in areas where the population prevalence is 2% or less (see Sect. 11.5.4).

Increased turnaround time. As mentioned above, a single PCR test can typically be performed in four to six hours. If pooling can be automated, using e.g. a pipetting robot, then one-stage (non-adaptive) pooled testing algorithms will not have a significantly longer turnaround time than individual testing. For multistage pooled testing algorithms, such as Dorfman's algorithm, conservative two-stage pooled testing algorithms based on (r, s)-regular designs, or the multistage algorithm piloted in Rwanda (see Sect. 11.5.3), the increase in turnaround time (relative to individual testing) will depend partly on the laboratory set-up. The impact of increased turnaround times clearly depends upon the damage done by letting infections go undetected for longer; in the case of screening healthcare workers or social care workers, who work with individuals highly vulnerable to severe illness if exposed to SARS-CoV-2, this impact is likely to be much greater than in the case of screening university students or factory workers (for example).

Laboratory infrastructure. Dorfman's algorithm does not require any sophisticated equipment to implement: the pooling of the samples can be done by hand. The pooling of the samples can even be done, as at the University of Cambridge (see Sect. 11.5.1), by the individuals to be tested, thus imposing no extra workload on laboratory staff. It only requires the laboratory to keep track of which individuals correspond to which pooled samples, and a capability to perform individual follow-up tests on the individuals whose pools test positive (or alternatively, for part-samples from each individual to be kept back during the first step, in case follow-up testing is needed on that individual in the second step). Some laboratories, e.g., those suffering from a shortage of well-trained personnel, or of equipment, would struggle to implement even Dorfman's algorithm (the simplest to implement): even keeping track of which samples belong to which individuals, once these have been divided in half, may prove challenging under conditions of extreme pressure and consequent disorganisation.

The grid-algorithm is most efficiently implemented using a pipetting robot with an arm that can move in two dimensions; this piece of equipment, while commercially available, may be too expensive for organisations operating with a low budget, and for poorer countries. An alternative is to do the pooling manually, if sufficient manpower is available.

The impact of dilution on test sensitivity.   When one infected sample is pooled with several others that are uninfected, the viral RNA is diluted, and this dilution leads to a decrease in the sensitivity of the PCR test on the pooled sample. The precise impact on sensitivity depends upon the laboratory protocol used, and also upon the distribution of viral loads in the samples being tested (which, in turn, depends upon the stage of the illness in the individuals being tested, as well as on individual biological factors). The following, however, gives a rough idea of the impact of dilution on sensitivity. Using a common protocol, and a set of 838 SARS-CoV-2 positive specimens, Bateman et al. (2021) found that dilution by a factor of five led (on average) to a 7% reduction in sensitivity, dilution by a factor of ten led (on average) to a 9% reduction in sensitivity, and dilution by a factor of 50 led (on average) to a 19% reduction in sensitivity. By contrast, a systematic review of the accuracy of individual PCR tests found false negative rates ranging from 2 to 29%, using repeated PCR-testing as the gold standard (for true positivity). Using repeated PCR testing as the gold standard for positivity is likely to underestimate the true rate of false negatives.

   More importantly, for tests taken in the field, is the fact that swabs are sometimes taken incorrectly (see the previous chapter of this book, by Dunbar and Tang, for further discussion of this issue). One community-based study of close contacts of confirmed COVID-19 cases in China, found an overall sensitivity of 71% for upper-throat swabbing (by trained medical personnel) followed by an individual PCR test, using repeated PCR tests as the gold standard (for true positivity). When individuals self-swab, sensitivity is likely to be lower, unless the individuals themselves are appropriately trained (e.g., healthcare workers). Compared to these factors, overall, the impact of dilution on sensitivity, is relatively minor.

   In the theoretical results we have described here, the number of samples per test s can get extremely large when the prevalence p is very small. In real applications, laboratories would be unwilling (due to dilution concerns) or unable (due to equipment capacity) to pool together very high numbers of samples. Thus for extremely low prevalences, the gains of pooled testing are unlikely to be as high as the theory suggests.

How to deal with inconsistent test results.   When the sensitivity or the specificity of the tests being used is less than 100%, a pooled testing algorithm can yield inconsistent results. For example, when Dorfman's algorithm is implemented, a particular pool can test positive due to the presence of one infected individual (a true positive), but then in the second stage of follow-up (individual) testing, all the individuals in that pool can test negative, due to the infected individual testing negative (a false negative). In this case, the testing authorities are faced with a dilemma: they could assume (wrongly, in this case), that the first (pooled test) was a false positive and that the follow-up tests were true negatives, or they could simply choose to declare the individuals in question free from infection (to avoid having to recall them for further testing), but they could also decide to repeat the second round of individual tests on that group (pool) of individuals, to hedge against the possibility that there was a false negative in the second round of individual tests. This could be regarded as a third, 'confirmatory', testing-step in the algorithm. (If extra samples cannot be taken from the individuals in question, a third, confirmatory, testing-step may in fact be impossible.)

Which option the authorities choose, may depend on the impact of letting an infection go undetected; if the individual in question is a school pupil or a member of the general community (in a mass-testing programme), the impact will be less than if the individual is a healthcare worker working with patients highly vulnerable to severe illness or death in the case of COVID-19 infection, or if the individual is a resident-facing social care worker. The extra resource-requirements, and the delay, of a third (confirmatory) round of testing, should it be judged necessary, would have to be taken into account when deciding whether to adopt the pooled testing strategy.

Prior prevalence estimates. In situations where surveillance is poor, or infection levels are changing very rapidly, a laboratory may have very little idea of the prevalence of infection in an incoming batch of samples to be tested for SARS-CoV-2 infection. In such circumstances, a suboptimal choice of the parameters in a pooled testing algorithm algorithm (due to an underestimate of the prevalence) can lead to an inefficient second step—less efficient, in fact, than individual testing. For example, if the prevalence is 25%, then using Dorfman's algorithm with pools of size 10 requires on average approximately 1.04 tests per individual, which is slightly worse than individual testing (in addition to having a longer turnaround time). However, if an upper bound on the prevalence is known with a reasonably high degree of certainty (and this upper bound is not too high), then the parameters in a pooled testing algorithm can still be chosen so as to achieve a significant resource-saving over individual testing, even though the parameters cannot be fined-tuned to the exact prevalence. For example, the number of tests per individual for Dorfman's algorithm with fixed pool size s, is increasing in p, so the 'worst case' is when p is maximal. Hence, if the prevalence is known to be at most 1%, then Dorfman's algorithm with pools of size 10, will require at most $\frac{1}{10} + 1 - (1-p)^{10} \leq \frac{1}{10} + 1 - (1-0.01)^{10} \approx 0.196$ tests per individual (on average). Thus we get at least a five-fold improvement on individual testing, as long as the prevalence does not rise above 1%.

Regulatory approval. All of the issues listed above may be obstacles to regulatory approval. An additional obstacle to regulatory approval is the complexity of pooled testing algorithms, compared to individual testing. Often, policymakers will need to give their approval, bearing in mind public opinion, and a procedure which cannot be understood by a large percentage of the public (or by policymakers without the requisite quantitative training), may be less likely to gain such approval. On the other hand, the simpler pooled testing algorithms, such as Dorfman's algorithm and the grid algorithm, almost certainly can be explained in such a way that policymakers (and the majority of the public) can understand them—and this may well be part of the reason why these two algorithms have seen the widest use, of all pooled testing algorithms, during the COVID-19 pandemic so far.

11.5 Uses of Pooled Testing in the COVID-19 Pandemic

Hitherto in the COVID-19 pandemic, Dorfman's algorithm has been the most widely-used pooled testing strategy. This is almost certainly because it is (i) easy to implement without necessarily needing large changes in laboratory equipment or infrastructure, (ii) relatively robust to changes in prevalence, (iii) simple and transparent enough for non-scientific decision makers or the public to understand and for regulators to approve, and (iv) of easily predictable and controllable sensitivity. At the same time, it can yield large efficiency gains over individual testing, as outlined above. In this section, we give some examples of places where pooled testing has been applied during the COVID-19 pandemic, focussing on examples where detailed information is available.

11.5.1 Dorfman's Algorithm at the University of Cambridge

During Autumn Term of 2020 (6 October–4 December 2020), University of Cambridge students in college accommodation were asked to participate in the University's asymptomatic COVID-19 screening programme (University of Cambridge 2020).

Students were divided into 'bubbles' of average size 8 and maximum size 10, with each bubble consisting of students sharing facilities (e.g., a bathroom, kitchen or living room). In a typical week, half of the students in each 'bubble' were requested to provide nasal swabs, which were then collected together into a single container by one of the students. This container was then sent to a local laboratory, where a pooled PCR test was performed on the pooled samples. If a pooled sample tested positive, each student in the corresponding bubble was informed, and instructed to take an individual PCR test at one of the National Testing sites. A simple rota determined which students were asked to provide swabs on which weeks; on average, a student was asked to provide a swab approximately once per fortnight. Students who were symptomatic did not take part on weeks when they were symptomatic, as all students experiencing symptoms were instructed to seek an individual test. Students who had recently tested PCR-positive did not take part either. Participation was voluntary, but consent rates were high, starting at 75% of all 15,479 eligible students during the first week of term, and steadily increasing to 82% of all 15,310 eligible students in the last week.

Students were requested not to socialise outside their bubbles. Assuming a high level of compliance, this would mean that positive cases were more likely to cluster within bubbles, leading to a resource-saving at the second stage (of follow-up individual testing): this is the 'best case' in Dorfman's algorithm, in terms of resource-use at the second step, when the infected individuals are distributed among as few pools as possible. In the sixth week of term, for example, 80 students individually tested positive across 59 positive-testing pools (Warne 2020), giving an average of approximately 1.3 infected students per positive testing pool. Had the infected individuals been more evenly distributed, there would have been only one positive-testing student per positive-testing pool.

11.5.2 The Grid and P-BEST Algorithms in Israel

In August 2020, Israel's Ministry of Health approved two single-step pooled testing protocols for use in clinical laboratories in the country: one based upon the 'grid algorithm', and one based upon the 'P-BEST' algorithm of Shental et al. (2020). The grid algorithm has been described earlier.

P-BEST uses an $(r = 6, s = 48)$-regular design with a code-based construction. This deals with $N = 384$ individuals in $T = Nr/s = 48$ pooled tests. A 'best guess' for the identifies of the infected individuals (from the results of the pooled tests), is obtained using the method described in Sect. 11.3.3, above.

It should be noted that the P-BEST algorithm does require software and computing resources to implement, in addition to a pipetting robot with an arm that can move in two dimensions. However, this equipment is affordable by most countries.

In trials where at most 5 out of 384 individuals are PCR-positive under individual PCR testing (corresponding to a prevalence of 1.3% or lower), the P-BEST algorithm usually correctly identified all infected and uninfected individuals. Problems can arise when a batch of samples is received with a much higher prevalence than 1.3%; in this case, even if there are no false positive and or false negative test results, it is often not possible to use P-BEST to determine which individuals are infected and which are not.

In Israel, in clinical laboratories where P-BEST (or the grid algorithm) has been employed, data analysis and machine-learning has been employed also, to predict which batches of samples are likely to have much higher prevalence rates than the national average (based on origin); such batches were typically dealt with using individual testing.

11.5.3 A Multi-stage (r, s)-Regular Algorithm in Rwanda

Starting in August 2020, a multi-stage algorithm was piloted in Rwanda, where infection prevalence was low but the supply of PCR tests was limited (Mutesa et al. 2021).

The stages are as follows. It is required to pick two integer parameters, a and r_2.

1. A Dorfman-like stage with $r_1 = 1$ pooled test for each individual and $s_1 = a^{r_2}$ samples in each test. A negative result shows that all s_1 individuals are noninfected; individuals in a positive pool go through to stage two.

2. An $(r_2, s_2 = a^{r_2-1})$-regular design, with a hypercube construction (see Sect. 11.2.5).
3. If the hypercube contains zero or one infected individuals, they can be identified. Otherwise, further stages of testing are used to disambiguate the results; we don't go into details here.

The parameters are chosen to be as efficient as possible while still ensuring that the chance of more than two stages being required is very small. One common choice is $a = 3, r_2 = 3$, so that the first stage is a $(1, 81)$-regular Dorfman-like design, and the second stage is a $(3, 9)$-regular hypercube design of 9 tests for 81 individuals.

11.5.4 Other Uses of Pooled Testing

Here is a brief (and far from exhaustive) list of some other examples of the use of pooled testing in the COVID-19 pandemic.

- PCR testing using Dorfman's algorithm in Wuhan, China (Fan 2020). Between 12 May and 1 June 2020, 9.9 million Wuhan residents were tested; the vast majority were asymptomatic. Dorfman's algorithm was reportedly used for approximately 25% of this testing, with pools of sizes between 5 and 10. Only 300 positive cases were identified.
- Screening of students on University campuses, using Dorfman's algorithm or the grid algorithm, during the Autumn/Fall term of 2020: Université de Liège, Belgium (saliva samples using Dorfman's algorithm with pools of size 8) (Université de Liège 2020); Duke University, USA (saliva samples using Dorfman's algorithm with pools of size between 5 and 10; participation mandatory for students on campus) (Denny et al. 2020); Michigan State University, USA (saliva samples using a grid algorithm) (Michigan State University 2020); Syracuse University, USA (saliva samples using Dorfman's algorithm with pools of size between 20 and 25) (Syracuse University 2020); Shenandoah University, (saliva samples using Dorfman's algorithm with pools of size 4 or 5) USA (Shenandoah University 2020).
- PCR testing using Dorfman's algorithm with pools of size up to 20, by Fundación Biomédica Galicia Sur, Galicia, Spain, to screen asymptomatic healthcare workers, social care workers, industrial workers and port workers in the province of Galicia from September 2020 onwards, raising screening-capacity to 100,000 screenings per month (with health and social care workers being screened twice per week). Pipetting robots have been used (La Voz de Galicia 2020).
- PCR testing using Dorfman's algorithm with pools of size up to 30, by Saarland University Hospital, Germany, for the regular screening of asymptomatic hospital patients and hospital staff, and care home residents in Saarland, from March 2020 onwards. Approximately 22,000 people screened (Universität des Saarlandes 2020).
- PCR testing using Dorfman's algorithm with pools of size 10, by Noguchi Memorial Institute for Medical Research, Ghana, to test contacts of confirmed cases.

Initially 10,000 people tested per day, from April 2020 onwards (World Health Organization 2020).

- PCR testing using Dorfman's algorithm with pools of size 5, by the states of Uttar Pradesh (Sharda 2020) and West Bengal (Yengkhom 2020), India, in areas with estimated prevalence of 2% or lower.

11.6 Applications of Pooled Testing for COVID-19: Some Conclusions

In this section, we conclude, by drawing some of the above analysis together and discussing our own personal perspective on the practical settings where pooled testing for COVID-19 is likely to be useful, or at least may merit serious consideration.

11.6.1 Pooled Testing for Asymptomatic Subpopulations

As stated in the introduction, we believe that pooled testing is most likely to be useful for the screening of asymptomatic people, for surveillance, and possibly for the testing of contacts of confirmed cases, provided the prevalence of infection among the group to be tested is sufficiently low. On the other hand, we believe that in most countries, pooled testing is unlikely to be useful for the testing of symptomatic people. (Here, we use the term 'asymptomatic' to denote someone who is not experiencing the recognised symptoms at the time of their test. This includes 'true asymptomatics', who never experience symptoms, and 'pre-symptomatics', who go on to develop symptoms after their test.)

There are two main reasons why we believe that, in most countries, pooled testing will be of limited use for the testing of symptomatic people. First, the prevalence of COVID-19 infection among those presenting symptoms is usually sufficiently high that the resource savings of pooled testing are modest compared to individual testing, and may be outweighed by the down-sides of pooled testing, such as increased turnaround time. Among those presenting COVID-like symptoms, prevalences of between 4 and 33% are realistic, depending upon the setting, the location, the symptoms used in the definition of 'symptomatic', and the prevalence of other respiratory viruses (which in turn depends on the time of year) (Menni 2020; Pueyo 2020). Second, many countries already have well-established testing programmes using individual testing for those presenting symptoms. In many countries, including the UK, an individual test on symptomatic people is mandated by the regulatory authorities to confirm infection, even if they can be proved positive solely via pooled tests.

On the other hand, in many countries, the prevalence of COVID-19 infection among the general population has for quite long periods been at levels low enough that pooled testing of asymptomatic people can yield large efficiency gains over individual testing. For example, the estimated prevalence of current SARS-CoV-2 infection in the general community in England, as estimated by the ONS Infection Survey (Office for National Statistics 2021), has ranged from 0.026% in early July, to 2.1% in early January (and 3.6% in London in early January). Hence, for the prevalence among asymptomatics in England, values of p between 0.02 and 1.6% are good estimates. Countries that have adopted more stringent non-pharmaceutical interventions (such as stricter lockdowns or strongly enforced quarantines for international arrivals) have experienced lower prevalence rates; for example, New Zealand probably eliminated COVID-19 infections in the general community between early May and mid-August 2020, except for international arrivals, who were quarantined (Baker et al. 2020).

It is often desirable to screen subpopulations where the prevalence of infection is likely to be significantly higher than in the general population: for example, patient-facing healthcare workers, resident-facing social care workers, and factory workers in high-risk environments such as meat-processing plants. But it may also be desirable to screen subpopulations where the prevalence of infection is likely to be similar to the general population: see some of the examples below.

We list here some of the settings where we believe pooled testing for COVID-19 is most likely to be useful. Of course, a careful cost-benefit analysis should be carried out for each potential application, with the decision to adopt or not depending on certain factors, including the prevalence level, laboratory resources and capabilities, the impact of increasing the turnaround time, and regulatory constraints.

- **University students**. As may be apparent from the relatively large number of examples of this in Sect. 11.5, screening of asymptomatic university students is one of the less controversial applications of pooled testing. Severe illness is very rare among those of student age, so the impact on students of an increase in the turnaround time associated with pooled testing is slight. If a significant amount of in-person teaching is taking place, there are higher risks to older members of staff in the event that they are infected, so the impact on them of increased turnaround time should be taken into account.
- **Key workers**—for example factory workers, warehouse workers and port workers (but excluding patient-facing healthcare workers and resident-facing social care workers) provided the prevalence is not too high among the workers in question. We recall from Sect. 11.5.4 that pooled testing has been used in Galicia for the screening of factory and port workers.
- **School pupils and staff**. The logistical challenges of the regular screening of asymptomatic school pupils are greater than in the case of university students (who can, if necessary, organise much of the process themselves; see Sect. 11.5.1). Schoolchildren—particularly younger schoolchildren—cannot do this, so the additional organisational burden is placed on schools, who are already overstretched. A further problem is that there is little incentive for low-income families to agree for their children to participate in regular screening, since if they test positive and this is reported, the parents are likely to have to take time off work. (Financial compensation for this may help.) Another concern is that school-aged children—

particularly primary-school aged children—may not tolerate regular swabbing, although saliva tests would not have this problem.

- **Members of sports teams**. Screening of asymptomatic members of sports teams is suggested in Mutesa et al. (2021).
- **Airline passengers**. This is also suggested in Mutesa et al. (2021).
- **Non-household contacts of confirmed cases**, provided the estimated prevalence among these is sufficiently low. We recall from Sect. 11.5.4 that this has been done in Ghana.

One potential application of pooled testing that many authors are circumspect about is the screening of asymptomatic healthcare workers and social care workers. Given the major risks to vulnerable patients and social care residents associated with any additional delay in finding positive cases among these workers, screening using individual testing is often thought to be preferable, if there are sufficient resources. Even in the presence of very severe resource constraints, a careful cost-benefit analysis should be performed to compare the impact of screening based on individual testing with that based on pooled testing, taking into account the increased turnaround time. Even Mutesa et al. (2021), who are in general strong advocates of the use of pooled testing for COVID-19, state explicitly that they do not advocate its use for the screening of healthcare workers. We do note, however, that pooled testing was used by Saarland University Hospital for this purpose (see Sect. 11.5.4).

11.6.2 Pooled Testing and Vaccination Programmes

At the time of writing (March 2021), vaccination programmes are proceeding rapidly in many developed countries and are having a large effect in reducing the number of COVID-19 cases (Aran 2021). It might be thought that, in such countries, there will soon be no need to consider pooled testing. We believe that such an assumption may be premature at this stage, mainly because we do not yet have reliable data on the extent to which the vaccines currently being distributed reduce the number of cases (symptomatic and asymptomatic) of new variants of COVID-19, particularly the South African and Brazilian variants (Mahase 2021). (See Aldridge and Ellis (2021) for a discussion of the evidence base on this to date.) Bearing in mind this uncertainty, and the risk of further new variants arising that are resistant to available vaccines, we believe it would be wise for decision-makers to bear in mind the possibility that it may be desirable to rapidly increase testing-capacity using pooled testing in the medium term. In poorer countries, vaccination programmes are likely to be long delayed. In such countries, pooled testing may still be a valuable tool to consider for the foreseeable future.

11.6.3 Pooled Testing for Surveillance

At low prevalence levels, pooled testing has the potential for very large resource-saving in national COVID-19 surveillance programmes. In some countries with a very large testing capacity, this may not be necessary—for example, the UK's ONS (Coronavirus) Infection Survey currently tests a random sample of approximately 400,000 members of the community population in England, once per fortnight, using individual testing (Office for National Statistics 2020); compared to the UK's Pillar II testing capacity of more than 200,000 tests per day, this is not too great a resource requirement. However, if it is desired to reduce the resource requirements of a nationwide surveillance programme, pooled testing provides a way of doing so. Using pools of size up to 100, the hypercube-based algorithm piloted in Rwanda by Mutesa et al. (2021) can estimate the prevalence fairly accurately, while achieving an approximately 100-fold reduction in number the tests used when the prevalence level is at 0.05%. The main concern here is the reduction in sensitivity caused by dilution, but Mutesa et al. (2021) report proof-of-concept experiments which suggest that, using an appropriate protocol, sensitivities of 98% or 92% (depending on the gene targeted), can be achieved at a 100-fold level of dilution. This suggests that their scheme can be used reliably to monitor prevalence. The extra turnaround time compared to individual testing is likely to be much less of an issue with surveillance than with case identification, particularly at low prevalence levels.

It is also plausible that pooled testing could be used to monitor the prevalence of new variants. This may become particularly important if new variants begin to seriously hinder the success of vaccination programmes. A new PCR-testing method (involving only a minor update to existing PCR tests) that can detect which variant of SARS-CoV-2 a patient is carrying is currently undergoing clinical trials by the biotechnology firm Novozymes (Merrifield 2021).

Acknowledgements The authors were supported in part by UKRI Research Grant EP/W000032/1.

References

M. Aldridge, Rates of adaptive group testing in the linear regime, in *2019 IEEE International Symposium on Information Theory (ISIT)* (2019), pp. 236–240. https://doi.org/10.1109/ISIT.2019.8849712

M. Aldridge, Conservative two-stage group testing. May 2020. arXiv:2005.06617

M. Aldridge, D. Ellis, Pooled testing and its applications in the COVID-19 pandemic. (Extended version.) (2021). arXiv:2105.08845

M. Aldridge, O. Johnson, J. Scarlett, Group testing: an information theory perspective. Found. Trends Commun. Inf. Theory **15**(3–4), 196–392 (2019)

D. Aran, Estimating real-world COVID-19 vaccine effectiveness in Israel using aggregated counts. Technical report, February 2021. medRxiv:2021.02.05.21251139

P. Ashcroft, J. Huisman, S. Lehtinen, J. Bouman, C. Althaus, R. Regoes, S. Bonhoeffer, Covid-19 infectivity profile correction. Swiss Med. Wkly. **150**, w20336 (2020)

M. Baker, N. Wilson, A. Anglemyer, Successful elimination of community transmission of SARS-CoV-2 in New Zealand. N. Engl. J. Med. **383**, e56 (2020)

L. Baldassini, O. Johnson, M. Aldridge, The capacity of adaptive group testing, in *2013 IEEE International Symposium on Information Theory Proceedings (ISIT)* (2013), pp. 2676–2680. https://doi.org/10.1109/ISIT.2013.6620712

A. Bateman, S. Mueller, K. Guenther, P. Shult, Assessing the dilution effect of specimen pooling on the sensitivity of SARS-CoV-2 PCR tests. J. Med. Virol. **93**, 1568–1572 (2021)

B. Böger, M. Fachi, R. Vilhena, A. Cobre, F. Tonin, R. Pontarolo, Systematic review with meta-analysis of the accuracy of diagnostic tests for COVID-19. Am. J. Infect. Control **49**(1), 21–29 (2021)

A. Broder, R. Kumar, A note on double pooling tests. Technical report, April 2020. arXiv:2004.01684

T. Denny et al., Implementation of a pooled surveillance testing program for asymptomatic SARS-CoV-2 infections. August 2–October 11, 2020. MMWR Morb. Mortal. Wkly. Rep. **69**(45), 1743–1747 (2020)

R. Dorfman, The detection of defective members of large populations. Ann. Math. Stat. **14**, 436–440 (1943)

D.-Z. Du, F. Hwang, *Combinatorial Group Testing and Its Applications*, 2nd edn. (World Scientific, Singapore, 2020)

W. Fan, Wuhan tests nine million people for coronavirus in ten days. Wall Street J. (2020), https://www.wsj.com/articles/wuhan-tests-nine-million-people-for-coronavirus-in-10-days-11590408910

P. Fischer, N. Klasner, I. Wegenera, On the cut-off point for combinatorial group testing. Discrete Appl. Math. **91**(1), 83–92 (1999)

F. Foucart, H. Rauhut, *A Mathematical Introduction to Compressive Sensing* (Springer, New York, 2013)

M. García-Fiñana, D. Hughes, C. Cheyne, G. Burnside, I. Buchan, C. Semple, Innova lateral flow SARS-CoV-2 antigen test accuracy in Liverpool pilot: preliminary data. Technical report, November 2020, https://assets.publishing.service.gov.uk/government/uploads/system/uploads/attachment_data/file/943187/S0925_Innova_Lateral_Flow_SARS-CoV-2_Antigen_test_accuracy.pdf

S. Ghosh et al., A compressed sensing approach to group-testing for COVID-19 detection. Technical report, May 2020. arXiv:2005.07895

X. He et al., Temporal dynamics in viral shedding and transmissibility of COVID-19. Nat. Med. **26**, 672–675 (2020)

W. Kautz, R. Singleton, Nonrandom binary superimposed codes. IEEE Trans. Inf. Theory **10**(4), 363–377 (1964)

La Voz de Galicia, Vigo estrena el laboratorio de microbiología para el sistema 'pooling' de cribado del covid-19. Technical report, September 2020, https://www.lavozdegalicia.es/noticia/vigo/vigo/2020/09/25/vigo-estrena-nuevo-laboratorio-microbiologia-sistema-pooling-cribado-covid-19/00031601031541411853966.htm

E. Mahase, Covid-19: point of care test reports 94% sensitivity and 100% specificity compared with laboratory test. BMJ **370**, m3682 (2020)

E. Mahase, Covid-19: where are we on variants and vaccines? BMJ **372**, n597 (2021)

C. Menni, Real-time tracking of self-reported symptoms to predict potential COVID-19. Nat. Med. **26**, 1037–1040 (2020)

R. Merrifield, New rapid coronavirus test can quickly reveal which variant a patient has. *The Daily Mirror*, February 2021, https://www.mirror.co.uk/news/world-news/new-rapid-coronavirus-test-can-23561606

Michigan State University, COVID-19 departure screening and testing. Technical report (2020), https://msu.edu/together-we-will/testing-reporting/departure-program-faq.html

L. Mutesa et al., A pooled strategy for identifying SARS-CoV-2 at low prevalence. Nature **589**, 276–280 (2021)

Office for National Statistics, COVID-19 Infection Survey (Pilot): methods and further information. September 2020, https://www.ons.gov.uk/peoplepopulationandcommunity/healthandsocialcare/conditionsanddiseases/methodologies/covid19infectionsurveypilotmethodsandfurtherinformation

Office for National Statistics, Coronavirus (COVID-19) infection survey pilot: England, Wales and Northern Ireland; UK statistical bulletins. Technical report (2021), https://www.ons.gov.uk/peoplepopulationandcommunity/healthandsocialcare/conditionsanddiseases/bulletins/coronaviruscovid19infectionsurveypilot/previousReleases

T. Pueyo, Coronavirus: learning how to dance. *Medium*, April 2020, https://tomaspueyo.medium.com/coronavirus-learning-how-to-dance-b8420170203e

S. Sharda, Pool testing to maximise COVID-19 screening in Uttar Pradesh. Technical report, April 2020, https://timesofindia.indiatimes.com/city/lucknow/pool-testing-to-maximise-covid-19-screening-in-uttar-pradesh/articleshow/75103433.cms

Shenandoah University, Shenandoah conducts pooled saliva testing for novel coronavirus. Technical report, October 2020, https://www.su.edu/blog/2020/10/shenandoah-conducts-pooled-saliva-testing-for-novel-coronavirus/

N. Shental et al., Efficient high-throughput SARS-CoV-2 testing to detect asymptomatic carriers. Sci. Adv. **6**(37), eabc5961 (2020)

J. Skitrall, M. Wilson, A. Smielewska, S. Parmar, M. Fortune, D. Sparkes, M. Curran, H. Zhang, H. Jalal, Specificity and positive predictive value of SARS-CoV-2 nucleic acid amplification testing in a low-prevalence setting. Clin. Microbiol. Infect. **27**(3), 469.e9–469.e15 (2021)

Syracuse University, Frequently asked questions about student testing. Technical report, July 2020, https://news.syr.edu/blog/2020/07/29/frequently-asked-questions-about-student-testing/

Universität des Saarlandes, Homburger Virologen steigern mit Poolverfahren die Kapazitäten für Coronavirus-Massentests. Technical report, April 2020, https://www.uni-saarland.de/universitaet/aktuell/artikel/nr/21848.html

Université de Liège, Dépistage Covid: une capacité de 30.000 à 60.000 tests par jour à l'U-Liège. Technical report, June 2020, https://www.news.uliege.be/cms/c_11967134/fr/depistage-covid-une-capacite-de-30-000-a-60-000-tests-par-jour-a-l-uliege

University of Cambridge, Asymptomatic COVID-19 screening programme. Technical report (2020), https://www.cam.ac.uk/coronavirus/stay-safe-cambridge-uni/asymptomatic-covid-19-screening-programme. Accessed 26 Feb 2021

B. Warne, University of Cambridge asymptomatic screening programme: week 6 (9–15 Nov. 2020). Technical report, November 2020, https://www.cam.ac.uk/sites/www.cam.ac.uk/files/documents/20115_uoc_screening_report_week_6.pdf

J. Watson, P. Whiting, Interpreting a COVID-19 test result. BMJ **369**, m1808 (2020)

S. Woloshin, N. Patel, A. Kesselheim, False negative tests for SARS-CoV-2 infection—challenges and implications. N. Engl. J. Med. **383**, e38 (2020). https://doi.org/10.1056/NEJMp2015897

World Health Organization, Pooling samples boosts Ghana's COVID-19 testing. Technical report, July 2020, https://www.afro.who.int/news/pooling-samples-boosts-ghanas-covid-19-testing

S. Yengkhom, West Bengal to start pool testing in low-risk zones. *Times of India*, April 2020, https://timesofindia.indiatimes.com/city/kolkata/bengal-to-start-pool-testing-of-samples-in-low-risk-zones/articleshow/75227413.cms

Open Access This chapter is licensed under the terms of the Creative Commons Attribution 4.0 International License (http://creativecommons.org/licenses/by/4.0/), which permits use, sharing, adaptation, distribution and reproduction in any medium or format, as long as you give appropriate credit to the original author(s) and the source, provide a link to the Creative Commons license and indicate if changes were made.

The images or other third party material in this chapter are included in the chapter's Creative Commons license, unless indicated otherwise in a credit line to the material. If material is not included in the chapter's Creative Commons license and your intended use is not permitted by statutory regulation or exceeds the permitted use, you will need to obtain permission directly from the copyright holder.

Chapter 12
Outlier Detection for Pandemic-Related Data Using Compositional Functional Data Analysis

Christopher Rieser and Peter Filzmoser

Abstract With accurate data, governments can make the most informed decisions to keep people safer through pandemics such as the COVID-19 coronavirus. In such events, data reliability is crucial and therefore outlier detection is an important and even unavoidable issue. Outliers are often considered as the most interesting observations, because the fact that they differ from the data majority may lead to relevant findings in the subject area. Outlier detection has also been addressed in the context of multivariate functional data, thus smooth functions of several characteristics, often derived from measurements at different time points (Hubert et al. in Stat Methods Appl 24(2):177–202, 2015b). Here the underlying data are regarded as compositions, with the compositional parts forming the multivariate information, and thus only relative information in terms of log-ratios between these parts is considered as relevant for the analysis. The multivariate functional data thus have to be derived as smooth functions by utilising this relative information. Subsequently, already established multivariate functional outlier detection procedures can be used, but for interpretation purposes, the functional data need to be presented in an appropriate space. The methodology is illustrated with publicly available data around the COVID-19 pandemic to find countries displaying outlying trends.

12.1 Introduction

The crisis caused by COVID-19 in almost all areas of life has also revealed that an accurate data collection is a challenge that cannot be easily resolved due to political or logistic problems. However, the availability of clean and reliable data is a key step in fighting a pandemic. On the one hand, knowing the real number of tested, newly infected and dead people allows to investigate the causes of the observed

C. Rieser · P. Filzmoser (✉)

Institute of Statistics and Mathematical Methods in Economics, TU Wien, Wiedner Hauptstr. 8-10, 1040 Vienna, Austria

e-mail: peter.filzmoser@tuwien.ac.at

C. Rieser

e-mail: christopher.rieser@tuwien.ac.at

© The Author(s) 2022

M. C. Boado-Penas et al. (eds.), *Pandemics: Insurance and Social Protection*,
Springer Actuarial, https://doi.org/10.1007/978-3-030-78334-1_12

developments and to take appropriate measures to stop the spread of an infection. On the other hand, insurance companies offering a protection linked to some specific events during a pandemic would like to have reliable data to avoid the possibility of moral hazard.

Many countries report the number of cases, deaths, tests, and further parameters (variables) related to the COVID-19 pandemic regularly over time, and the data are accessible in public data repositories. Rather than treating the data with tools from time series analysis, it is common to consider them as functional data, so that the measurements are represented by smooth functions over time. One could then analyse the multivariate information contained in the functions for the different variables, and compare the countries with respect to this information. Thus, countries for which the multivariate information differs from the main trend given by the majority of the countries are possible outliers. Instead of directly considering the reported number (represented by the functions), one could also focus on analysing relative information. This can be done by taking (log-)ratios between the variables. Thus, the source of information for the analysis would not consist in the number of cases, death, tests, etc., for a particular day in a particular country, but in the (log-)ratios between these numbers. This is what is done in compositional data analysis, and outlier detection in this context will focus on atypical behaviour in the multivariate information of such (log-)ratios. For example, if the development of the number of cases over time is similar in some countries, but in one country the number of deaths develops more rapidly, this could be much better visible in a (log-ratio) than in the reported values. Thus, treating COVID-19 data as compositional data and analysing relative rather than absolute information can be very beneficial for outlier detection.

In this paper we consider a new method for the detection of outliers in the compositional functional data setting. The detection of outliers in the p-dimensional multivariate data case has been intensively investigated throughout the years and many methods have been developed. Denote by $\mathbf{x}_k \in \mathbb{R}^p$, for $k = 1, ..., K$, the observed samples. A popular approach considers an outlier of these samples as a point $\mathbf{x}_{k_0}$ for which the robustified version of the Mahalanobis distance, $\sqrt{(\mathbf{x}_{k_0} - \mathbf{m})' \mathbf{C}^{-1} (\mathbf{x}_{k_0} - \mathbf{m})}$, where $\mathbf{m}$ respectively $\mathbf{C}$ are robust estimators for the mean and the covariance matrix, is above a certain threshold and thus far away from the centre $\mathbf{m}$ with respect to the covariance structure $\mathbf{C}$; see Rousseeuw (1985), Rousseeuw and Driessen (1999) and Hubert and Debruyne (2010). The idea of defining an outlier as a point being far away from the centre has been extended to more general measures related to statistical depth, see Tukey (1975), Serfling (2006) and Mosler (2012).

In recent years, many methods of multivariate statistics have been generalised to Functional Data Analysis (FDA). In FDA one considers data points to be whole functions, i.e. in the notation above, data points $\mathbf{x}_k : I \to \mathbb{R}^p$ are multivariate functions; for an overview of FDA we refer to Ramsay (2004), Ferraty and Vieu (2006) or Kokoszka and Reimherr (2017). Accordingly, the concept of outliers has been extended from the multivariate to the FDA setting, see Fraiman and Muniz (2001), Febrero et al. (2008), Sun and Genton (2011) and Hubert et al. (2015b).

In this paper we consider extending the ideas of outlyingness to functional data with image in the compositional data space. Thus, Sects. 12.1.1 and 12.1.2 provide a short introduction to the concepts of compositional data analysis and functional data, respectively. Further, in Sect. 12.2 we consider smoothing for functional data with image in the compositional space. In Sect. 12.3 we look at how one can detect outliers for the latter setting. That is, we extend the methods of detecting outliers from the non-compositional FDA case to the compositional one. Furthermore, Sect. 12.4 contains an application of the method presented. The data is comprised of COVID-19 data of different countries over time. Each country represents a functional data point. We finish in Sect. 12.5 with a summary and some conclusions.

12.1.1 Compositional Data Analysis Concepts

Assume we have given a D-dimensional random vector $\mathbf{x}$ for which each entry is strictly positive, i.e. $\mathbf{x} \in \mathbb{R}^D_+$, where $\mathbb{R}^D_+$ denotes the D-dimensional real number space with strictly positive entries. In the framework of compositional data analysis (CODA) it is assumed that the ratios $\frac{x_j}{x_k}$, for any $j, k \in \{1, ..., D\}$, $j \neq k$, carry the relevant information, and thus only relative information is essential. As ratios do not change when multiplying $\mathbf{x}$ with a strictly positive scalar $\lambda > 0$, it holds that $\lambda \mathbf{x} =: \mathbf{y}$ carries the same information as $\mathbf{x}$. This motivates defining the equivalence relation

$$\mathbf{x} \sim \mathbf{y} \quad \Longleftrightarrow \quad \exists \lambda > 0 \quad \lambda \mathbf{x} = \mathbf{y} \text{ for any } \mathbf{x}, \mathbf{y} \in \mathbb{R}^D_+$$

which partitions the space $\mathbb{R}^D_+$ into equivalence classes. Choosing for each equivalence class the representative $\mathbf{x} = (x_1, ..., x_D)'$ satisfying $\sum_{j=1}^D x_j = 1$, leads to the set of equivalence classes called the D-part simplex

$$\mathcal{S}^D := \left\{ \mathbf{x} = (x_1, ..., x_D)' \in \mathbb{R}^D_+, \sum_{j=1}^D x_j = 1 \right\}.$$

The space $\mathcal{S}^D$ is turned into a Hilbert space—called the Aitchison geometry on the simplex, see Aitchison (1982)—by defining addition (perturbation), multiplication with a scalar (powering), an inner product and a norm for $\mathbf{x} = (x_1, ..., x_D)'$, $\mathbf{y} = (y_1, ..., y_D)' \in \mathcal{S}^D$ and $\alpha \in \mathbb{R}$:

- Perturbation: $\mathbf{x} \oplus \mathbf{y} := (x_1 y_1, ..., x_D y_D)'$
- Powering: $\alpha \odot \mathbf{x} := (x_1^\alpha, ..., x_D^\alpha)'$
- Inner product:

$$\langle \mathbf{x}, \mathbf{y} \rangle_A := \frac{1}{2D} \sum_{j=1}^D \sum_{k=1}^D \log\left(\frac{x_j}{x_k}\right) \log\left(\frac{y_j}{y_k}\right)$$

- Norm: $\|\mathbf{x}\|_A := \sqrt{\langle \mathbf{x}, \mathbf{x} \rangle_A}.$

Furthermore, the Aitchison geometry is (bijectively) isometric to $\mathbb{R}^{D-1}$. To show this, firstly define the centred log-ratio (clr)

$$\text{clr} : \mathcal{S}^D \to \mathbb{R}^D, \quad \text{clr}(\mathbf{x}) := \left(\log \left(\frac{x_1}{\sqrt[D]{\prod_{j=1}^D x_j}} \right), ..., \log \left(\frac{x_D}{\sqrt[D]{\prod_{j=1}^D x_j}} \right) \right)'$$

(12.1)

which satisfies the properties of being invariant under the above operations and the norm, i.e.

$$\text{clr}(\mathbf{x} \oplus \mathbf{y}) = \text{clr}(\mathbf{x}) + \text{clr}(\mathbf{y}) \tag{12.2}$$

$$\text{clr}(\alpha \odot \mathbf{x}) = \alpha \, \text{clr}(\mathbf{x}) \tag{12.3}$$

$$\langle \mathbf{x}, \mathbf{y} \rangle_A = \langle \text{clr}(\mathbf{x}), \text{clr}(\mathbf{y}) \rangle_E, \tag{12.4}$$

see Filzmoser et al. (2018). However, as for any $\mathbf{x} \in \mathcal{S}^D$, the entries of $\text{clr}(\mathbf{x})$ sum up to zero, $\sum_{i=1}^D \text{clr}(\mathbf{x})_i = 0$, it follows that the clr mapping does not satisfy the property of being one-to-one onto $\mathbb{R}^D$. To obtain a bijective mapping, choose a $D-1$ dimensional basis $\mathbf{V} = (\mathbf{v}_1, ..., \mathbf{v}_{D-1})$, where $\mathbf{v}_j \in \mathbb{R}^D$, for $j = 1, ..., D-1$, are clr coefficients, and define the isometric log-ratio (ilr) mapping as

$$\text{ilr}_\mathbf{V} : \mathcal{S}^D \to \mathbb{R}^{D-1}, \quad \text{ilr}_\mathbf{V}(\mathbf{x}) := \mathbf{V}' \text{clr}(\mathbf{x}). \tag{12.5}$$

The latter is a one-to-one mapping fulfilling (12.2), (12.3) and (12.4), see Filzmoser et al. (2018). As there are infinitely many possibilities to choose a basis $\mathbf{V}$, ilr coefficients are frequently considered to express all relative information of a composition appropriately in the usual Euclidean geometry, for which the common statistical tools have been designed. If an interpretation is desirable, the relative information is often re-expressed in terms of clr coefficients by $\text{clr}(\mathbf{x}) = \mathbf{V} \, \text{ilr}_\mathbf{V}(\mathbf{x})$, because they relate to the original compositional parts in terms of relative information of the part to an "average" (geometric mean), see (12.1).

12.1.2 Functional Data

In FDA we consider observations to be multivariate smooth functions $\mathbf{f} : [t_1, t_N] \to \mathbb{R}^D$. In practice, such observations often originate as time series, measured at certain time points t_i, with $i = 1, ..., N$, and thus they are not necessarily forming smooth functions. In this case, a preprocessing step is needed to find an estimate $\hat{\mathbf{f}}$ for $\mathbf{f}$ given $(t_i, \mathbf{y}_i)$, with $\mathbf{y}_i \in \mathbb{R}^D$, $i = 1, ..., N$, being noisy samples of $\mathbf{f}(t_i)$. We assume in the following Gaussian centred uncorrelated noise with equal variance. Although many methods exist to recover smooth functions, it is common that $\hat{\mathbf{f}}$ is estimated by smoothing spline methods. The literature on spline methods is vast and we refer to

Reinsch (1967), Wood (2017) and Yee (2015) for a good overview. The main idea is that given multivariate data $(t_i, \mathbf{y}_i)$ we find an estimate $\hat{\mathbf{f}}$ which is, on the one hand, sufficiently smooth but, on the other, also a good approximation to the data. It is common to look at the following vector valued smoothing problem

$$\hat{\mathbf{f}} := \arg\min_{\mathbf{f}} \sum_{i=1}^{N} \|\mathbf{y}_i - \mathbf{f}(t_i)\|_E^2 + \lambda \int_{t_1}^{t_N} \|\mathbf{f}''(t)\|_E^2 \, dt, \qquad (12.6)$$

where $\lambda > 0$ is a fixed smoothing parameter, and $\|\cdot\|_E$ denotes the Euclidean norm. The idea is that with increasing λ, the second derivative $\mathbf{f}''$ is forced to zero, i.e. towards a linear function. From Problem (12.6) it can be deduced that the solution is of the form $\mathbf{f}(t) := \sum_{i=1}^{N} \mathbf{a}_i b_i(t)$, see Yee (2015), with b_i being basis functions of the cubic spline space, and $\mathbf{a}_i$ being fixed vectors in $\mathbb{R}^D$. Plugging this basis expansion into (12.6) shows that the penalty function acts as regularisation penalty on $\mathbf{a}_i$ restraining the flexibility of the latter. In reality, one never uses the full basis expansion as given above, but rather a different and equally flexible expansion with less basis functions to save coefficients and avoid unnecessary computation in the case of a lot of data, for example a B-spline basis. Plugging in a specific basis expansion $\mathbf{f}(t) := \sum_{i=1}^{N} \mathbf{a}_i b_i(t)$ we can see that the problem is a convex problem, and solving this vector valued problem is discussed in Yee (2015).

12.2   Smoothing for CODA Time Series

In this section we consider functional observations with image in $\mathcal{S}^D$, i.e. functions $\mathbf{u} : [t_1, t_N] \to \mathcal{S}^D$. As before, we assume that only a set of discrete samples $(t_i, \mathbf{x}_i)$ is given, with $i = 1, \ldots, N$ and $\mathbf{x}_i \in \mathcal{S}^D$, where $\mathbf{x}_i$ is a sample of $\mathbf{u}(t_i)$. To construct a smooth estimate $\hat{\mathbf{u}}$ of $\mathbf{u}$, we firstly define derivatives and smoothing splines in a compositional context. For a function $\mathbf{u} : [t_1, t_N] \to \mathcal{S}^D$, its derivative at a time point t is defined as

$$\mathbf{u}'(t) := \lim_{h \to 0} \frac{1}{h} \odot \mathbf{u}(t + h) \ominus \mathbf{u}(t). \qquad (12.7)$$

Accordingly, one can define higher order derivatives inductively, e.g. $\mathbf{u}''(t) := (\mathbf{u}')'(t)$. For a reference on compositional calculus we refer to Pawlowsky-Glahn and Buccianti (2011). In accordance with the previous section, define $\hat{\mathbf{u}}$ as

$$\hat{\mathbf{u}} := \arg\min_{\mathbf{u}} \sum_{i=1}^{N} \|\mathbf{x}_i \ominus \mathbf{u}(t_i)\|_A^2 + \lambda \int_{t_1}^{t_n} \|\mathbf{u}''(t)\|_A^2 \, dt, \qquad (12.8)$$

where $\lambda > 0$ is again a fixed smoothing parameter controlling the smoothness.

Using the continuity of ilr_V and (12.2), it follows that

$$
\begin{aligned}
\mathrm{ilr}_V(\mathbf{u}')(t) &= \mathrm{ilr}_V\left(\lim_{h\to 0}\left\{\frac{1}{h}\odot \mathbf{u}(t+h)\ominus \mathbf{u}(t)\right\}\right)\\
&= \lim_{h\to 0}\mathrm{ilr}_V\left(\frac{1}{h}\odot \mathbf{u}(t+h)\ominus \mathbf{u}(t)\right)\\
&= \lim_{h\to 0}\frac{\mathrm{ilr}_V(\mathbf{u}(t+h)) - \mathrm{ilr}_V(\mathbf{u}(t))}{h}\\
&= \mathrm{ilr}_V(\mathbf{u})'(t)
\end{aligned}
$$

holds. With the same arguments, the equation $\mathrm{ilr}_V(\mathbf{u}'')(t) = \mathrm{ilr}_V(\mathbf{u})''(t)$ follows.

Therefore, defining $\mathbf{f} := \mathrm{ilr}_V(\mathbf{u})$, Problem (12.8) can be reformulated using the latter, as well as the properties (12.2) and (12.4):

$$
\arg\min_{\mathbf{u}}\sum_{i=1}^{N}\|\mathbf{x}_i\ominus \mathbf{u}(t_i)\|_A^2 + \lambda\int_{t_1}^{t_n}\|\mathbf{u}''(t)\|_A^2\,dt \tag{12.9}
$$

$$
\Longleftrightarrow\ \arg\min_{\mathbf{u}}\sum_{i=1}^{N}\|\mathrm{ilr}_V(\mathbf{x}_i) - \mathrm{ilr}_V((\mathbf{u}(t_i))\|_A^2 + \lambda\int_{t_1}^{t_n}\|\mathrm{ilr}_V((\mathbf{u}''(t))\|_A^2\,dt \tag{12.10}
$$

$$
\Longleftrightarrow\ \arg\min_{\mathbf{f}}\sum_{i=1}^{N}\|\mathrm{ilr}_V(\mathbf{x}_i) - \mathbf{f}(t_i)\|_E^2 + \lambda\int_{t_1}^{t_n}\|\mathbf{f}''(t)\|_E^2\,dt. \tag{12.11}
$$

The latter is a vector valued smoothing problem in $\mathbb{R}^{D-1}$ for the data $(t_i, \mathrm{ilr}_V(\mathbf{x}_i))$, see Problem (12.6), and it can be solved accordingly.

Given a solution $\hat{\mathbf{f}}$ to (12.11), a solution to (12.8) is then $\hat{\mathbf{u}} = \mathrm{ilr}_V^{-1}(\hat{\mathbf{f}})$ per definition of $\mathbf{f}$. In the case that different solutions to (12.11) exist, e.g. $\hat{\mathbf{f}}_1$ and $\hat{\mathbf{f}}_2$, we know from the equivalence chain before and from the fact that ilr_V is isometric, that also $\mathrm{ilr}_V^{-1}(\hat{\mathbf{f}}_1)$ and $\mathrm{ilr}_V^{-1}(\hat{\mathbf{f}}_2)$ are different solutions to Problem (12.8). Equally, having two different solution of (12.8) leads to different solutions of (12.11). This means that if (12.11) is uniquely solvable for a chosen V, we get that $\hat{\mathbf{u}}$ is also uniquely determined. Therefore, the choice of V is irrelevant. With the exception of some very degenerate settings, Problem (12.11) is uniquely solvable in most applications.

12.3 Outlier Detection in Compositional FDA

In the univariate case we can think of outliers as observations being very far away from the main mass of the data set, thus far away from the data centre with respect to the scale (Maronna et al. 2006).

The outlyingness of a multivariate observation $\mathbf{x} \sim P_{\mathbf{X}}$, where $P_{\mathbf{X}}$ denotes the distribution of a p-dimensional random vector $\mathbf{X}$ and $\mathbf{x}$ a realisation, can be built on the univariate case by means of projection onto a line defined by $\mathbf{r} \in \mathbb{R}^p$, with $\|\mathbf{r}\| = 1$, thus $\mathbf{r}'\mathbf{X}$. As discussed in Donoho et al. (1992), the outlyingness of an observation $\mathbf{x}$ of the projection $\mathbf{r}'\mathbf{x}$ can be measured by

$$\frac{|\mathbf{r}'\mathbf{x} - \text{median}(\mathbf{r}'\mathbf{X})|}{\text{mad}(\mathbf{r}'\mathbf{X})}, \tag{12.12}$$

where "mad" denotes the median absolute deviation, i.e. the median of $|\mathbf{X} - \text{median}(\mathbf{X})|$. Taking the supremum of (12.12) over all $\mathbf{r}$ with $\|\mathbf{r}\| = 1$ yields a measure of outlyingness for any $\mathbf{x}$ independent of the direction $\mathbf{r}$. Adjusting (12.12) for skewness—see Hubert and Vandervieren (2008) for adjusted boxplots of skewed distributions in the univariate case—the adjusted outlyingness (AO) is defined as

$$AO(\mathbf{x}, P_{\mathbf{X}}) := \begin{cases} \sup_{\|\mathbf{r}\|=1}\left(\frac{\mathbf{r}'\mathbf{x}-\text{median}(\mathbf{r}'\mathbf{X})}{w_2(\mathbf{r}'\mathbf{X})-\text{median}(\mathbf{r}'\mathbf{X})}\right) & \text{if } \mathbf{r}'\mathbf{x} > \text{median}(\mathbf{r}'\mathbf{X}) \\ \sup_{\|\mathbf{r}\|=1}\left(\frac{\text{median}(\mathbf{r}'\mathbf{X})-\mathbf{r}'\mathbf{x}}{\text{median}(\mathbf{r}'\mathbf{X})-w_1(\mathbf{r}'\mathbf{X})}\right) & \text{if } \mathbf{r}'\mathbf{x} \leq \text{median}(\mathbf{r}'\mathbf{X}), \end{cases}$$

where w_1 and w_2 are functions that allow to adjust for the skewness of the univariate distributions, see Hubert et al. (2015b) for an exact definition of these two functions.

To obtain a measure of outlyingness in the FDA case, e.g. for the data $(\mathbf{f}: [t_1, t_N] \to \mathbb{R}^p) \sim P_{\mathbf{F}}$, Hubert et al. (2015b) propose to use the functional adjusted outlyingness of a FDA point $\mathbf{f}$:

$$FAO(\mathbf{f}, P_{\mathbf{F}}) := \int_{t_1}^{t_N} AO(f(t), P_{F(t)})dt,$$

where $P_{f(t)}$ denotes the marginal distribution of $\mathbf{F}$ for fixed t.

In a compositional functional data context, where the compositions are functions of the form $\mathbf{u}: [t_1, t_N] \to \mathcal{S}^D$, with distribution $P_{\mathbf{U}}$, we propose to define the compositional functional adjusted outlyingness as

$$CFAO(\mathbf{u}, P_{\mathbf{U}}) := \int_{t_1}^{t_N} AO(\text{ilr}_{\mathbf{V}}(u(t)), P_{\text{ilr}_{\mathbf{V}}(U(t))})dt. \tag{12.13}$$

For Definition (12.13) to be a valid measure of outlyingness it needs to be checked that it is well defined, i.e., this measure needs to be independent of the choice of the basis matrix $\mathbf{V}$. As $\mathbf{V}\,\text{ilr}_{\mathbf{V}}(\mathbf{x}) = \text{clr}(\mathbf{x})$ holds by definition for a matrix with orthonormal columns $\mathbf{V}$, we have for a different matrix $\widetilde{\mathbf{V}}$ with orthonormal columns $\text{ilr}_{\mathbf{V}}(\mathbf{x}) = \mathbf{V}'\,\text{clr}(\mathbf{x}) = \mathbf{V}'\widetilde{\mathbf{V}}\,\text{ilr}_{\widetilde{\mathbf{V}}}(\mathbf{x})$, see Filzmoser et al. (2018). As the matrix $\mathbf{V}'\widetilde{\mathbf{V}} \in \mathbb{R}^{(D-1)\times(D-1)}$ is of full rank $D - 1$, we get, for any fixed t

$$AO(\text{ilr}_{\mathbf{V}}(u(t)), P_{\text{ilr}_{\mathbf{V}}(U(t))}) = AO(\mathbf{V}'\widetilde{\mathbf{V}}\,\text{ilr}_{\widetilde{\mathbf{V}}}(u(t)), P_{\mathbf{V}'\widetilde{\mathbf{V}}\,\text{ilr}_{\widetilde{\mathbf{V}}}(U(t))}) \quad (12.14)$$

$$= AO((\mathbf{V}'\widetilde{\mathbf{V}})(\text{ilr}_{\widetilde{\mathbf{V}}}(u(t))), P_{(\mathbf{V}'\widetilde{\mathbf{V}})(\text{ilr}_{\widetilde{\mathbf{V}}}(U(t)))}) \quad (12.15)$$

$$= AO(\text{ilr}_{\widetilde{\mathbf{V}}}(u(t)), P_{\text{ilr}_{\widetilde{\mathbf{V}}}(U(t))}) \quad (12.16)$$

where the last equality follows from the affine invariance property of AO, see Hubert and Van der Veeken (2008); affine invariance means that $AO(\mathbf{x}, P_{\mathbf{X}}) = AO(\mathbf{Ax} + \mathbf{b}, P_{\mathbf{AX}+\mathbf{b}})$ holds for any regular matrix $\mathbf{A} \in \mathbb{R}^{p\times p}$ and $\mathbf{b} \in \mathbb{R}^p$ for $\mathbf{x} \in \mathbb{R}^p$ with $\mathbf{x} \sim P_{\mathbf{X}}$. As CFAO is defined as an integral over (12.16) it follows that the latter is equally invariant and thus well defined.

To visually find outliers in the FDA setting, Hubert et al. (2015a) introduced a functional outlier map (FOM). Assume that the evaluation of K multivariate functional data points $\mathbf{f}_1, \ldots, \mathbf{f}_K$ is given at time points $t_1, \ldots, t_n$, and denote P_K the sample distribution of the functional data points, and P_{t_i} the sample distribution of the evaluations at time point t_i. The FOM is defined as a two dimensional graph, plotting $FAO(\mathbf{f}_k, P_K)$ on the horizontal axis against

$$\frac{\sigma_{i=1,\ldots,N}((AO(\mathbf{f}_k(t_i), P_{t_i}))_i)}{(1 + FAO(\mathbf{f}_k, P_K))} \quad (12.17)$$

on the vertical axis, for $k = 1, \ldots, K$, where σ denotes the standard deviation. The motivation behind this map is that when a data point $\mathbf{f}_k$ is a shift outlier, its according point in the FOM plot will be higher on the horizontal axis. If a data point $\mathbf{f}_k$ displays an outlying high variability in time, this will result in a high value on the vertical axis in the FOM plot. The denominator in (12.17) is necessary to correct for the effect that when a data point is shifted further, this is reflected in the standard deviation accordingly, see Hubert et al. (2015a).

Given the evaluation of the compositional functional data $\mathbf{u}_1, \ldots, \mathbf{u}_K$, $k = 1, \ldots, K$, at time points $t_1, \ldots, t_N$, we suggest equivalently to plot $CFAO(\mathbf{u}_k, P_K)$ on the horizontal axis, against

$$\frac{\sigma_{i=1,\ldots,N}((AO(\text{ilr}_{\mathbf{V}}(\mathbf{u}_k(t_i)), P_{t_i}))_i)}{(1 + CFAO(\mathbf{u}_k, P_K))} \quad (12.18)$$

on the vertical axis. Again, the latter is independent of the choice of $\mathbf{V}$, because AO as well as $CFAO$ are affine invariant, see the reasoning for (12.16) and its conclusion.

12.4 Application to COVID-19 Data

In this section we use data from `https://covid.ourworldindata.org`, which are publicly available. This page contains for most countries of the world daily information related to the COVID-19 pandemic. Here we focus on European countries only, and on the following information:

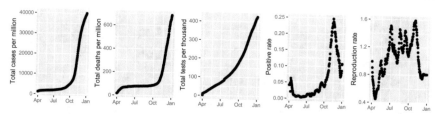

Fig. 12.1 COVID-19 data from Austria in the period April 1 until December 31, 2020. The plots show daily data for the 5 variables used for the analysis

- Total number of COVID-19 infections per million inhabitants.
- Total number of COVID-19 deaths per million inhabitants.
- Total number of COVID-19 tests per million inhabitants.
- Positive rate, i.e. share of total COVID-19 tests that were positive.
- Reproduction rate, referring to the expected number of cases directly generated by one case.

We select the time period from April 1 until December 31, 2020, because from April onwards the information was consistently collected in the data base. However, for some of the European countries the information on some of the variables was not available, so that finally only 35 European countries could be used. Still, for some countries there were missing values (or shorter time periods with missings), which have been imputed by a weighted moving average imputation method, implemented as function na_ma() in the R package imputeTS (Moritz and Bartz-Beielstein 2017).

As an example, Fig. 12.1 shows the data for Austria, and the data structure is similar in many of the other countries. Still, there might be countries with deviations in the multivariate data structure, and the task is to identify such countries. The focus here is on relative information in terms of log-ratios between the different variables.

Figure 12.1 reveals that the total number of cases starts to grow quickly in October 2020, and the same is true for the total number of deaths (per million). The number of tests grows steadily over the time period. The positive rate decreases at the beginning of this selected time period, but it increases drastically in October, followed by a decline in November/December. The reproduction rate fluctuates more, and has higher values than one in the summer and fall.

Multivariate functional outlier detection is here first applied to the data expressed in relative information, i.e. as ilr coordinates. In a second stage we also compare with an analysis based on absolute information, as reported in Fig. 12.1 for Austria. Naturally, the different treatment of the data will very likely lead to different results. As an example for relative versus absolute information, we may consider just the number of cases and the number of deaths (per million). For most countries, an increase of cases also implies an increase of deaths, probably with a different time delay. If one looks at relative information in terms of a log-ratio, however, differences between the countries might get more clearly pronounced. We will come back to this issue later.

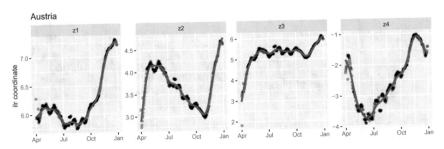

Fig. 12.2 Ilr coordinates of the data from Austria, together with the lines after smoothing. The smoothed lines (for every country) are the input for compositional functional outlier detection

For every country, the data are first ilr-transformed, resulting in time series of the ilr coordinates. Since the specific choice of the ilr coordinates is not relevant here, we use so-called pivot coordinates, where the first coordinate expresses all relative information of the first part to the remaining parts in the composition, see Filzmoser et al. (2018). Figure 12.2 shows the resulting ilr coordinates for the Austrian data; since there are 5 variables available, see Fig. 12.1, we end up with 4 ilr coordinates. Figure 12.2 also shows the lines after smoothing the data in ilr coordinates, thus after solving Problem (12.11). The information of these lines form the compositional functional data as they are used for multivariate outlier detection. Since we used pivot coordinates, only the first coordinate (denoted here by z_1) has a clear interpretation in terms of all relative information of the total cases to the remaining variables. This coordinate is in fact proportional to the first clr coefficient (Filzmoser et al. 2018). We will show and discuss the corresponding clr coefficients later in Fig. 12.5.

Once the smooth functions are estimated for every country, compositional functional outlier detection can be performed. Figure 12.3 shows the compositional functional outlier map (CFOM). Every point in the plot corresponds to a country, and the line indicates the outlier cutoff. It can be seen that one (red) point (Iceland) slightly exceeds the cutoff, and another point (Belarus) is just below the cutoff. The sorted compositional functional adjusted outlyingness is again shown in Fig. 12.4 (left), with the corresponding country names added. The values for Iceland and Belarus clearly stick out, and the next biggest value originates from the data from Luxembourg. These countries are not particularly outlying in their variability in time, since their values in Fig. 12.4 (right) are not unusual.

Figure 12.5 is an attempt to identify the reason for outlyingness. The plots show the smoothed functional data in clr coefficients, which are simply obtained by a transformation from the functions in ilr coordinates, see Eq. (12.5). The function for Iceland is shown in red, and that for Belarus in blue. For example, the clr coefficients for the total cases (left plot) mainly show a strongly increasing trend at the beginning, and again at the end of the considered time period. This means that the cases have grown rapidly, relative to the remaining variables (on average). The function for Belarus (blue) shows a quite different behaviour, with very high values especially around May. This means that the total cases are very much dominating over the values of the other variables. The reason for this is not because of high values of

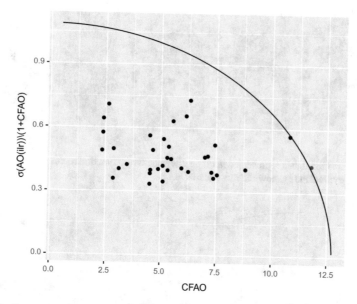

Fig. 12.3 Compositional functional outlier map: the points represent the countries, and the line is the outlier cutoff. Iceland exceeds the cutoff value, Belarus is just below the cutoff, see also Fig. 12.4

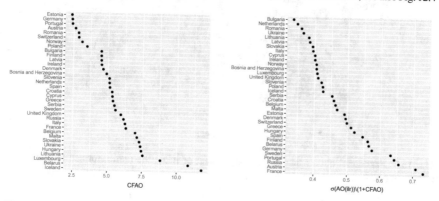

Fig. 12.4 Sorted compositional functional adjusted outlyingness (left), and sorted values from the vertical axis in Fig. 12.3 (right)

cases, but because of exceptionally low (reported) values of the remaining variables. Also the values for Iceland (red curves) are seen as atypical. For example, the clr coefficients of the total cases started to be the lowest in April, but then increased to be the highest in August. In a ratio, it can either be the change in the numerator or in the denominator, or in both, to get this behaviour, but in any case it turns out to be quite different compared to the other countries.

As a comparison, the following analysis is based on absolute information. Thus, the smoothed curves are directly estimated from the raw input data without any trans-

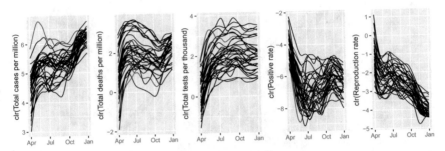

Fig. 12.5 Functional data represented in clr coefficients. Every function represents the time series of one country; Iceland is shown in red, Belarus in blue

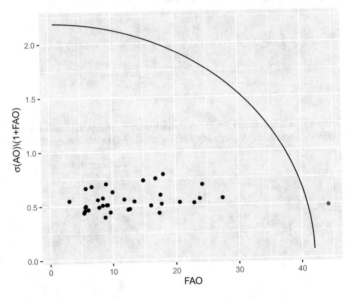

Fig. 12.6 Functional outlier map (FOM) as a result of using the untransformed absolute data information

formation, see Eq. (12.6). Then multivariate functional outlier detection is applied, which results in the functional outlier map presented in Fig. 12.6. Here, one point clearly exceeds the outlier cutoff value, and this point is Luxembourg.

Details are presented in Fig. 12.7, where the left plot are the sorted values from the horizontal axis, and the right plot the sorted values of the vertical axis from the FOM of Fig. 12.6. Indeed, Luxembourg appears with an exceptionally high value of FAO, and neither Iceland nor Belarus are atypical in any of these plots.

Finally, Fig. 12.8 shows the raw functional data. The outlier Luxembourg is shown by green curves, Iceland in red, and Belarus in blue. Luxembourg shows a very clear difference in the total tests, which might be the reason for the multivariate outlyingness. The countries Iceland and Belarus, which were clearly different in the

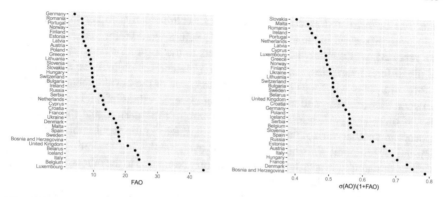

Fig. 12.7 Horizontal axis (left) and vertical axis (right) from Fig. 12.6 for functional outlier detection based on the untransformed absolute information

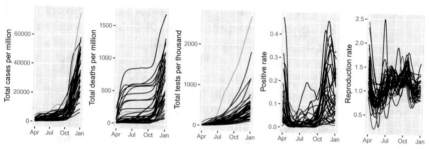

Fig. 12.8 Smoothed curves for the untransformed (absolute) data, with Luxembourg in green, Iceland in red, and Belarus in blue

compositional analysis, follow the main data structure well and do no longer appear as atypical. This shows that both types of analysis indeed focus on different data aspects, and it will be based on the task and research question to determine which of the analysis is more appropriate.

12.5 Summary and Conclusions

Outlier detection has been a relevant task in data analysis already since the beginning of data collection, and it continues being important also for more complex data structures. The identified outliers may point at atypical events, and depending on the context even at possible cases of fraud; see, e.g., van Capelleveen et al. (2016) or Nian et al. (2016). Outlier detection methods are also useful for pandemic-related data, as they may guide policy makers to draw appropriate conclusions.

Here we have used publicly available time series data related to COVID-19, as they are reported from different countries. The multivariate information, here in terms of the number of cases, deaths, tests, the positive rate, and the reproduction rate, has been treated as compositional data, where relative rather than absolute values are processed in the analysis. Absolute values would refer to the data as they are reported, while relative information refers to the log-ratios between the values of the different variables. An outlier detection method which makes use of relative information thus will focus more on the differences of the developments over time between the variables, and not necessarily on extreme values in single variables. In fact, if there is a peak in one variable in a certain time period, and the peak also appears in another variable in the same period, the log-ratio would not show up as unusual. A temporal shift of the peaks, however, creates big log-ratios, and if the position or magnitude is different for one country compared to the others, this country will appear as a potential outlier.

The time trends of the COVID-19 data have been treated here as functional data. Functional data which are processed with tools from compositional data analysis commonly have a constant sum constraint, such as probability density functions or particle-size curves, see van den Boogaart et al. (2014) or Menafoglio et al. (2014). Here we considered the single variables of the multivariate data information as parts of a composition, and since the information is derived continuously over a domain (here time), such data are regarded as multivariate compositional functional data. As functional data are supposed to be smooth functions, the concepts from compositional data analysis already need to be taken into account when generating the compositional functional data. Thus, the original data information, which usually needs to be smoothed in order to represent functions, has to be presented in the appropriate geometry. Since we deal with multivariate information, smoothing also needs to be done in a multivariate context. Here we have used isometric log-ratio coordinates to move the data from the simplex to the standard Euclidean geometry, and we have shown that the specific choice of these coordinates is not relevant for obtaining the smooth functions.

Once the multivariate compositional functional data are available and expressed in the appropriate geometry, standard tools for multivariate functional outlier detection can be used. The application of the methodology to the COVID-19 data revealed that the outlyingness values for the two countries Iceland and Belarus were clearly higher compared to the other investigated countries. Diagnostics in clr coefficients, again referring to relative information, has shown that some of the functions for these countries indeed deviated clearly, at least in certain time periods. Because clr coefficients refer to log-ratios of a specific variable to the geometric mean, deviations can be caused either by atypical values of this variable, or by atypical values of the geometric mean, representing an "average behaviour" of all analysed variables. The analyst would then have to compare this information to that from the other countries, or even go back to the original data source for such a comparison. There could be many reasons for outlyingness: data reporting is done differently (probably only for some of the variables), the policy of the restrictions in the context of the pandemic is very different, the behaviour of the people to deal with the pandemic is very different, etc.

We have also compared such an analysis with multivariate functional outlier detection using the absolute information, where outliers are, for example, countries with extreme values of a function in a certain time period. This analysis led to different outliers, and it finally will depend on the underlying task and research question which type of analysis is most appropriate.

There are many further methodological challenges, which are revealed when considering real data applications as, for instance, the full COVID-19 data set provided from the source mentioned in the paper: zero values, missings, poor data quality, some countries do not provide information for some of the characteristics, etc. These issues are relevant already for estimating the multivariate smooth functions, and subsequently also for the purpose of outlier detection. Our future research will be devoted to such tasks.

Acknowledgements This research was supported by the Austrian Science Fund (FWF) under the grant number P 32819 Einzelprojekte.

References

J. Aitchison, The statistical analysis of compositional data. J. R. Stat. Soc. Ser. B (Methodological) **44**(2), 139–160 (1982)

D.L. Donoho, M. Gasko et al., Breakdown properties of location estimates based on halfspace depth and projected outlyingness. Ann. Stat. **20**(4), 1803–1827 (1992)

M. Febrero, P. Galeano, W. González-Manteiga, Outlier detection in functional data by depth measures, with application to identify abnormal NOx levels. Environmetrics Off. J. Int. Environmetrics Soc. **19**(4), 331–345 (2008)

F. Ferraty, P. Vieu, *Nonparametric Functional Data Analysis: Theory and Practice* (Springer Science & Business Media, 2006)

P. Filzmoser, K. Hron, M. Templ, *Appl. Compos. Data Anal.* (Springer Nature, Switzerland, 2018)

R. Fraiman, G. Muniz, Trimmed means for functional data. Test **10**(2), 419–440 (2001)

M. Hubert, M. Debruyne, Minimum covariance determinant. Wiley Interdiscip. Rev. Comput. Stat. **2**(1), 36–43 (2010)

M. Hubert, S. Van der Veeken, Outlier detection for skewed data. J. Chemom. J. Chemom. Soc. **22**(3–4), 235–246 (2008)

M. Hubert, E. Vandervieren, An adjusted boxplot for skewed distributions. Comput. Stat. Data Anal. **52**(12), 5186–5201 (2008)

M. Hubert, P. Rousseeuw, P. Segaert, Rejoinder to 'multivariate functional outlier detection'. Stat. Methods Appl. **24**(2), 269–277 (2015a)

M. Hubert, P.J. Rousseeuw, P. Segaert, Multivariate functional outlier detection. Stat. Methods Appl. **24**(2), 177–202 (2015b)

P. Kokoszka, M. Reimherr, *Introduction to Functional Data Analysis* (CRC Press, 2017)

R. Maronna, D. Martin, V. Yohai, *Robust Statistics: Theory and Methods* (Wiley, Chichester, 2006)

A. Menafoglio, A. Guadagnini, P. Secchi, A kriging approach based on Aitchison geometry for the characterization of particle-size curves in heterogeneous aquifers. Stoch. Environ. Res. Risk Assess. **28**, 1835–1851 (2014)

S. Moritz, T. Bartz-Beielstein, imputeTS: time series missing value imputation in R. R. J. **9**(1), 207–218 (2017)

K. Mosler, *Multivariate Dispersion, Central Regions, and Depth: The Lift Zonoid Approach*, vol. 165 (Springer Science & Business Media, 2012)

K. Nian, H. Zhang, A. Tayal, T. Coleman, Y. Li, Auto insurance fraud detection using unsupervised spectral ranking for anomaly. J. Financ. Data Sci. **2**(1), 58–75 (2016)

V. Pawlowsky-Glahn, A. Buccianti, *Compositional data analysis: Theory and applications* (John Wiley & Sons, 2011)

J.O. Ramsay, Functional data analysis. *Encyclopedia of Statistical Sciences*, vol. 4 (2004)

C. Reinsch, Smoothing by spline functions. Numerische Mathematik **10**, 177–183 (1967)

P.J. Rousseeuw, Multivariate estimation with high breakdown point. Math. Stat. Appl. **8**(283–297), 37 (1985)

P.J. Rousseeuw, K.V. Driessen, A fast algorithm for the minimum covariance determinant estimator. Technometrics **41**(3), 212–223 (1999)

R. Serfling, Depth functions in nonparametric multivariate inference. DIMACS Ser. Discret. Math. Theor. Comput. Sci. **72**, 1 (2006)

Y. Sun, M.G. Genton, Functional boxplots. J. Comput. Graph. Stat. **20**(2), 316–334 (2011)

J. Tukey, Mathematics and picturing data, in *Proceedings of the 1974 International Congress of Mathematicians*, vol. 2 (1975), pp. 523–531

G. van Capelleveen, M. Poel, R. Mueller, D. Thornton, J. van Hillegersberg, Outlier detection in healthcare fraud: a case study in the Medicaid dental domain. Int. J. Account. Inf. Syst. **21**, 18–31 (2016)

K. van den Boogaart, J. Egozcue, V. Pawlowsky-Glahn, Bayes Hilbert spaces. Aust. N. Z. J. Stat. **56**, 171–194 (2014)

S. Wood, *Generalized Additive Models: An Introduction With R* (Chapman and Hall/CRC, Boca Raton, USA, 2017)

T.W. Yee, *Vector Generalized Linear and Additive Models: With an Implementation in R* (Springer, 2015)

Open Access This chapter is licensed under the terms of the Creative Commons Attribution 4.0 International License (http://creativecommons.org/licenses/by/4.0/), which permits use, sharing, adaptation, distribution and reproduction in any medium or format, as long as you give appropriate credit to the original author(s) and the source, provide a link to the Creative Commons license and indicate if changes were made.

The images or other third party material in this chapter are included in the chapter's Creative Commons license, unless indicated otherwise in a credit line to the material. If material is not included in the chapter's Creative Commons license and your intended use is not permitted by statutory regulation or exceeds the permitted use, you will need to obtain permission directly from the copyright holder.

Chapter 13
The Legal Challenges of Insuring Against a Pandemic

Rachel Hillier

Abstract COVID-19 has raised, and continues to raise, questions about the traditional approach to insurance cover. For instance, business interruption insurance covering "pandemics" under all risks insurance policies are likely to be a thing of the past. With tensions between businesses and the insurance industry on the rise, what can be done to offer businesses some protection at a premium they can afford, without emptying insurers' reserves? In this chapter we talk about legal challenges related to traditional insurance against the risk of losses caused by a pandemic, and whether parametric insurance is the solution.

13.1   Introduction

The insurance industry has always been adept at developing insurance products to meet the changing needs of society, from the development of death benefit and personal injury insurance during the Industrial Revolution to motor vehicle insurance in the late 1890s,[1] to aircraft insurance in the 1920s, to drone insurance in the 2010s. The ability to adapt insurance to meet new perils arising from changing human activity is a common thread in the history of insurance.

The COVID-19 pandemic as a new peril is no different. Living with the risk of loss associated with COVID-19 has become a reality and the insurance industry is working to provide financial protection to those affected. However, some of our most established and oldest types of insurance policies have been found wanting during the pandemic, particularly in relation to small businesses hit hard by financial losses during repeated lockdowns and trading restrictions imposed across the world in an attempt to curb the spread of the virus. This chapter looks at the weaknesses and strengths of the legal structure of traditional insurance policies, and how parametric

---

[1] The first motor policies in the UK were written around the time of the original London-to-Brighton car run, which took place on 14th November 1896.

R. Hillier (✉)
Partner at Capital Law, Cardiff & London, UK
e-mail: r.hillier@capitallaw.co.uk

© The Author(s) 2022
M. C. Boado-Penas et al. (eds.), *Pandemics: Insurance and Social Protection*,
Springer Actuarial, https://doi.org/10.1007/978-3-030-78334-1_13

insurance, with the aid of new technologies could assist in providing new insurance to meet the needs of a COVID-19 world.

As a lawyer trained and qualified in England and Wales, I have concentrated in this chapter on challenges to the legal system in England and Wales. However, many other jurisdictions have the same or a similar insurance contracts and therefore similar challenges.

13.2 Summary of the Traditional Approach to Insurance

13.2.1 The Origins of Insurance

Fire insurance was the start of commercial business insurance in England and can be traced directly to the Great Fire of London in 1666. A company called the "Fire Office" was set up by Nicholas Barbon (a building entrepreneur who made his money rebuilding London after The Great Fire). From 1680, the Fire Office also had a "company of men" to put out the fires. By the 17th Century, marine and life insurance policies were commonly purchased. By this time, the English courts were regularly hearing cases involving insurance disputes. The legal construct of an insurance contract has in many cases not changed significantly since the 17th Century and is predominantly based on the indemnity principle.

13.2.2 The Insurance Indemnity Principle

A definition for a "contract of insurance" is conspicuously missing from the central piece of legislation that regulates the insurance industry in the United Kingdom, the Financial Services and Markets Act 2000 (commonly known as FSMA). This allows the industry to adapt and develop new models of insurance to meet new types of risk. In this section, we will explore why the indemnity principle has become so entrenched in insurance law and why, in many cases, it is not fit for purpose in a pandemic.

Modern insurance has been defined in broad terms by the English courts as *"an agreement to confer upon the insured a contractual right which, prima facie, comes into existence immediately when loss is suffered by the happening of an event insured against, to be put by the insurer into the same position in which the insured would have been had the event not occurred, but in no better position"*.[2] This definition includes the principle of "indemnity"—that payment under an insurance contract puts the policyholder back in the position they would have been had the event not occurred, but in no better position.

---

[2] Callaghan v. Dominian Insurance Co. [1997]2 Lloyds Rep. 541 per Sir Peter Webster.

The indemnity principle has served the insurance industry well for many years. It is a good tool to ensure that insurers pay only the value of what is lost, so that an insurance policy is not gambling, or merely a security entered into for the purpose of profit.

The fundamental purpose of insurance is to mitigate risk. Risk can be mitigated in many ways. For example, doing business through a limited company can limit the risk of its shareholders to the nominal value of shares owned. Another example is the presence of fire proof doors and smoke alarms in a building to reduce the risk of damage by fire. Mitigation of risk through insurance is achieved through the transfer of risk to another (to insurers and reinsurers). The insurance company ensures it has sufficient reserves from premiums collected from many insureds to protect the few who suffer a loss, so that those few have their losses reinstated, at the expense of all the insureds who paid the insurer their premiums. Thus, risk is transferred to insurers. If the insurer has not collected sufficient premiums to cover the insured losses, it is the insurer that must pay the difference. The insurer must therefore calculate not only the likelihood of a claim occurring, but the likely value of each claim. The indemnity principle provides a good method for insurers to learn from its previous claims as well as to collect information about insured property to build sophisticated models to work out how much each insured is likely to lose on average on an indemnity basis, whilst ensuring at the same time that it does not pay out more than it needs to.

Whilst often convenient to insurers, in the UK, there is no statutory requirement for the indemnity principle to be the basis on which claims are valued, as long as a policyholder has an "insurable interest" in what is being protected. That is, an insured must *"benefit by the safety or due arrival of insurable property, or may be prejudiced by its loss, or by damage thereto, or by the detention thereof, or may incur liability in respect thereof"*.[3]

Non-indemnity insurance (including parametric insurance) does exist (and has existed for some years), but such policies are small inlets compared to an ocean of indemnity policies. Section 27(3) of the Marine Insurance Act 1906 provides that an insurance policy may be "valued or unvalued" and that where it is valued, *"Subject to the provisions of this Act [...], and in the absence of fraud, the value fixed by the policy is, as between the insurer and assured, conclusive of the insurable value of the subject intended to be insured, whether the loss be total or partial."* An example of such insurance is income protection insurance. The amount the insured will be able to claim each month if they are unable to work is fixed in advance, usually with reference to the insured's income at the time the insurance is taken out. There may be some indexed rise in that fixed amount built into the policy, but the agreed payment will be made whether the income actually lost is less or more than the pre-agreed fixed amount.

In the face of new threats to the modern world, such as cyber-crime, climate change and COVID-19, alongside the ability of modern computing power, machine learning and artificial intelligence to provide more accurate model future risks, the

---

[3] Marine Insurance Act 1906, Section 5.

indemnity principle, if not obsolete, may start to lose favour as the primary measure of claims payments in the insurance industry.

13.3   The Effect of COVID-19 on the Insurance Industry

On 10th January 2020, the World Health Organization reported that Chinese authorities had determined that an outbreak of a pneumonia type disease, first reported in December in the Wuhan province, was caused by a novel coronavirus, see WHO (2020). The virus spread rapidly into and across Europe. On 23rd March 2020, the UK prime minister, Boris Johnson, announced that people in the United Kingdom must "stay at home" and that non-essential businesses must close.

The closure of manufacturing plants, bars, restaurants, retail establishments, entertainment venues and other places of business resulted in significant business interruption losses and an influx of claims under commercial business insurance policies. But it was not all bad news for insurers. People used their cars less and accidents on the roads reduced significantly, reducing claims under vehicle insurance. Meanwhile, deaths due to COVID-19 spiralled (in June 2020, more than 1,000 deaths a day were recorded due to Coronavirus for 22 consecutive days). The majority of deaths were older people, and many were above the age range for life insurance cover. Yet, the population as a whole was faced with its own mortality, and uptake of life insurance for those who were within normal age ranges for life insurance soared.

13.3.1   The Effect of COVID-19 on Business Interruption Insurance Policyholders

When Coronavirus hit, those businesses that had purchased specialist pandemic insurance, for example, the All England Lawn Tennis Club (host of the world famous Wimbledon Tennis Championship) who had learned from the SARS epidemic and taken out specialist pandemic insurance in 2013, did receive payments under insurance policies. However, as they were indemnity based policies, calculating loss was a long and complicated process. Richard Lewis, Chief Executive of the All England Lawn Tennis Club said "It's a wide-ranging policy but part of the reason it takes so long to work through is that everything is looked at, so there's no blanket payout.", Gangcuangco (2020). A Daily Mail report quoted Lewis as explaining "It's looked at line by line quite literally, every cost, expenditure, bit of income, revenue, whether it has to be repaid, all that sort of thing".

Many small and medium sized businesses (SMEs) thought they were covered under their all risks commercial business insurance policies. However, most insurers believed that these policies were not designed to cover a pandemic and there was no cover under them. This dichotomy of understanding between insured and insurer and the resulting court cases that ensued across the world have often been in the press during 2020/21. The insurance industry has not covered itself in glory by refusing to pay claims under these policies and taking claims through to the bitter end of the court process in many countries.

In the UK, SMEs complained to the Financial Conduct Authority (FCA) that insurers were refusing the vast majority of claims under all risk business insurance policies. This led to the FCA seeking permission from the UK courts to submit test cases relating to common policy wordings, following widespread concern over the lack of clarity and certainty for businesses. The FCA hoped to obtain some certainty for hard pressed SMEs. The Supreme Court heard the case. It considered a representative sample of 21 types of policy wordings from eight different insurers as test cases. The FCA applied to the court for a decision as to whether the business interruption sections of each policy covered losses due to business interruption flowing from COVID-19 closures. The Court's decisions were broadly in favour of business owners. However, despite that decision, insurers continue to decline claims as policy wordings are often slightly different to those wordings considered by the court. Similar test cases have been considered by courts in other countries across the globe—France,[4] Germany,[5] Ireland,[6] Australia.[7]

Even where insurers have accepted a policy covers economic loss due to the disease, many policy terms have exclusions and conditions that mean claims were legitimately rejected. For example, many policies only cover losses flowing from closure due to the presence of disease on the premises. During the first wave of the pandemic "test and trace" in the UK was not in place, and in the second wave was found to be wanting. Insureds therefore often have difficulty proving that COVID-19 was actually present on the premises and claims were rejected. Even where liability has been accepted, in many cases payments have not been made. Assessment of loss under the indemnity principle is leading to long delays in payment, because not only does the policy wording have to cover the loss event, and causation have to be proved, but loss must also be proved. Seven weeks after the Supreme Court judgement, the UK regulator reported that Hiscox (whose policy wording was considered by the court) had paid just 151 of its customers affected and was still deciding more than 4,500.

Insurers are vilified in the press, but as Hiscox said in response to criticism, "These are complicated claims that require comprehensive financial information and discussions with customers in order to settle them fairly.", see BBC News (2021). Quantifying economic loss on an indemnity basis is a time-consuming, administra-

---

[4] On 22 May 2020, for example, Paris's commercial court ordered Axa SA unit Axa France IARD S.A. to pay €45,000 to restaurant group Maison Rostang SAS over the closure of one of its restaurants.

[5] On 1 October 2020 the Munich regional court ordered insurer Versicherungskammer Bayern to pay €1.01 million to the operator of the Augustinerkeller beer garden.

[6] In October 2020, the Irish High Court heard business interruption cases between insurer FBD Holdings PLC and four pubs.

[7] On 18 November 2020, the NSW Court of Appeal handed down its much-anticipated decision in HDI Global Specialty SE v Wonkana No. 3 Pty Ltd [2020] NSWCA 296 ("HDI Global").

tively burdensome and costly task, particularly when a once in a life time constantly evolving world-wide pandemic, waxes and wanes, and the same businesses go in and out of lockdown in response pandemic waves (more on the difficulties of quantifying economic loss below). Expecting SMEs to provide detailed evidence of loss in an ever-changing environment of Government led opening and forced closure, whilst they tackle additional and evolving requirements for COVID safe environments and ongoing staff absence due to COVID-19 illness or being required to self-isolate, is creating even further stress on business owners. Many businesses are becoming insolvent and failing before they can prove their losses, even when they have a valid claim.

Ongoing delays and disputes concerning claims for business interruption losses resulting from COVID-19 are happening across the world. For example, in the United States, by August 2020 over 1,000 COVID-19-related insurance coverage lawsuits had reportedly been filed with early outcomes suggesting different judicial interpretations of key issues and limited potential for any consolidation of proceedings, see Covington (2020).

Meanwhile, ongoing loss to SMEs and increased premiums at renewal (as insurers adjust premiums to recoup claims paid and ensure adequate reserves) means many businesses are now finding themselves unable to afford increased premiums without agreeing to large deductibles. Those that can afford cover are often unable to procure insurance against future COVID-19 related losses due to policy terms offered.

13.3.2 The Effect of COVID-19 on Insurers

For insurers (and reinsurers), COVID-related business interruption claims have meant many potential losses and uncertainty as to whether policies are valid, affecting levels of reserves required. As of March 2021, publicly available data puts an estimate of total losses due to the pandemic worldwide at $34.523 billion, reported in Reinsurance News (2021). However, the use of the indemnity principle as a measure to pay claims means that this figure is an estimate, and the true figures will take years to be known as evidence of economic loss claims is painstakingly gathered and claims are slowly paid. Greater uncertainty as to both liability and quantum means greater uncertainty in traditional claims modelling, resulting in insurers having to hold more cash to top up the reduction in the value of assets held, to meet reserves and regulatory capital requirements.

In my view, the legal costs, and the costs of payments of claims is nothing compared to the reputational damage the insurance industry has suffered as a result of high-profile court cases across the world. Many SMEs have paid substantial premiums for many years, never making a claim, and now feel let down by their insurers in their time of need. This has led to a worldwide loss of faith in the insurance industry by the small business community. Insurance is built on trust, and that trust has been seriously diminished.

We are now starting to see that many insurers are adding wide ranging exclusions to their all risks business insurance so that SMEs have no cover against further waves of COVID-19. It remains to be seen what the appetite of businesses for renewal of these policies will be, particularly where premiums are increased from the previous year as insurers seek to recover loss.

13.4 Life Insurance Versus Business Interruption Insurance

2020 brought an interesting comparison between the fortunes of non-investment life insurance providers and business interruption insurers.

Whilst insurers of business interruption policies have become embroiled in expensive litigation, claims payments have been slow, renewal prices have increased and many insurers have withdrawn from insuring the risk, in comparison, life insurers have been largely unaffected by the pandemic, despite the unprecedented loss of life in 2020. There is no evidence of insurers withdrawing from the life insurance market. Despite high numbers of claims, insurers have not sought to exclude death due to COVID-19, premiums have remained fairly static, many claims have been paid swiftly and life insurance remains a relatively cheap insurance.

Why is that? We all know of the devastating number of deaths due to COVID during 2020. The demographics of those deaths in the UK, according to the Office of National Statistics, is as follows, see Office for National Statistics (2021):

Age range	Deaths in the UK as of 27/12/2020 (%)
0–19	0.06
20–39	0.63
40–59	6.76
60–79	37.98
80+	54.57

The majority of UK life insurance policies are purchased when individuals take a mortgage and are designed to expire upon mortgage repayment. Although the age at which people repay their mortgages is increasing and of course there are life insurance products available for those in their 80s, the majority of people have repaid their mortgage and their life insurance policies have ended by the time they reach 80 years. This means the demographics of the COVID-19 death rates have been kind to the life insurance industry. Also, whilst there is no doubt that the number of claims under life insurance has increased, at the same time, life insurers have seen a significant increase in the sale of life insurance, which has meant healthy cash reserves to set off against claims.

There is also a stark difference in the mechanics of life insurance products compared to all risks commercial business insurance. Life insurance is a relatively simple product, not based on the indemnity principle. Policyholders decide at the outset of the policy what the payment should be upon the event that is insured—their death. The trigger for payment is simple and easily proved by production of a death certificate. Whilst claims may be investigated where fraud is expected and death due to pre-existing conditions are often excluded, payments are usually made quickly on production of a death certificate. There is nowhere near the administrative burden relating to claims payments, which also keeps premiums lower. There is no ongoing loss—there is either a death, or not a death.

13.5 Why Existing Indemnity Based Pandemic Insurance Products Are Not Working

Advances in epidemic risk analytics, including monitoring and modelling tools, have helped insurers better understand their risks and improve response strategies for indemnity based claims. Technology has enabled better risk measurement, monitoring, mitigation and claims management. There are innovative Insurtechs offering tech-based solutions in several areas of indemnity based insurance (UK examples are, Distribind which provides an automated bordereaux system, and Concirrus which uses onboard sensors to provide real time analytics of marine shipping risk). But technology cannot entirely mitigate the time it takes to predict and evidence loss for a new indemnity based risk.

Specialist pandemic insurance was available to businesses prior to COVID-19. In 2018, Marsh McClennan launched a specialist pandemic insurance which provided indemnity based cover, using triggers such as mortality or infections in a defined area.[8] The policy could be tailored to the policyholder to provide coverage for specific expenses, geographies, types of disease, or portions of a calendar year, but until COVID-19 it reportedly had very little uptake, see Collins (2020). I suspect that the slow uptake was due to the perceived gap in the cost of the insurance compared to the risk of suffering losses due to a pandemic. Hindsight is a wonderful thing, and it must be remembered that the last global pandemic was the influenza pandemic of 1918. Whilst Ebola and SARS were modern epidemics, they were seen in the Western World as foreign diseases of little relevance or risk to communities in the West.

Even for specialist pandemic insurance using the latest technology, the methodology used to determine future loss was not well developed. Insurers had some knowledge of epidemics such as SARs and Ebola, but both were different to each other and each was different to COVID-19. The unpredictability of epidemiology meant that actuaries considering risk models for pandemic insurance were involved in more art than science. Furthermore, modelling of the economic input was based on sparse evidence and on economic simulations rather than empirical data. Unlike natural catastrophes and other crises, pandemics and epidemics typically do not cause

---

[8] https://www.marsh.com/ca/en/press-centre/marsh-to-help-businesses-minimize-financial-loss-from-pandemics.html.

immediate physical damage, and so they are difficult to model because every business will have bespoke economic loss, dependent on their circumstances at the time.

Then there is the challenge of anticipating behaviour of both governments (severity of lockdowns, enforcement of restrictions, availability of vaccines) and of individuals (super spreaders, mask wearing, social distancing, ability and willingness to work from home).

Most businesses did not hold specialist pandemic insurance, but did hold all risks commercial business insurance which covers any peril in relation to a particular category of risk, unless excluded in the policy. To give an example, an all risks business insurance policy may cover any loss arising from theft, but exclude theft by way of deception, so a break-in would be covered, but fraud would not. This approach to insurance has led to long and complicated insurance policies with many exclusions[9] and there is often a mismatch of understanding between insured and insurer as to what these policies cover.

Insurers viewed commercial business insurance policies as protection against losses flowing from physical damage to insured premises, subject to restrictions and limitations and that only very few of such policies covered economic loss flowing from COVID-19. For policyholders, they read their commercial business policies as including cover for economic losses relating to closure of their business due to disease and had the expectation that, having paid insurance premiums for years, they would be covered for losses to their businesses flowing from COVID-19.

This mis-match of expectations stems from the nature of all risks insurance policies which started life as simple insurance cover against physical damage to property resulting from fire. Many additions of cover and exclusions over time mean these policies are often repetitive and inconsistent. Add to the mix some unfortunate poor drafting, and it is unsurprising that these types of insurance policies have become a hot bed for litigation in the last year.

Looking at some common policy wordings, you can see why commercial business insurance policyholders had the expectation that they were covered for economic loss flowing from COVID-19:

We shall indemnify you in respect of . . .

– interruption of or interference with the business arising from any notifiable disease

– any occurrence of a notifiable disease within a radius of 25 m from the premises

– the discovery of an organism at the premises which is likely to give rise to the occurrence of a notifiable disease

– in respect of interruption of or interference with the business arising from any infectious disease

– loss resulting from prevention of access to the premises due to the actions or advice of a government or local authority due to an emergency which is likely to endanger life or property

---

[9] For example, the RSA business combined policy is 72 pages long.

- loss of income arising from the closure or restriction in use of the premises by a competent local authority due to defects in the drains or other sanitary arrangements or discovery of vermin or pests at the premises

- damage to Property or premises within one mile of the boundary of Your Premises which causes a loss of Income directly due to a reduction in customers visiting the area."

Reading the extracts above, you could be forgiven for siding entirely with the policyholders making claims for losses during the COVID-19 pandemic. But all-risks policies will also have exclusions, for example,

- Any pandemic coronavirus or strain identified by the World Health Organization.

- Any loss or damage directly or indirectly caused by or contributed to by or arising from contamination.

One of the biggest areas of dispute in COVID-19-related business interruption claims is whether COVID-19 has caused damage to business properties from which economic loss flows. This is because most indemnity based business insurance policies have, at their core, the principle that the insurance covers loss flowing from physical damage. However, over time additional cover against economic loss has been added, such as economic loss due to closure by a local authority and in amongst the list, often loss arising from interruption of trade due to closure of the premises on the occurrence of disease, either on or near the premises. Policyholders argue that COVID-19 has damaged their business and their premises and so they are covered when they have to close as a result.

Where claims were made on the basis that COVID-19 had physically damaged the premises, the insurers' arguments against this were varied and included:

- The presence of invisible microbes on surfaces or objects, however, potentially injurious to human health do not amount to physical damage to the property, and the policy only covers loss flowing from such physical damage.
- COVID-19 microbes cannot be said to alter the physical state or condition of surfaces or objectives whether at surface level or molecular level, so there cannot be physical damage.
- Even if COVID-19 were to be held to be physical damage, it can be removed by deep cleaning, and even if left individual microbes dissipate after between 18 and 100 h, so it is not permanent.
- A deposit of something on property (i.e. the COVID-19 virus) which causes no physical alteration to the property itself does not equate to damage to that property.

Where the words "physical damage" are used, insurers felt they had a strong case and this has been born out in recent case law.[10] However, some policies refer only to

---

[10] TKC London Limited v Allianz Insurance plc [2020] EWHC 2710 (Comm).

"damage", and so the argument then is, whether COVID-19 has caused "damage" to the business.

Where policies clearly and separately include cover for loss flowing from business interruption due to closure because of disease, insurers have argued as to the meaning of "disease", "notifiable disease", and "human disease". Many wordings refer to "notifiable disease", but it is often not defined. Medical practitioners must report "notifiable diseases" to the government under statutory duties in the Public Health (Control of Disease) Act 1984 and the Health Protection (Notification) Regulations 2010, and on 5 March 2020, COVID-19 was added to the list of notifiable diseases. Where notifiable disease is not defined, is that what was meant? Some policies are even more vague—referring only to "human disease" or just "diseases", particularly with reference to local authorities shutting businesses down.

Cover often includes losses due to "closure by an authority" because of disease. Insurers have argued that an "authority" relates to a local authority, and it was the UK Government, Welsh Government and Scottish Government that initially "advised" and then required businesses to close, so there is no cover.

For good measure, where policies do not specifically exclude loss due to a pandemic, insurers argue that commercial business insurance simply is not designed to cover pandemics and that policyholders must have realised that when they purchased the insurance, because of the price of the premium. Whilst the concept of applying a construction of a policy that makes commercial sense for both parties is established law, in my view raising such arguments both in court and in the press, does nothing but harm the reputation of the insurance industry.

Where primary liability is proved by the insured, under the indemnity principle, policyholders then have to prove causation between the insured event that has occurred and the economic loss suffered. This principle has led to further disputes between the insurer and the insured, particularly in relation to economic damage suffered by SMEs during lockdown due to the presence of COVID-19 on or near the insured premises. Insurers have argued that in many cases, the economic loss of businesses has been caused by a national lockdown and an economic downturn, not by the presence of COVID-19 on the premises. Insurers have required evidence from their insureds of COVID-19 physically being present on the premises, which, particularly during the first wave of COVID-19 is impossible for most as testing was not widely available and there was no means of tracing where infected people had been. Even where COVID-19 on the premises could be proved, policyholders had to show that this caused the closure of the premises and that such a closure would not have happened irrespective of the COVID-19 instance on their premises. In many cases this is an impossible hurdle to climb.

A perfect storm has brewed between insurers and their policyholders. It is not that the insurers are the corporate bad guys, and nor do the policyholders have unrealistic expectations of cheap property based insurance. The problem is the structure of the indemnity policy, which does not meet the fundamental needs of risk transfer in a pandemic. It is little wonder that insurers are choosing to change policy wordings on renewal to remove all cover for losses related to disease, COVID-19 or otherwise.

This is not a UK only issue. Businesses across the world have been hit by the pandemic. In the United States, Charles Chamness, president of the US insurer's trade association, National Association of Mutual Insurance Companies (NAMIC) has been quoted as saying "Pandemics simply are not insurable risks; they are too widespread, too severe and too unpredictable for the insurance industry to underwrite", see Insurance Journal (2020).

13.6 Proposals Across the World for Resolving the Business Interruption Insurance Deficit

Across the world, governments and insurers have established working groups to consider proposals for future insurance of business interruption losses resulting from a pandemic. Considerations have included incentivising risk prevention measures, using different models of risk transfer between insurers, reinsurers and governments, see EIOPA (2020), and charging a flat rate levy in all insurance products to create a central fund to be paid to commercial businesses in the event of a World Health Organization declared pandemic, see GDV (2020).

Parametric solutions are also being considered:

1. The French, Fédération française de l'assurance, have devised a parametric solution providing coverage for business interruption losses resulting from various pre-determined catastrophies (e.g. terrorist attack, pandemic, natural disaster). Cover will be triggered by state action to close businesses. Payments will be fixed lumps sums, not indemnity based claims, but will be "calibrated" to replace gross business disruption costs net of salaries and profits, see FFA (2020).

2. In the United States, insurers and the Federal Government are considering the development of a national parametric solution. A formulaic payout is proposed to be made to businesses, triggered by a presidential declaration of viral emergency. Businesses would choose the desired level of protection for three months relief of up to 80% of their payroll. Businesses would purchase the insurance via state-regulated insurance entities. Aid would come from the Federal Government, see Insurance Journal (2020).

13.7 What Is Parametric Insurance?

Parametric insurance is not a new phenomenon. For around 20 years, insurers have used parametric triggers in relation to catastrophic events such as flooding, hurricanes and earthquakes, and it has been used successfully to write policies relating to Ebola and Zika outbreaks. Parametric insurance (also called index-based insurance) is a "pre-valued" policy. Pre-agreed payments are made upon a trigger event occurring.

Trigger events depend on the nature of the parametric policy and can include environmental triggers such as wind speed and rainfall measurements, see Molini et al. (2007). Previously, it has been used to protect the agriculture industry against losses due to catastrophic weather events, because the level of actual losses affecting often large areas of the world are not viable for indemnity insurance. Payments are linked to a pre-agreed index, which is linked to loss of production. For example, rainfall in a particular area. Where rainfall falls below a certain level (drought conditions) or over a certain level (flood), pre-set payments are "triggered". Payments are automatically paid upon the trigger event occurring at a pre-agreed amount.

Smart contracts are often used. These are usually associated with block chain technology, but are not exclusive to it. Smart contracts are self-executing contracts, so the agreed amount is paid automatically in accordance with written lines of computer code.

To work well, parametric insurance requires the following:

(1) a recognised and trusted set of data which both parties to the contract agree to rely on;

(2) a pre-agreed payment figure, which may or may not be linked to various levels of data; and

(3) a clearly defined event which acts as a trigger;

(4) an acceptance by the insured in clear terms that payments may not (and probably will not in most cases) put the insured back in the position they had been before the trigger event.

13.7.1 Working Examples of Parametric Insurance

13.7.1.1 Case Study 1—Flooding

The insurance need: flood damage to commercial premises in flood prone areas. The UK is experiencing increasing flood events. According to a study from Heriot-Watt University, flooding in the UK could increase by an average of 15–35% by 2080, see Ellis et al. (2021). Many businesses who have previously suffered flood damage or are located in an area at risk of flooding find it hard and sometimes impossible to find indemnity based insurance to protect their property from flooding at a premium they can afford.

An Insurtech parametric solution: FloodFlash[11] uses an internet connected flood depth sensor which is fixed to the premises. The insured chooses a depth of flooding (for example, 0.5 m) which will trigger payment under the insurance policy. The insured also chooses the sum they want to receive when the sensor notifies Floodflash

---

[11] Flood Flash—rapid payout flood insurance for any business.

that the trigger is met. The premium for insurance is calculated using the depth trigger and the agreed settlement amount. Payment of the pre-agreed settlement figure is made to the insured automatically within hours of the trigger being met. Several triggers can be set, for example, a payment of £50,000 is triggered when the water depth reaches 0.5 m, then a payment of a further £50,000 is triggered when the water depth reaches 1.2 m.

Why it works: The payments are swift, but are usually less than the cost of recovery from the flood as the insured will set the flood depth trigger and the settlement amount to the amount of premium it can afford. Fast payment often means that businesses can use the settlement as a "first responder" pot of money to get the business trading again. For the insurer, meteorological predictions have become more sophisticated so that flood events can be predicted more accurately, and the level of payment is already set, meaning risk can be better measured, and significant claims and loss adjuster costs are negated. The insurer can offer smaller settlement amounts on risks it would reject for insurance on an indemnity basis.

13.7.1.2 Case Study 2—Earthquakes

The insurance need: Insurance for households against losses due to earthquakes in areas of high earthquake risk areas. In the United States, damage caused by earthquake is not commonly included in household insurance policies. Where it is available, traditional indemnity based insurers require large deductibles, because otherwise the premiums are too high to be affordable. When there is an earthquake there are many claims at once and it takes time to process. Where there is catastrophic damage, insurance often pays for replacement accommodation whilst rebuilding takes place, but for lower level damage insurance payments are slow.

An insurtech meeting the needs of customers through parametric insurance: Jumpstart Recovery[12] uses its tech platform to monitor data from the United States Geological Survey (USGS) Shake Maps. When peak ground hits a velocity of 30cm per second in a certain area, the USGS turns its Shake Map red in that area. This triggers the Jumpstart platform to automatically send a text to its policyholders whose properties are insured in the area that turns red, asking if they have suffered a loss. The insured texts "yes" in response, and $10,000 is automatically sent to the insured's bank account.

Why parametric insurance is a good solution: This is a great example of parametric insurance providing a fast effective solution to a large amount of low level losses arising from one catastrophic event. It is a first responder type insurance which provides fast payment to cover immediate losses. It is not a replacement for more comprehensive indemnity based insurance against the cost of rebuilding, but either compliments it, or provides lower level risk transfer for those who cannot afford the "bells and whistles" of full indemnity based insurance.

---

[12] https://www.jumpstartrecovery.com/.

13.7.2 Challenges

There are challenges to the use of parametric insurance for business interruptions policies.

13.7.2.1 Triggers and Quality of Data

Parametric insurance is hugely dependent on the basis of the trigger used for payment. There needs to be confidence on the part of both the insured and the insurer in the veracity of the data. By veracity I mean the accuracy and truthfulness of the data being used, as well as the ability of the insured, and potentially a court, to verify the truth of the data and that the trigger has been fired in the correct circumstances.

Some parametric insurance terms rely on more than one data point. For example, the World Bank Pandemic Emergency Financing Facility's parametric catastrophe bond[13] relies on publicly available data to determine how much money the facility would release to the poorest countries in the world. According to the World Bank, the triggers are based on outbreak size (the number of cases of infections and fatalities), outbreak growth (over a defined time period), and outbreak spread (with two or more IBRD/IDA countries affected by the outbreak). Pay-out occurs only after a slew of conditions are met in connection with a country: (1) a rolling daily average of at least 250 cases; (2) the virus exists for at least 84 days; (3) total confirmed deaths to be greater than 250 cases (for class B issuances) or 2,500 cases (for class A issuances); (4) an exponential growth rate; and (5) geographic spread of the virus. The World Bank was criticised by the Financial Times, see Financial Times (2020), for making the parametric triggers so high that it was nearly 40 days after the World Health Organization officially announced that COVID-19 was a pandemic, before the first payments were made. The more data points relied on, the more chance there is of the trigger being in dispute.

Statistical parameters could be used, relying on third parties to determine whether a trigger has been satisfied. Such triggers are typically utilised in catastrophe bonds and insurance-linked securities (instruments connected to insurance-related risks that provide issuers funding for specific events), where scientific measurements of the severity of tornadoes or hurricanes are used. Third party analytics of infection rates or death rates in a particular area could be used as a trigger for pandemic insurance.

The epidemic data analytics firm Metabiota has also developed the Pathogen Sentiment Index, Metabiota (2018), that measures the effects of changes to behaviour in a pandemic, giving "fear" score based on disease attributes, symptoms and mortality, disease transmission and availability of treatment, to provide a score to reflect reduction in consumption. Others are based on an index of available data to determine when the policy pay-outs can begin. Some pandemic insurance has used the rate of

---

[13] A financing mechanism used at the World Bank (https://www.worldbank.org/) is designed to provide an additional source of financing to help the world's poorest countries respond to cross-border, large-scale outbreaks.

hotel bookings in comparison to year-on-year averages or the measures of footfall in pedestrian areas.

My view is that for a pandemic business interruption solution for SMEs, a simple single trigger should be used and would mean a faster rollout of the product and more transparency for the policyholder. While using several third party behaviour-based metrics are useful for international organisations where a more sophisticated trigger for difference areas of the business is needed, for SMEs a single parametric trigger relying on one datapoint would be preferable where there are many smaller claims.

Civil authority triggers could be a solution. However, the challenge for insurers is to pick which civil authority trigger to use in a rapidly changing political response to a novel situation. In the UK, the Health Protection (Coronavirus, Restrictions) (All Tiers) (England) Regulation 2020 came into force came into force on 2nd December 2020. This legislation introduced the concept of "tiers" and has been updated and amended several times during as the pandemic has evolved and our knowledge of how the disease spreads has increased. These regulations have been made under Section 45 of the Public Health (Control of Disease) Act 1984, which allows UK Government to make a regulation "imposing or enabling the imposition of restrictions or requirements on or in relation to persons, things or premises in the event of, or in response to, a threat of public health".[14] A specific trigger regarding the implementation of regulations restricting access under this Act could be used, but then the devolved nature of the United Kingdom means that the Welsh and Scottish Government have devolved powers to make these regulations, so this would only work for England unless the Welsh and Scottish regulations were added to the trigger. There is also the issue of "Guidance". In the early stage of the pandemic, the UK Government issued guidance that people should stay at home, severely impacting the entertainment and leisure industry, despite there being no regulation in place.

An alternative trigger would be infection rates in a particular area. There are potential difficulties with how these figures are arrived at: in the UK they are based on the percentage of people who have tested positive for COVID-19 at a point in time, but an alternative trigger might be the incidence rate—a measure of only the new infections in a given period of time. This trigger relies on a good testing regime. We know that at the start of the pandemic in the UK, testing was not well developed, and it is widely accepted that infection rates were far higher in reality compared with the figures recorded.

13.7.2.2 Pricing Challenges

For insurers, it will be a continuing challenge to build a risk model that can price premiums that are affordable and get the ratio right between premiums and claims payments. Can AI, machine learning help? Do we trust it enough? There are developments in this area which could help. US Insurtech, Thimble, joined the 5th cohort of Lloyds lab accelerator programme in 2020, Lloyds Lab (2021). Thimble is using

---

[14] Section 45C(3)(c) of the Public Health (Control of Disease) Act 1984.

AI to devise a parametric business interruption insurance policy that has low limits before payment is made, and pays out incrementally and instantly, for example where businesses in a certain zip code are forced to close. There is certainly more that can be done in this area and Insurtechs continue to work on solutions.

13.7.2.3 Legal And/or Regulatory Uncertainty

Few countries have specific laws or regulations about parametric insurance. Whilst parametric insurance is an accepted concept in many countries, it can cause uncertainty as to lawfulness in jurisdictions where the indemnity principle sits at the core of insurance law and regulation. For example, in India and South Africa parametric insurance is often purchased by farmers in relation to drought and flood. The law in India around contingency based agreements means that the insured must prove a loss to the insurer. This requirement means that when a trigger is met, there is then the same slowness in claim payments under parametric insurance, as for traditional insurance. South Africa similarly has legal requirements for the insured to prove loss, but has approached the issue in a different way, by not requiring the insured to prove the extent of loss, only that some loss has occurred. Countries where parametric insurance is now well established includes the United States, many African states, the UK, France and the Caribbean.

13.7.3 Opportunities

13.7.3.1 Combination Policies

There is room for a combination parametric/indemnity policy. This would be a means of indemnifying against loss to a certain extent, but it would also provide an immediate parametric based payment as a first response. I am not aware of a single insurance policy that provides both options at the time of writing.

13.7.3.2 Combining Parametric Insurance with Government Support

Parametric insurance could be a simple method of providing a combination of Government support and insurance claim payments. In the UK, Government has worked with insurers in relation to flooding loss (Flood Re). Flood Re is a joint initiative between the UK Government and insurers. All home insurance companies in the UK pay a levy to Flood Re, which is a reinsurer set up for the purpose. All those insurers can then use Flood Re to reinsure the flood risk part of a policy where a home is in an area that is at higher risk of flooding, so providing a transfer of risk where insurance might otherwise be refused. A scheme could be implemented at Government level where payments triggered by parametric technology are be partly insurance funded, partly government funded.

13.7.3.3 Capital Investment

Insurance has not typically been attractive to capital markets as part of an investment portfolio. Insurance linked securities (ILS) are traded, but are not a main stream investment. The clean transparent nature of parametric insurance is likely to be more attractive, as investors only need to understand the event trigger, not potential quantum (as payments are always a pre-valued fixed sum). It could be that as parametric insurance grows in popularity insurers/reinsurers can hedge against risk aggregated exposure to parametric triggers with parametric ILS securities, thereby transferring the risk of a pandemic to some extent away from government and insurers/reinsurers to capital markets and institutional investors.

13.7.4 Could Parametric Insurance Be the Answer for SMEs During a Pandemic?

Parametric pandemic insurance could be the solution as a "first defence" against business interruption during a pandemic. For larger businesses it may well be that both a parametric insurance product and a more specialist traditional indemnity based pandemic insurance policy are the way forward. However, for the many SMEs in the UK and the rest of the world, for whom a huge protection gap has emerged, parametric insurance could be the only affordable option.

In the same way as a paramedic being first on the scene of an accident can make the difference between life and death in the aftermath of an accident, payment within 48 h of forced closure to a pandemic, could mean the difference between business survival and closure. The paramedic is very quickly available to give life-saving aid with enough equipment and enough training to keep those involved alive. The treatment does not cure the victims of their injuries, but it is enough to keep them alive, to relieve some of the pain, and to give comfort until the ambulance arrives with more equipment and the injured is taken to hospital for more thorough treatment and hopefully a full recovery. Parametric business insurance can work in the same way, providing fast, automatic payment with no requirement for proof of loss, upon a mutually agreed trigger associated with the pandemic that affects that business. Furthermore, parametric insurance allows those fast automatic payments to be made repeatedly, every time a prescribed trigger event occurs.

The payment is unlikely to indemnify all losses, but will be enough to keep the business alive in the immediate crisis so that it does not fail, but lives to trade another day. That initial payment may be enough to purchase necessary PPE and change the premises to meet social distancing requirements, or may pay for fixed cost commitments such as rent/salaries during lockdown or reduced trade due to social distancing measures.

Causation is taken out of the equation in a parametric solution. All the problems that business owners have experienced under indemnity business interruption policies are gone: complicated terms, having to prove their losses are caused by COVID-19 being on or near the premises, lengthy waits for payment. It does not matter whether losses are due to a general downturn, poor management or COVID-19, the agreed sum is paid automatically.

Parametric insurance is not only good for the insured. For insurers, it significantly reduces the administrative claims burden and allows for smaller pay-outs in return for payments policyholders can afford. An army of loss adjusters and claims handlers are not required. Payments are made automatically when data feeds from the agreed source hit the trigger. There is no requirement to assess or prove loss.

References

BBC News, D. Ascher, It's do or die time for my insurer to pay up, 29 March 2021, https://www.bbc.co.uk/news/business-56535583

S. Collins, Insurers wary of meeting growing demand for specialist pandemic cover. Commer. Risk, 9 April 2020, https://www.commercialriskonline.com/insurers-wary-meeting-growing-demand-specialist-pandemic-cover/

Covington. Developments in Coronavirus coverage litigation and legislation, Covington & Burling LLP, 13 August 2020, https://www.cov.com/en/news-and-insights/insights/2020/08/developments-in-coronavirus-coverage-litigation-and-legislation

EIOPA. Issues Paper on resilience solutions for pandemics. European Insurance and Occupational Pensions Authority (2020), https://www.eiopa.europa.eu/content/issues-paper-resilience-solutions-pandemics_en

C. Ellis, A. Visser-Quinn, G. Aitken, L. Beevers, Quantifying uncertainty in the modelling process; future extreme flood event projections across the UK. Geosciences, 11(1), 33 (2021), https://doi.org/10.3390/geosciences11010033

Fédération Française de l'Assurance (FFA). La Fédération Française de l'Assurance présente sa contribution au débat sur la création d'un régime de catastrophes exceptionnelles: le dispositif CATEX, Fédération Française de l'Assurance, 2020. https://www.ffa-assurance.fr/actualites/la-federation-francaise-de-assurance-presente-sa-contribution-au-debat-sur-la-creation-un

Financial Times, A. Gross. World Bank pandemic bonds to pay $133m to poorest virus-hit nations, 19 Apr 2020, https://www.ft.com/content/c8556c9f-72f7-48b4-91bf-c9e32ddab6ff

T. Gangcuangco, Outgoing boss lifts the lid on Wimbledon's pandemic insurance. Bus. Insur. UK, 29 June 2020

Gesamtverband der Deutschen Versicherungswirtschaft (GDV). Green paper: Supporting the economy to better cope with the consequences of future pandemic events (2020), https://www.en.gdv.de/resource/blob/59854/079826b589006ed3bd4fc7a09e64cf1a/pandemiefonds-vorschlag-download-green-paper-data.pdf

Insurance Journal. Insurers, Agents Propose Pandemic Business Relief Plan; Plaintiff's Offer BIG Compromise, May 22, 2020. https://www.insurancejournal.com/news/national/2020/05/22/569611.htm

Lloyds Lab. Cohort 5: InsurTech solutions to support COVID-19 response, recovery and future-resilience (2021). https://lloydslab.com/insurtechs/cohort-5/

Metabiota. Infectious disease outbreaks can inflict enormous social and economic disruption (2018). https://metabiota.com/sites/default/files/presentation_files/Metabiota%20Pathogen%20Sentiment%20Index%20Flyer_Apr%202018.pdf

V. Molini, M. Keyzer, B. van den Boom, W. Zant. Creating safety nets through semi-parametric index-based insurance: a simulation for Northern Ghana, in *European Association of Agricultural Economists. 101st Seminar, Berlin Germany*, 5–6 July 2007. https://EconPapers.repec.org/RePEc:ags:eaa101:9263

Office for National Statistics (ONS). Latest data and analysis on coronavirus (COVID-19) in the UK and its effect on the economy and society, https://www.ons.gov.uk/peoplepopulationandcommunity/healthandsocialcare/conditionsanddiseases

Reinsurance News. COVID-19 loss reports and reserves reported by insurance or reinsurance companies. Reinsur. News, April 2021. https://www.reinsurancene.ws/covid-19-insurer-reinsurer-loss-reports/

WHO. Timeline: WHO's COVID-19 response, https://www.who.int/emergencies/diseases/novel-coronavirus-2019/interactive-timeline#event-7

Open Access This chapter is licensed under the terms of the Creative Commons Attribution 4.0 International License (http://creativecommons.org/licenses/by/4.0/), which permits use, sharing, adaptation, distribution and reproduction in any medium or format, as long as you give appropriate credit to the original author(s) and the source, provide a link to the Creative Commons license and indicate if changes were made.

The images or other third party material in this chapter are included in the chapter's Creative Commons license, unless indicated otherwise in a credit line to the material. If material is not included in the chapter's Creative Commons license and your intended use is not permitted by statutory regulation or exceeds the permitted use, you will need to obtain permission directly from the copyright holder.

Chapter 14
An Actuary's Opinion: How to Get Through a Pandemic

Frank Schiller

Abstract We discuss in this chapter how the insights and methods presented in the previous chapters can be effectively and practically implemented to manage and mitigate pandemics. The findings are not only analysed for the current COVID-19 crisis, but we also present some insights that could be gained for future pandemics and other extreme events. Coming from an actuarial background, the main focus lies on practical and technical aspects of the presented articles, namely on data, models and possible risk mitigation through (re)insurance, capital markets and similar approaches.

14.1 Questions to Be Tackled from an Actuary's Perspective

The COVID-19 pandemic and the widespread shutdown of social and economic life in 2020 hit the insurance industry hard. In addition to losses in investments due to falling share prices and lower interest rates in reinvestment, non-life insurers and reinsurers in particular suffered high claims costs especially in travel insurance, business closure and event cancellation. Even if the situation of the industry is not existentially threatening from today's perspective, the COVID-19 pandemic has impressively shown the negative effects extreme events can have on the financial and solvency situation of companies (see also Frank et al. (2020) and Actuarial Association of Europa (2020)).

To make matters worse, the course of the COVID-19 pandemic is a "natural disaster in slow motion", with no end in sight even at present. With all the uncertainties of further waves and the effectiveness of the vaccination program the further course of the pandemic and the economic consequences are still difficult to predict.

Management of the crisis for both the private and public sectors is based on the analysis of data and appropriate models and metrics. Therefore, in Sect. 14.2 we first discuss the learnings and their practical application from COVID-19 on models, risk modelling and assessment. We especially share some thoughts on the general insurability of a pandemic in Sect. 14.2.1 including a first assessment of how the

F. Schiller (✉)
Deutsche Aktuarvereinigung (DAV), Hohenstaufenring 47â£"51, 50674 Cologne, Germany

© The Author(s) 2022
M. C. Boado-Penas et al. (eds.), *Pandemics: Insurance and Social Protection*,
Springer Actuarial, https://doi.org/10.1007/978-3-030-78334-1_14

risk appetite and the capacity for covering a pandemic might have changed due to COVID-19 and what innovative solutions for a pandemic cover could be expected for the future. In 2020 we especially gained more experience on how to manage an extreme event like a world-wide pandemic and we elaborate in Sect. 14.2.2 on aspects of data quality and information uncertainty, how the massively increased volatility during such a crisis can be handled and, based on this experience, which changes of the general risk management framework Solvency II can be expected for the future. Finally, in Sect. 14.2.3 we summarize some potential future consequences which COVID-19 could have on existing products and how this can be monitored and reflected in their prices and reserves. In Sect. 14.3 we discuss how to improve measures for mitigating risks from a governmental perspective. Again, we first discuss in Sect. 14.3.1 what can be learnt for the management of the crisis, now from the perspective of a government. Finally, we present in Sect. 14.3.2 some thoughts on how the economy can generally be made more resilient for such a crisis, how pandemic risks can be covered and better spread on several risk carriers, what role the capital market and especially the World Bank could have and why we have to especially solve the issue of basis risk to implement an effective risk transfer for a pandemic. In a brief summary in Sect. 14.4 we conclude with the main takeaways and what are relevant learnings after such a pandemic.

14.2 Managing a Pandemic as a (Re)insurer

The COVID-19 pandemic is testing the models, methods and processes (re)insurers have used so far. Let us first consider in general terms how we can deal with new knowledge and uncertainties in assessing risks during an extreme event such as a pandemic. For assessing prices and risks three types of models are typically used: calculation or pricing models, valuation models, and risk models.

The COVID-19 pandemic may have an impact on all three types in different ways. A conscious approach to the respective characteristics and requirements and the questions derived from them for the individual model types are important:

- Calculation/pricing: Are pricing assumptions still valid? Is there a sufficient margin/return on investment (also in conjunction with a possible adjustment of the allocation of capital costs from the risk model)?
- Reserving: For which potential losses should provisions already be formed and reported, even under uncertainty? How do assumptions already need to be adjusted, if necessary, also for the expected payments and future projections?
- Risk: Are the model assumptions still appropriate? Was the exposure fully and correctly considered in the valuation? Are there other dependencies/correlations in extreme events between individual models?

In the following sections we will consider effects and learnings of COVID-19 on all three types of models and the respective processes and measures when applying these models.

14.2.1 *Insurability of a Pandemic*

The main purpose of (re)insurance is to provide compensation for losses from risks. Insurance is ery effective when covering, e.g., losses stemming from accidents with limited exposures. In this case the main risk is simply volatility and large enough portfolios help to diversify effectively against high aggregate losses. Even local catastrophic events like floods, storms, earthquakes or epidemics are (re)insurable, if on a global level these risks have comparable low frequency, limited size, and, hence, can be diversified against other rare risks in a global portfolio. For such covers, it might happen that a (re)insurer faces rather high losses in a single year but these can be compensated by reserves and equity buffers build up in the previous 5–10 years— some companies even accept longer periods for certain risks and under Solvency II (re)insurance companies have to hold equity to compensate for a 200-years event.

Risk equalisation in the collective or over time no longer works if the risks are systemic or world-wide events have to be considered. A pandemic is a very relevant example for such an event. Obviously, in such a situation the diversification over a world-wide portfolio would not work as effectively as it is necessary for a collective diversification. Assets and liabilities of several lines of business from life, health to non-life are effected. And losses could be so extreme that it is simply not possible to built up sufficient reserves and equity to compensate for the total loss that needs to be (re)insured in such an event (see also Frank et al. (2020) for a more detailed discussion).

Risk appetite and capacity for pandemic risks

With COVID-19 we saw quite impressively that during a pandemic collective risk sharing over different risks within a (re)insurance portfolio does not work. Both sides of the balance sheet, assets and liabilities, were affected, in some cases heavily. And also especially for the liabilities, claims in many lines of business were triggered simultaneously, starting from the obvious life and health business[1] to non-life portfolios with main exposures in business closure, business interruption and event cancellation, and other lines of business like, e.g., credit insurance (see also Actuarial Association of Europa (2020) for a more detailed overview and current data can be observed on Roser et al. (2020)).

The problem of the massive potential exposure during a pandemic can be better illustrated by an example from Germany (more detailed data can be found in Destatis (2020), GDV (2019) and GDV (2020)). When only considering a business closure insurance for the hospitality industry sector[2] during lockdown we could face a loss

---

[1] For COVID-19 in 2020 only to a very limited extend in many countries, since portfolio exposures are comparably small for the older population—for other pandemics like flu other age groups might be affected.

[2] In 2019 the annual turnover for this sector in Germany was €94.7bn, https://www.dehoga-bundesverband.de/zahlen-fakten/umsatz/.

of more than €90bn.[3] In contrast to that in 2018 the annual premium income of the German non-life insurance sector was only €70.7bn, the claims expenditures €52.5bn and the own funds were only €110bn. For such rare events like a pandemic the premium income would increase not more than 2–3% of exposure, i.e., less than €3bn. All in all after such an extreme event the German non-life insurers would be insolvent, if they provided such a cover without a relevant limitation.

In several chapters of this book insurability of a pandemic is discussed coming from different perspectives. Obviously, improving the models does not help, as the problem lies with the simultaneous trigger on different lines of business and both, assets and liabilities, and in aggregate a too large total exposure. Improved data and model quality would only improve the prize of the cover, which is not relevant during the realisation of such rare events. What is important, however, is clarity on the claims trigger itself. Treaty wording has to be strong and clear and the definition of the claims trigger simple but adequate for transparently transferring the risk as promised from the client. In such cases, parametric (re)insurance might be a very valuable too, as discussed in the chapter on "The Legal Challenges of Insuring Against a Pandemic" by R. Hillier. The payout is quicker and supports more effectively the safeguarding of the whole economy. However, the main problem with parametric triggers is always their potential basis risk, i.e. the real claims costs might be materially larger or smaller than the amount paid through the parametric (re)insurance. We discuss this issue in more detail in Sect. 14.3.2.

New solutions and innovations after the pandemic

Greater demand for pandemic cover from clients on the one hand, but lower risk appetite among reinsurers on the other, will drive innovation and new solutions for this cover.

During the crisis (re)insurance companies have developed rather a more limited risk appetite for pandemics. This means that relevant exposure of carve-out or even stand-alone covers for a pandemic will be not available in (re)isurance. To the contrary, the focus will more and more lie on how to enhance the diversification of portfolios, to limit exposures for extreme events, and to better stabilize the whole balance sheet also during an extreme event. This could even mean that several (re)insurers will rather further limit or even try to exclude a pandemic cover for their new products.

On the other hand, the demand for mitigating pandemic risks is now higher than before the crisis for many decision makers in the concerned sectors. Solutions for mitigating or transferring such risks cannot be provided alone by the private (re)insurance sector and solutions will have to include the public sector. This topic is dealt with excellently in the chapter on "Risk Sharing and Stochastic Premia in the Presence of Systematic Risk: The case study of the UK COVID-19 economic losses" by

---

[3] The actual reduction in turnover in 2020 was €35.2bn or 36.5%, https://www.dehoga-bundesverband.de/zahlen-fakten/umsatz/.

H. Assa and T. J. Boonen. A more detailed discussion from an actuarial point of view on such solutions is given in Sect. 14.3.2.

14.2.2 Risk Management During a Pandemic

The information uncertainty during the pandemic poses particular challenges to management and risk management of a (re)insurance company but also to the management of the crisis under social economic and general economic aspects, and we can learn from the current pandemic how to improve these measures for this crisis and also future extreme events. For (re)insurance it is important to better understand:

- How can the period of increased uncertainty be managed?
- How to adjust requirements during the pandemic, if necessary, even with uncertain information,
- and especially taking into account errors and uncertainties in modelling and assessing the risk situation?

Data quality and information uncertainty

Especially at the beginning of the COVID-19 crisis we had to cope with limited statistical comparability between countries due to different methods used to measure rates of infection and death, as well as different approaches to testing, testing capacity and criteria applied for test eligibility. As a consequence, it was extremely difficult to assess and compare the success of different approaches taken by the countries to contain the pandemic. Also for this reason some research institutes even decided not to look into the data of reported cases at all but only analyse the reported deaths, as they appeared to be more reliable (see for example Imperial College COVID-19 response team (2021)). Even for a single country the data originating from the first till half year are not really comparable to the second wave starting in October for the same reasons, and this still leads to relevant uncertainties in interpreting the observations.

Now, after one year, testing quality and capacity has increased significantly and so the data quality of the reported cases. Still, we have to cope with quality issues during the reporting process and it might be valuable to apply methods as described in "Outlier Detection for Pandemic-related Data Using Compositional Functional Data Analysis" by Ch. Rieser and P. Filzmoser to smoothen the data. However, a major problem remains the attribution of deaths that were directly caused by COVID-19 and not only died with a COVID-19 infection (a more detailed discussion can be found in Ealy et al. (2020)). To solve this ambiguity of potential wrong or missing attributions to deaths caused by COVID-19, it is helpful to analyse the observed excess mortality instead. Reporting and understanding of this data has proven to be of adequate quality for most countries, and hence this is also a robust source for a deeper analysis of pricing and reserving adjustments required due to the pandemic. A valid source for Europe can be found online at EuroMOMO, EuroMOMO (2021).

Managing volatility

The COVID-19 related fluctuations on the capital markets since March 2020 have shown that short-term, temporary volatility can be much higher than long-term volatility. As of mid-August 2020, share prices have already recovered significantly from the previous low in mid-March and are partly back at the level of year-end 2019. The temporary increase in spreads for government and corporate bonds after the lockdown also reduced the market values of interest-bearing securities on the asset side and led to a loss of own funds. The effect of the (temporary) market distortions on own funds was partially mitigated by the instrument of volatility adjustment (VA) under Solvency II, which serves to avoid pro-cyclical behaviour by market participants. The symmetric adjustment factor ("equity dampener") has mitigated equity stress in the standard formula by up to 10% for the same reasons.

In many (re)insurance companies, the impact and uncertainty of the COVID-19 pandemic have triggered special analyses and sometimes also ad-hoc ORSA[4] reporting, depending on the requirements as defined in the internal ORSA guidelines and the actual relevance of COVID-19 on the companies' economic balance sheet. In addition, the supervisors have taken an interest in the current risk situation of the companies and the impact of the COVID-19 pandemic on the risk profile of the companies and have carried out special queries.

Since the development and duration of the pandemic was uncertain from the beginning and is still difficult to predict, scenario analyses in all phases of the pandemic are proven instruments for assessing risks and determining the risk profile. In this process, several scenarios are run, each with a different course of the pandemic, and the effects of the macroeconomic developments associated with these courses on the company's capital investment, insured portfolio and ultimately equity and solvency capital requirements are estimated.

In this context it is important to provide robust tools to analyse data and transparent models to calibrate and run such scenarios and use the results for deciding what measures are adequate to manage through the current status of the pandemic. Aspects of improving data analysis and models are covered in several chapters of this book, as in "Some Investigations with a Simple Actuarial Model for Infections such as COVID-19" by A. D. Wilkie, "A Mortality Model for Pandemics and Other Contagion Events" by G. Venter or "Epidemic Compartmental Models and Their Insurance Applications" by R. Feng et al. With this input a cascading approach with a crisis intervention team for balance sheet management can provide the relevant guidance for business decisions:

- Risk drivers assessed as highly relevant are identified with high frequency. For a pandemic, new drivers can be added, such as new business or reported losses of highly exposed segments. And these drivers would then also be able to be taken into account in parallel in the assessment of the risks by running the above mentioned scenarios.

---

[4] Own Risk Solvency Assessment, as required by Solvency II.

- Relevant limits are set for these risk drivers and if they are exceeded, measures are initiated by crisis team. The crisis team should also plan regular meetings with high frequency and have been mandated by the Executive Board with the corresponding decision-making powers.
- The crisis team regularly reports to the board of directors on the development, proposes adjustments to the risk appetite and corresponding measures, and thereby also obtains an adjusted mandate.

In addition to the crisis team for economic and balance sheet risk management issues, it may also make sense to set up another team to deal with purely operational issues such as business continuity, IT and human resources. A representative of risk management, e.g. the RMF,[5] should also participate in this group and build a bridge between the two groups.

Adjustments to risk models and processes

We have experienced so far that the principle based Solvency II framework enabled (re)insurers quite effectively to manage through this crisis. Currently, the Solvency II Review 2020 is in its final phase and it is a good opportunity to directly reflect learnings of the crisis in adapting the framework accordingly (cf. EIOPA (2020b)). Also in the future, Solvency II will be reviewed with a predefined frequency and new insights and requirements can be reflected. So far, no relevant changes to the standard formula due to COVID-19 are expected. Only topics on extended or improved reporting, e.g. of the ORSA, are under discussion in some of the latest EIOPA consultation papers (cf. EIOPA (2020a)). Again, the chapters on modelling and scenario analysis in this book mentioned above might provide helpful input for how to improve and extend the ORSA report with relevant scenarios for pandemic events.

14.2.3 Reflecting Potential Future Consequences of COVID-19

For the capital market and many non-life insurance covers, most of the effects of COVID-19 were rather immediate and directly observable. In the global economy and in life and health (re)insurance we have to consider long-term effects. Some of these effects are discussed in this book, e.g. , behavioural changes of the population due to COVID-19, others might be relevant to be reflected in pricing and reserving updates of existing products, as, e.g., different demand and consumption with effects on the economy, effects on the health of recovered with longer term impacts and potential triggering of claims at a later point in time (e.g. the chronic fatigue syndrome, effects on the respiratory system, the heart, or neurological illnesses as

---

[5] Risk Management Function, one of the second line of defence risk management functions required by Solvency II.

late effects of COVID-19), or a worsened general health status because of poorer health care during the pandemic. Again, all these effects might only be observed in the future and we will need a close cooperation between data scientists, modelling, actuarial and medical experts. Methods to monitor and model such developments are again discussed in this book in several chapters and from different perspectives, as again "A Mortality Model for Pandemics and Other Contagion Events" by G. Venter and "Outlier Detection for Pandemic-related Data Using Compositional Functional Data Analysis" by Ch. Rieser and P. Filzmoser.

14.3 Managing Pandemics from a Governmental Perspective

For governments and public bodies it might be relevant to learn from the current pandemic

- how to optimally impose measures like a lockdown, what are effective test strategies and how to implement a vaccination program,
- how to manage through the sometimes conflicting objectives of minimising both the number of casualties of the pandemic and the economic loss during the pandemic crisis,
- and finally how to prepare better in advance to mitigate some of the risks by providing obligatory or facultative (re)insurance or risk pools?

14.3.1 Deciding on the Right Measures During a Pandemic

What we discuss on how to implement an adequate risk management during a pandemic in Sect. 14.2.2 is also true for governments. Managing uncertainty is very relevant especially at the beginning of the pandemic and politicians and public servants need to consult closely with experts from all different areas to understand potential options for mitigating risks during a pandemic and to closely manage their effects, both positive and negative. This book features several chapters on strategies how to better and quicker test for infected, optimal strategies to implement a lockdown in the different stages of the pandemic, and already some thoughts on the effectiveness of vaccination, see "Pooled Testing in the COVID-19 Pandemic" by M. Aldridge and D. Ellis and "Diagnostic Tests and Procedures During a Pandemic" by S. Dunbar and Y. Tang, "Changes in Behaviour Induced by COVID-19: Obedience to the Introduced Measures" by N. Badenes-Plà and "COVID-19 and Optimal Lockdown Strategies: The Effect of New and More Virulent Strains" by J. P. Caulkins et al.

After the SARS epidemic in the late 2000s the Robert Koch Institute prepared an analysis of potential future pandemics for the German Bundestag in 2013 (cf. Deutscher Bundestag (2013)). In this work, as in later analyses of the topic, consideration of lockdown as one of the possible solutions to contain a pandemic was completely absent. Also further risk analysis like the Risk Radar of the CRO-Forum in 2019 (cf. CRO-Forum (2019)) still had no reference to a major risk stemming from a potential lockdown during a pandemic. All in all, the research results presented in this book and the open and transparent discussion of potential measures during a pandemic will help the (re)insurance industry and also other sectors like trade and hospitality to better reflect and prepare for measures taken by the government.

14.3.2 Mitigating Economic Risks for Future Pandemics

After COVID-19 we all have to ask ourselves what we can do to become more resilient in future extreme situations to come. Simply applying the same measures as during COVID-19 might not be good enough, as the next crisis might probably be completely different. The next pandemic might affect a completely different age profile (as already did the Spanish flu one hundred years ago), and we might have to manage completely different events like a global black-out, global cyber attacks or, not to forget, the climate change.

Risk cover for future pandemics

As already discussed in Sect. 14.2.1 the private (re)insurance sector will not be able to provide sufficient cover to protect the whole economy against the next pandemic. For such extreme exposures only pool solutions with limitation of the maximal exposure for the entire insurance industry are possible as private-sector insurance solutions for pandemics. Beyond this maximal exposure and also to limit the moral hazard of governments implementing a lockdown without having to compensate for any economic losses, states must provide the major part of the cover. Due to the systemic nature of pandemic risks, the diversification of risks via pools on a global level will only work to a limited extent, unlike natural catastrophe, nuclear power plant and terrorism risks, where already pool solutions exist in Europe, as e.g. the Insurance Compensation Consortium in Spain, EXTREMUS in Germany and an earthquake pool in Switzerland.

A more detailed discussion on this topic, the potential design of such a pool solution using parametric triggers and the role of the World Bank can be found in three chapters of this book: "Risk Sharing and Stochastic Premia in the Presence of Systematic Risk: The case study of the UK COVID-19 economic losses" by H. Assa and T. J. Boonen, "The Legal Challenges of Insuring Against a Pandemic" by R. Hillier and "All-Hands-On-Deck!—How International Organisations Respond to the COVID-19 Pandemic" by M. C. Boado-Penas et al.

The potential role of the capital market

Often the capital market is seen as one possibility to off-load parts of the pandemic exposure of a (re)insurance company. First concepts for such solutions have been tested in the mid- to end-2000s as mortality catastrophe bonds. The experience shows that the appetite in the capital market is rather limited and this solution cannot provide a relevant relief for the total exposure. The main reasons for this limited interest—in contrast to other catastrophe bonds for, e.g., earthquakes or hurricanes—is the expected and, again during COVID-19, observed high correlation of a pandemic loss and losses in the capital market. Only well diversifying alternative investment products have shown a relevant demand from investors.

However, especially for the world's poorest countries, epidemic or pandemic covers are not affordable at all. To provide a certain economic protection for such countries the World Bank launched in 2016 the Pandemic Emergency Financing Facility (PEF) covering a maximum exposure of US$195.84 m for 64 of the poorest countries. It was actually designed for covering and providing quick support for expenditures during local epidemics like Ebola. During COVID-19 the full amount was paid out (cf. The World Bank (2021)). Until now, the cover has not been renewed by the World Bank, but it might be a very useful tool for providing and effectively protecting the development aid for these countries and hence it might be of political interest to launch a second bond.

Why basis risk sometimes prevents effective risk transfer

The PEF's parametric criteria are often criticised for being too slow and complicated. The facility, which is based on a set of disbursement triggers, only releases funds once there have already been a certain number of cases, deaths and countries affected by an outbreak. In particular, when infections occur only singularly in a country, the PEF does not provide funds even though there is an obvious need: a classic case of basis risk.

Basis risk has to be considered from two perspectives: a pure economic for effectively hedging against adverse cash flows which should be mitigated, and a regulatory for criteria if and to what extend risk mitigation instruments can be considered under Solvency II:

1. Economically, for a risk transferring instrument it has to be ensured that in most of the cases when a compensating cash flow should be provided by the instrument, this cash flow should also be triggered and paid out in the amount expected. If not then this has to be considered as *basis risk*.
2. For reflecting *basis risk* under Solvency II, EIOPA issued a guideline[6] to be considered when and how the treatment of risk mitigation techniques in the calculation of the Solvency Capital Requirement with the standard formula needs to be assessed.

---

[6] https://www.eiopa.europa.eu/content/guidelines-basis-risk_en.

As parametric triggers cannot directly use internal portfolio information of the insurer and must approximate the desired effect by publicly available data, it is obvious that there will be always situations where the instrument was originally intended to provide a compensating cash flow but actually does not. In case of a pandemic and an instrument based on parametric triggers covering the mortality risk of a portfolio, this can especially happen, if

- the age profile of the portfolio and the public data base,
- the social status and hence the availability and access to medical treatment in the different portfolios,
- the gender mix, health status or other relevant risk drivers for the mortality rate during the pandemic event.

do not match. As different events might be driven by different risk drivers, it is almost impossible to find a perfect fit for all conceivable situations. And then the parametric trigger should also be transparent, easy and quick to calculate. Effectively achieving both goals is impossible and we have to cope with a certain but hopefully limited residual basis risk when trying to mitigate risks of extreme events with instruments using a parametric design.

14.4 Conclusion: Our Learnings

For us, one of the most important lessons learned was that it is impossible to predict and (re)insure such rare and extreme events as a pandemic. Each new crisis will be different from the last, and we certainly do not have the right risk models and mitigation techniques to safely manoeuvre through this event. For example, in the event of a blackout following a global cyber-attack, it will be the newly established home office environment that will be the Achilles' heel, rather than the perfect solutions, as was the case with COVID-19. Therefore, we must always remain critical and quickly learn to adapt our measures during a crisis based on new data, robust statistical methods and experts' insights. Solvency II, as a principles-based risk management framework that builds a more resilient risk culture in companies, helped us stay in control during COVID-19. And the same approach, with appropriate adjustments, will work during other extreme events.

However, we have also learned that the capacity for these extreme events in the (re)insurance market, but also in the financial market, is limited and may not be sufficient to cover all macroeconomic risks. In such extreme events, governments need to step in and stand by the affected companies.

In the event of a crisis, we must act together in a determined and concerted manner. Global disasters cannot be dealt with by individual states or industries acting alone. Especially for the coming climate crisis, we will have to think about how we can tackle the problem across sectors and as a community of states.

References

Actuarial Association of Europa, *Position Paper on COVID-19* (2020), https://actuary.eu/wp-content/uploads/2020/05/202005-COVID-19.pdf

CRO-Forum, *Emerging Risks Initiative, Major Trends and Emerging Risk Radar*, May 2019 Update (2019), https://www.thecroforum.org/wp-content/uploads/2019/07/CRO-ERI_Emerging-Risk-RadarTrends_May-2019-Update.pdf

Destatis, Turnover in accommodation and food services in 2020 estimated to be 38% lower in real terms (2020), https://www.destatis.de/EN/Press/2021/01/PE21_024_45213.html

Deutscher Bundestag. Unterrichtung durch die Bundesregierung Bericht zur Risikoanalyse im Bevölkerungsschutz 2012. Drucksache 17/**12051** (2013), https://dipbt.bundestag.de/dip21/btd/17/120/1712051.pdf

H. Ealy, M. McEvoy, D. Chong, J. Nowicki, M. Sava, S. Gupta, D. White, J. Jordan, D. Simon, P. Anderson, *COVID-19 Data Collection: A Historical Retrospective* (Comorbidity & Federal Law, 2020)

EIOPA, COVID-19 measures, (2020a), https://www.eiopa.europa.eu/browse/covid-19-measures_en

EIOPA, 2020 review of Solvency II (2020b), https://www.eiopa.europa.eu/browse/solvency-ii/2020-review-of-solvency-ii_en

EuroMOMO, Graphs and maps (2021), https://www.euromomo.eu/graphs-and-maps

D. Frank, R. Fürhaupter, F. Schiller, Bestandsaufnahme der Coronapandemie und aktueller Handlungsbedarf für künftige globale Kumule (Pandemien). **2020**(2), 10–14 (2020)

GDV, Insurance in figures (2019), https://www.en.gdv.de/en/issues/our-news/insurance-in-figures-52050

GDV, *Statistical Yearbook of German Insurance 2019* (2020). https://www.en.gdv.de/resource/blob/52084/8586ea0d4ff8aba4982b18792111967a/statistical-yearbook-2019---broschuere-data.pdf

Imperial College COVID-19 response team, Short-term forecasts of COVID-19 deaths in multiple countries (2021). https://mrc-ide.github.io/covid19-short-term-forecasts/index.html

M. Roser, H. Ritchie, E. Ortiz-Ospina, J. Hasel, *Coronavirus Pandemic (COVID-19)* (2020). https://ourworldindata.org/coronavirus

The World Bank, *The World Bank: Fact Sheet: Pandemic Emergency Financing Facility* (2021). https://www.worldbank.org/en/topic/pandemics/brief/fact-sheet-pandemic-emergency-financing-facility

Open Access This chapter is licensed under the terms of the Creative Commons Attribution 4.0 International License (http://creativecommons.org/licenses/by/4.0/), which permits use, sharing, adaptation, distribution and reproduction in any medium or format, as long as you give appropriate credit to the original author(s) and the source, provide a link to the Creative Commons license and indicate if changes were made.

The images or other third party material in this chapter are included in the chapter's Creative Commons license, unless indicated otherwise in a credit line to the material. If material is not included in the chapter's Creative Commons license and your intended use is not permitted by statutory regulation or exceeds the permitted use, you will need to obtain permission directly from the copyright holder.

Printed in the United States
by Baker & Taylor Publisher Services